Best IDEAS: International Database Engineered Applications Symposium

Best IDEAS: International Database Engineered Applications Symposium

Guest Editor

Peter Z. Revesz

 Basel • Beijing • Wuhan • Barcelona • Belgrade • Novi Sad • Cluj • Manchester

Guest Editor
Peter Z. Revesz
School of Computing
University of Nebraska-Lincoln
Lincoln, NE
USA

Editorial Office
MDPI AG
Grosspeteranlage 5
4052 Basel, Switzerland

This is a reprint of the Special Issue, published open access by the journal *Information* (ISSN 2078-2489), freely accessible at: www.mdpi.com/journal/information/special_issues/1D5WDEWMG5.

For citation purposes, cite each article independently as indicated on the article page online and using the guide below:

Lastname, A.A.; Lastname, B.B. Article Title. *Journal Name* **Year**, *Volume Number*, Page Range.

ISBN 978-3-7258-2754-1 (Hbk)
ISBN 978-3-7258-2753-4 (PDF)
https://doi.org/10.3390/books978-3-7258-2753-4

© 2024 by the authors. Articles in this book are Open Access and distributed under the Creative Commons Attribution (CC BY) license. The book as a whole is distributed by MDPI under the terms and conditions of the Creative Commons Attribution-NonCommercial-NoDerivs (CC BY-NC-ND) license (https://creativecommons.org/licenses/by-nc-nd/4.0/).

Contents

About the Editor . . . . **vii**

Preface . . . . **ix**

Peter Z. Revesz
Best IDEAS: Special Issue of the International Database Engineered Applications Symposium
Reprinted from: *Information* **2024**, *15*, 713, https://doi.org/10.3390/info15110713 1

Victor Casamayor Pujol, Andrea Morichetta, Ilir Murturi, Praveen Kumar Donta and Schahram Dustdar
Fundamental Research Challenges for Distributed Computing Continuum Systems
Reprinted from: *Information* **2023**, *14*, 198, https://doi.org/10.3390/info14030198 5

Ali Abbasi Tadi, Saroj Dayal, Dima Alhadidi and Noman Mohammed
Comparative Analysis of Membership Inference Attacks in Federated and Centralized Learning †
Reprinted from: *Information* **2023**, *14*, 620, https://doi.org/10.3390/info14110620 15

Francisco Enguix, Carlos Carrascosa and Jaime Rincon
Exploring Federated Learning Tendencies Using a Semantic Keyword Clustering Approach
Reprinted from: *Information* **2024**, *15*, 379, https://doi.org/10.3390/info15070379 41

Panagiotis Filippakis, Stefanos Ougiaroglou and Georgios Evangelidis
Prototype Selection for Multilabel Instance-Based Learning †
Reprinted from: *Information* **2023**, *14*, 572, https://doi.org/10.3390/info14100572 68

Shruti Daggumati and Peter Z. Revesz
Convolutional Neural Networks Analysis Reveals Three Possible Sources of Bronze Age Writings between Greece and India †
Reprinted from: *Information* **2023**, *14*, 227, https://doi.org/10.3390/info14040227 90

Peter Z. Revesz
Archaeogenetic Data Mining Supports a Uralic–Minoan Homeland in the Danube Basin †
Reprinted from: *Information* **2024**, *15*, 646, https://doi.org/10.3390/info15100646 109

Giacomo Bergami, Samuel Appleby and Graham Morgan
Quickening Data-Aware Conformance Checking through Temporal Algebras †
Reprinted from: *Information* **2023**, *14*, 173, https://doi.org/10.3390/info14030173 128

Giacomo Bergami
Streamlining Temporal Formal Verification over Columnar Databases
Reprinted from: *Information* **2024**, *15*, 34, https://doi.org/10.3390/info15010034 188

Joseph Ajayi, Yao Xu, Lixin Li and Kai Wang
Enhancing Flight Delay Predictions Using Network Centrality Measures
Reprinted from: *Information* **2024**, *15*, 559, https://doi.org/10.3390/info15090559 210

Muhammad Alfian, Umi Laili Yuhana, Eric Pardede and Akbar Noto Ponco Bimantoro
Correction of Threshold Determination in Rapid-Guessing Behaviour Detection
Reprinted from: *Information* **2023**, *14*, 422, https://doi.org/10.3390/info14070422 221

Reza Shahbazian and Irina Trubitsyna
DEGAIN: Generative-Adversarial-Network-Based Missing Data Imputation
Reprinted from: *Information* **2022**, *13*, 575, https://doi.org/10.3390/info13120575 233

About the Editor

Peter Z. Revesz

Dr. Peter Z. Revesz earned a B.S. summa cum laude with a double major in computer science and mathematics from Tulane University and a Ph.D. in computer science from Brown University. He was a postdoctoral fellow at the University of Toronto before joining the University of Nebraska-Lincoln, where he is a professor at the School of Computing. He is an expert in computational linguistics, databases, bioinformatics, and geoinformatics. He is the editor of Computational Linguistics and Natural Language Processing (MDPI, 2024) and the author of Introduction to Databases: From Biological to Spatio-Temporal (Springer, 2010). He has held visiting appointments at the IBM T.J. Watson Research Center, INRIA, the Max Planck Institute for Computer Science, the University of Athens, the University of Hasselt, the University of Helsinki, the U.S. Air Force Office of Scientific Research, and the U.S. Department of State. He is a recipient of an AAAS Science & Technology Policy Fellowship, a J. William Fulbright Scholarship, an Alexander von Humboldt Research Fellowship, a Jefferson Science Fellowship, a National Science Foundation CAREER award, a Faculty International Scholar of the Year award by Phi Beta Delta, and was ranked in the Stanford/Elsevier World's Top 2% Scientists list.

Preface

The aim of this Special Issue is to present the latest research on database engineered applications, including applications in the areas of distributed computing systems, federated learning, learning instance selection, data analytics and data mining, temporal databases, data prediction and imputation, and the detection of user behavior. The authors are leading international experts in these areas. Most of the authors already presented preliminary versions of their works at various International Database Engineered Applications Symposia in the past few years, but this Special Issue contains a collection of their full journal articles. The readers can apply these authors' cutting-edge ideas and techniques to various specific applications. In addition, several articles describe many challenging open problems for researchers who would like to find novel topics in database engineering.

Peter Z. Revesz
Guest Editor

Editorial

Best IDEAS: Special Issue of the International Database Engineered Applications Symposium

Peter Z. Revesz [1,2]

[1] School of Computing, College of Engineering, University of Nebraska-Lincoln, Lincoln, NE 68588, USA; peter.revesz@unl.edu
[2] Department of Classics and Religious Studies, College of Arts and Sciences, University of Nebraska-Lincoln, Lincoln, NE 68588, USA

1. Introduction

Database engineered applications cover a broad range of topics including various design and maintenance methods, as well as data analytics and data mining algorithms and learning strategies for enterprise, distributed, or federated data stores. The exponentially growing amounts of commercial, governmental, and non-government organizational data provide a continued challenge for many database engineered applications. The collection of papers in this Special Issue makes several fundamental contributions to this research area.

This Special Issue is primarily based on extended versions of selected papers from the 27th International Database Engineered Applications Symposium (IDEAS) held in 2023 in Heraklion, Crete, Greece, as well as selected papers from prior IDEAS conferences. We also invited additional papers on the conference theme, and they also underwent a rigorous review process.

These invited papers included the paper "Fundamental Research Challenges for Distributed Computing Continuum Systems" by Schahram Dustdar, Professor and Head of the Distributed Systems group at the Vienna University of Technology (TU Wien), and his coworkers [1]. Schahram Dustdar was invited to contribute to this Special Issue because he has served as one of the invited speakers at several IDEAS conferences. This paper [1] lays out a bold vision for the future of distributed computing systems.

The other papers in the Special Issue cover a range of topics, as follows.

2. Federated Learning and Learning Instance Selection

In 2010, Revesz and Triplet [2] introduced the concept—though not the term—of federated learning, in which multiple entities collaborate to train a model without sharing the data due to privacy concerns. Revesz and Triplet [2] gave the example of a set of hospitals who may not share information about their cardiology patients because of patient privacy restrictions. Revesz and Triplet [2] proposed that each hospital train its own classification model on their local data, and then they share the classification models instead of the raw data. Revesz and Triplet [2] also presented several classification integration methods based on constraint databases [3]. Bonawitz et al. [4] termed this type of collaborative learning 'federated learning' and applied it to a set of mobile devices training a neural network. Federated learning has become a very active research area since then [5]. It should not be confused with federated databases [6], which also cooperate in answering queries, but not in learning.

Two papers in this Special Issue deal with the topic of federated learning. The paper "Comparative Analysis of Membership Inference Attacks in Federated and Centralized Learning" by Abbasi Tadi et al. [7] describes several methods that can be used to prevent potential attackers from inferring sensitive data by intercepting updates transmitted between training parties and a central server which maintains the common learned model.

The paper "Exploring Federated Learning Tendencies Using a Semantic Keyword Clustering Approach" by Enguix, Carrascosa, and Rincon [8] considers identifying current trends and emerging subareas within a research area. The authors propose an automatic semantic keyword clustering method. They apply their method to the set of federated learning research papers published since 2017 and identify the fastest growing subareas.

The paper "Prototype Selection for Multilabel Instance-Based Learning" by Filippakis, Ougiaroglou, and Evangelidis [9] considers the problem of reducing the size of the training set in the case of multilabel instance-based classification learning. Here, the term "multilabel" means that each instance can belong to several classes. While there are several well-known algorithms for reducing the size of the training set in the case of single-label instance-based classification learning, the multilabel case was an open problem. Filippakis et al. [9] propose several solutions to this open problem.

We would like to point out that the authors of [9] had the highest and the authors of [7] had the second highest ranked paper at the IDEAS 2023 conference, and their journal articles are also excellent contributions to this Special Issue.

3. Data Analysis and Data Mining

Learning is closely related to data analysis and data mining. In fact, the paper "Convolutional Neural Networks Analysis Reveals Three Possible Sources of Bronze Age Writings between Greece and India" by Daggumati and Revesz [10] included training a set of convolutional neural networks (CNNs) to recognize eight Bronze Age scripts as a first step. In a second step, Daggumati and Revesz passed each script's signs to each other script's trained CNN. As each CNN recognized each of the foreign scripts' signs as a local sign, a table of sign correspondences was found. Two scripts could be identified as being related if their sign correspondence table showed a one-to-one function. Based on that idea, the eight Bronze Age scripts were found to form three groups: (1) Sumerian pictograms, the Indus Valley script, and the proto-Elamite script; (2) Cretan hieroglyphs and Linear B; and (3) the Phoenician, Greek, and Brahmi alphabets. The CNN-based script similarity method of Daggumati and Revesz [10] improves on an earlier computational script similarity method based on feature vectors [11]. A better understanding of script similarities helps in the decipherment of ancient inscriptions [12,13].

The paper "Archaeogenetic Data Mining Supports a Uralic–Minoan Homeland in the Danube Basin" by Revesz [14] applies data mining to the rapidly growing archaeogenetic data. The available archaeogenetic data are often incomplete and therefore more difficult to analyze than regular genetic data. By using some novel data mining algorithms, it was possible to show that the Minoans, who formed the first Bronze Age civilization in Europe, mostly originated from the lower Danube Basin. A better understanding of the origin of the Minoans helps to narrow down the set of languages to be considered as likely cognates with the Minoan language. This could avoid resorting to brute-force methods of cryptanalysis where all possible ancient languages are considered from the Mediterranean and Black Sea areas [15]. The lower Danube Basin is a good candidate for a Proto-Uralic language area in the Neolithic.

4. Temporal Logic and Verification

Linear Temporal Logic over finite traces (LTL_f) can be used to express a set of temporal specifications Φ. Verifying that a system satisfies an LTL_f specification is a computationally difficult task. Therefore, an extended LTL_f ($xtLTL_f$) is proposed by Bergami, Appleby, and Morgan [16] in the paper "Quickening Data-Aware Conformance Checking through Temporal Algebras". They describe systems by a set of traces of observed and completed labeled activities expressing one possible run of a process. Verifying that such system descriptions satisfy an $xtLTL_f$ specification can be efficiently checked if the set of traces are first converted to a columnar data storage [16].

The paper "Streamlining Temporal Formal Verification over Columnar Databases" by Bergami [17] takes this idea further by considering the following four new operators:

ChainResponse(A,B), ChainPrecedence(A,B), AltResponse(A,B), and AltPrecedence(A,B). For example, ChainResponse(A,B) is true if the activation of A is immediately followed by the target B. Bergami [17] shows that expressions including these operators can also be checked efficiently if the traces are converted to columnar data storage.

5. Prediction, Detection and Imputation

The paper "Enhancing Flight Delay Predictions Using Network Centrality Measures" by Ajayi et al. [18] aims at improving the accuracy of predicting airplane flight delays. The authors improve the prediction accuracy by introducing a novel method based on network centrality measures that are sensitive to the structure of the flight network.

The paper "Correction of Threshold Determination in Rapid-Guessing Behaviour Detection" by Alfian et al. [19] concerns detecting whether a student is only guessing answers on a multiple-choice test. The traditional method of detecting whether a student is guessing is based on setting a fixed threshold response time, say K seconds, where K is a small number like 3 or 5 depending on the overall difficulty level of the test. If the student's response time is less than K seconds, then the student is assumed to have guessed the answer. Alfian et al. [19] criticize this K-seconds approach because the difficulty of the questions could vary on a test. They show that the accuracy of detecting guessing is improved when the threshold is a variable depending on the difficulty level of the questions.

Greco, Molinaro, and Trubitsyna already considered the challenging topic of incomplete databases in an earlier IDEAS paper [20]. Now, Shahbazian and Trubitsyna [21] address the issue again in the paper "DEGAIN: Generative-Adversarial-Network-Based Missing Data Imputation". They propose handling missing data in incomplete databases by means of data imputation, where the missing values are estimated based on the rest of the data. Generative Adversarial Imputation Nets (GAINs) can be used to generate synthetic data that are like the real data [22]. The main idea is to have a generator of fake data and a discriminator that tries to tell whether a datum is real or fake. However, Shahbazian and Trubitsyna [21] argue that there is a strong correlation among real data. Hence, a deconvolution process is needed to reduce these correlations, and then the generator and discriminator network will work more effectively. Combining deconvolution and GAIN gives rise to the name DEGAIN. We hope that DEGAIN will gain widespread acceptance in data imputation in the future.

Acknowledgments: I would like to thank the many reviewers of the papers submitted to this Special Issue. Their detailed comments and thoughtful recommendations regarding acceptance or rejection helped to maintain a high standard for this Special Issue. I also would like to thank the Section Managing Editor at MDPI for this Special Issue, for her outstanding help in every aspect of the organization work, including her help in finding some reviewers. I thank Haridimos Kondylakis for serving as editor of [14] and Xin Ning for serving as editor of [10] to avoid any conflicts of interest in the review process. Finally, I express my great appreciation for the many talented contributors to this Special Issue, as well as the authors and editors that contributed to the IDEAS 2023 conference and earlier IDEAS conferences. It was great working with the authors throughout the publication process and learning about their exciting results. I wish all of them much success in their future research.

Conflicts of Interest: The author declares no conflicts of interest.

References

1. Casamayor Pujol, V.; Morichetta, A.; Murturi, I.; Donta, P.K.; Dustdar, S. Fundamental Research Challenges for Distributed Computing Continuum Systems. *Information* **2023**, *14*, 198. [CrossRef]
2. Revesz, P.Z.; Triplet, T. Classification integration and reclassification using constraint databases. *Artif. Intell. Med.* **2010**, *49*, 79–91. [CrossRef] [PubMed]
3. Kanellakis, P.C.; Kuper, G.M.; Revesz, P.Z. Constraint query languages. *J. Comput. Syst. Sci.* **1995**, *51*, 26–52. [CrossRef]
4. Bonawitz, K.; Ivanov, V.; Kreuter, B.; Marcedone, A.; McMahan, H.B.; Patel, S.; Ramage, D.; Segal, A.; Seth, K. Practical secure aggregation for privacy-preserving machine learning. In Proceedings of the 2017 ACM SIGSAC Conference on Computer and Communications Security, Association for Computing Machinery, New York, NY, USA, 30 October–3 November 2017; pp. 1175–1191.

5. Kairouz, P.; McMahan, H.B.; Avent, B.; Bellet, A.; Bennis, M.; Bhagoji, A.N.; Bonawitz, K.; Charles, Z.; Cormode, G.; Cummings, R.; et al. Advances and open problems in federated learning. *Found. Trends Mach. Learn.* **2021**, *14*, 1–210. [CrossRef]
6. Sheth, A.P.; Larson, J.A. Federated database systems for managing distributed, heterogeneous, and autonomous databases. *ACM Comput. Surv.* **1990**, *22*, 183–236. [CrossRef]
7. Abbasi Tadi, A.; Dayal, S.; Alhadidi, A.; Mohammed, N. Comparative Analysis of Membership Inference Attacks in Federated and Centralized Learning. *Information* **2023**, *14*, 620. [CrossRef]
8. Enguix, F.; Carrascosa, C.; Rincon, J. Exploring Federated Learning Tendencies Using a Semantic Keyword Clustering Approach. *Information* **2024**, *15*, 379. [CrossRef]
9. Filippakis, P.; Ougiaroglou, S.; Evangelidis, G. Prototype Selection for Multilabel Instance-Based Learning. *Information* **2023**, *14*, 572. [CrossRef]
10. Daggumati, S.; Revesz, P.Z. Convolutional Neural Networks Analysis Reveals Three Possible Sources of Bronze Age Writings between Greece and India. *Information* **2023**, *14*, 227. [CrossRef]
11. Revesz, P.Z. Establishing the West-Ugric Language Family with Minoan, Hattic and Hungarian by a Decipherment of Linear A. *WSEAS Trans. Inf. Sci. Appl.* **2017**, *14*, 306–335.
12. Revesz, P.Z. A Translation of the Arkalochori Axe and the Malia Altar Stone. *WSEAS Trans. Inf. Sci. Appl.* **2017**, *14*, 124–133.
13. Hughes-Castleberry, K. Could AI Language Models Like ChatGPT Unlock Mysterious Ancient Texts? *Discover Magazine*. 11 April 2023. Available online: https://www.discovermagazine.com/technology/could-ai-language-models-like-chatgpt-unlock-mysterious-ancient-texts (accessed on 15 April 2023).
14. Revesz, P.Z. Archaeogenetic Data Mining Supports a Uralic–Minoan Homeland in the Danube Basin. *Information* **2024**, *15*, 646. [CrossRef]
15. Nepal, A.; Perono Cacciafoco, F. Minoan Cryptanalysis: Computational Approaches to Deciphering Linear A and Assessing its Connections with Language Families from the Mediterranean and the Black Sea Areas. *Information* **2024**, *15*, 73. [CrossRef]
16. Bergami, G.; Appleby, S.; Morgan, G. Quickening Data-Aware Conformance Checking through Temporal Algebras. *Information* **2023**, *14*, 173. [CrossRef]
17. Bergami, G. Streamlining Temporal Formal Verification over Columnar Databases. *Information* **2024**, *15*, 34. [CrossRef]
18. Ajayi, J.; Xu, Y.; Li, L.; Wang, K. Enhancing Flight Delay Predictions Using Network Centrality Measures. *Information* **2024**, *15*, 559. [CrossRef]
19. Alfian, M.; Yuhana, U.L.; Pardede, E.; Bimantoro, A.N.P. Correction of Threshold Determination in Rapid-Guessing Behaviour Detection. *Information* **2023**, *14*, 422. [CrossRef]
20. Greco, S.; Molinaro, C.; Trubitsyna, I. Algorithms for computing approximate certain answers over incomplete databases. In Proceedings of the 22nd International Database Engineering and Applications Symposium, Villa San Giovanni, Italy, 18–20 June 2018; ACM Press: New York, NY, USA, 2018; pp. 1–4.
21. Shahbazian, R.; Trubitsyna, I. DEGAIN: Generative-Adversarial-Network-Based Missing Data Imputation. *Information* **2022**, *13*, 575. [CrossRef]
22. Yoon, J.; Jordon, J.; Schaar, M. GAIN: Missing data imputation using generative adversarial nets. In Proceedings of the International Conference on Machine Learning, Stockholm, Sweden, 10–15 July 2018; pp. 5689–5698.

Disclaimer/Publisher's Note: The statements, opinions and data contained in all publications are solely those of the individual author(s) and contributor(s) and not of MDPI and/or the editor(s). MDPI and/or the editor(s) disclaim responsibility for any injury to people or property resulting from any ideas, methods, instructions or products referred to in the content.

Article

Fundamental Research Challenges for Distributed Computing Continuum Systems

Victor Casamayor Pujol *, Andrea Morichetta, Ilir Murturi, Praveen Kumar Donta and Schahram Dustdar

Distributed Systems Group, TU Wien, 1040 Vienna, Austria; a.morichetta@dsg.tuwien.ac.at (A.M.); i.murturi@dsg.tuwien.ac.at (I.M.); p.donta@dsg.tuwien.ac.at (P.K.D.); dustdar@dsg.tuwien.ac.at (S.D.)
* Correspondence: v.casamayor@dsg.tuwien.ac.at

Abstract: This article discusses four fundamental topics for future Distributed Computing Continuum Systems: their representation, model, lifelong learning, and business model. Further, it presents techniques and concepts that can be useful to define these four topics specifically for Distributed Computing Continuum Systems. Finally, this article presents a broad view of the synergies among the presented technique that can enable the development of future Distributed Computing Continuum Systems.

Keywords: distributed computing continuum systems; system representation; system model; lifelong learning; business model

Citation: Casamayor Pujol, V.; Morichetta, A.; Murturi, I.; Kumar Donta, P.; Dustdar, S. Fundamental Research Challenges for Distributed Computing Continuum Systems . *Information* **2023**, *14*, 198. https:// doi.org/10.3390/info14030198

Academic Editor: Hamid R. Arabnia

Received: 14 January 2023
Revised: 7 March 2023
Accepted: 16 March 2023
Published: 22 March 2023

Copyright: © 2023 by the authors. Licensee MDPI, Basel, Switzerland. This article is an open access article distributed under the terms and conditions of the Creative Commons Attribution (CC BY) license (https:// creativecommons.org/licenses/by/ 4.0/).

1. Introduction

The expansion from Cloud computing to Edge computing has brought a new paradigm called the Distributed Computing Continuum [1–4]. This combines the virtually unlimited resources of the Cloud with the heterogeneity and proximity of the Edge. To do so, the Distributed Computing Continuum combines the underlying infrastructure of all other computing tiers. Hence, the infrastructure becomes a first-class citizen compared to current Internet-distributed systems.

Current research on Edge computing and Distributed Computing Continuum focuses on solving specific problems, which produce particular solutions with narrow applicability. A few examples include: in [5], the approach is tailored to an ultradense network; in [6], the authors present a solution for a static description of the system; or in [7], the authors present an orchestration for the edge-cloud that requires centralization on the Cloud. In this article, we aim to show pointers to generalized solutions; we organize it through the highlighting of four key aspects that require in-depth analysis, as well as a high degree of agreement among the scientific community and the other stakeholders to make tangible progress on the development of the Distributed Computing Continuum.

First, the Distributed Computing Continuum needs a novel **representation** beyond the classical architecture of computer systems. Distributed Computing Continuum Systems are built of a large variety of heterogeneous devices and networks. The system's functional requirements can either naturally evolve during their lifetime, dynamically change the running services [8], or, more critically, suffer unexpected events. These changes will affect the underlying infrastructure configuration, out-dating previous architectural representations. For instance, the Edge infrastructure requires a dynamic adaption to new devices and network connections, leading to a completely new system from the perspective of its design phase. This behavior contrasts with Cloud computing, where changes in the underlying infrastructure can be updated, but the application is not affected.

Another challenge for Distributed Computing Continuum Systems is their **model**. First, we need to clarify the difference between the representation and the model. Our view is that representation is a description of the system, its components, its relations, and its characteristics. Whereas we address the dynamic behavior of the system and its components by the model. Nevertheless, both concepts require a certain level of agreement as, ideally, one aims to have a compatible representation and model of the system. The complexity of the system, coupled with its openness (i.e., many external and spontaneous actors can affect the system), endangers the correctness of adaptation strategies; in the Cloud, this is usually solved by only considering a single elasticity strategy per component [9]. Further, it is complex to assess the impact of the adaptation on the entire system, i.e., using a different set of Edge devices might imply moving data through another network, which can affect privacy/security constraints. Hence, tools to describe the new behavior of the infrastructure are needed.

The third key element is a **lifelong learning framework**. The dynamicity of the environment, the user's variety of behaviors, the evolution of functional requirements, and the long-term usage of the underlying infrastructure require developing a learning framework to keep high-quality standards during the system's entire life cycle. This is aligned with the idea presented in [10] about lifelong learning for self-adaptive systems.

The last facet that requires agreement among the community is the **business model**. A key enabler for the Cloud tier has been its successful business model. Nevertheless, the multi-tenant and multi-proprietary characteristics of the underlying infrastructure produce a more challenging set of stakeholders for Distributed Computing Continuum Systems. However, to attract the needed collaborations and investments to develop such an ambitious computing tier, it is of utmost importance to develop certain agreements that can enable the best business model. In this regard, we have witnessed, in the context of the Mobile World Congress 2023, how the big telecommunication companies want to be part of the Distributed Computing Continuum Systems by providing an API to application developers to tailor their networks to the application's needs, e.g., see "GSMA Open Gateway", from Future Networks (https://www.gsma.com/futurenetworks/?page_id=35 168—accessed 1 March 2023).

The main goal of this work presented as a vision for the emerging Distributed Computing Continuum Systems, is to highlight the need for a holistic perspective. This type of system is far from being a reality, and we believe that common grounds are required to advance their development. Hence, we provide what we think are fundamental research challenges to be solved and what are our research road-maps for each of them. We seek to spark discussion and creativity in the research community to enable these future systems.

To sum up, Distributed Computing Continuum Systems require a broad agreement on a representation, a model, a lifelong learning framework, and a business model to enable its development. This article presents a few ideas on how to start building these required blocks. In the following section, a technique or concept is presented for each of the presented aspects to shape our vision of Distributed Computing Continuum Systems. Then, we discuss the overall merging of all presented concepts and techniques, and we finish this article with a conclusion and future work.

2. Vision

In this section, we present key elements for each of the previously introduced aspects that, from our perspective, will be key for developing Distributed Computing Continuum Systems.

2.1. Representation

The characteristics of the Distributed Computing Continuum require new representations for Internet-distributed systems enabling dynamic systems and topologies. This contrasts with the usage of the concept of system architecture, where the word architecture refers to a static structure of the system. As previously discussed, the complex and dynamic behavior of Distributed Computing Continuum calls for other techniques to represent these systems, which can accommodate the dynamic behavior of the underlying infrastructure and the system's environment.

Further, a fundamental concept requires an in-depth discussion: the definition and scope of the system and its relation to each component. Simply put, the Distributed Computing Continuum needs to be understood as an ecosystem in which there are different abstraction layers, where components are described and aggregated differently. Interestingly, when one thinks about an ecosystem, the synergies and dependencies between the components blur the definition of self, i.e., the boundaries of an autonomic component are flexible and dependent on the purpose. Components interact with others both horizontally and vertically with respect to their abstraction levels. In this regard, Distributed Computing Continuum Systems have multi-level and multi-scale structures, and their components show a dual tendency; from one side, they aim to be autonomous, and from the other side, they need to be integrated with others to provide a complete view of the system. Hence, they require defining what is the *self*: the entire system, a single autonomic component, or all things simultaneously. Hence, we need to analyze and provide arguments for each case to understand which is the best solution for the system's representation.

Our initial intuition is that we need a holistic view of the system, considering it as a system of systems and providing compatible tools at any level. Hence, we envision the **Markov Blanket** as a key element to represent Distributed Computing Continuum Systems given its nesting and filtering capacity [11].

Markov Blanket

The Markov Blanket, in Bayesian statistics, refers to the set of variables that contain all needed information to determine a target variable. Simply put, the Markov Blanket concept can be used to determine which variables influence another. Formally, if x is a random variable, and Y is the set of random variables of the Markov Blanket of x, then $P(x|Y) = P(x|Y,Z)$, where Z represents any other random variable [12].

In large-scale Distributed Computing Continuum Systems, the Markov Blanket concept can be seen as causal filtering, given that it allows working only with the subset of variables affecting the target. This is key in terms of scalability, e.g., the problem of selecting the best device for each service in an application is NP-Hard with exponential time complexity [13]; hence dealing with only the required subset of components can drastically alleviate the difficulty. Further, the definition of a random variable within the Markov Blanket scope is flexible, which means that regardless of the abstraction level in which the variable exists, it will be possible to build its Markov Blanket. Hence, we can foresee this as a nesting capability in which the higher-level abstraction Markov Blanket can be decomposed as a set of other Markov Blankets at lower-level abstractions. If we bring this to the Distributed Computing Continuum Systems, it is possible to define the entire system with a Markov Blanket describing the main components that affect the system, and it is possible to go deeper in detail and describe smaller components also in terms of a Markov Blanket. Hence, both scalability and the self's definition are addressed by using the Markov Blanket abstraction. In conclusion, the Markov Blanket concept is needed to represent Distributed Computing Continuum Systems.

2.2. Model

Two main challenges are identified specific to Distributed Computing Continuum Systems to develop models of their behavior. (1) Its decentralization precludes developing a model of the entire system as a single entity; conversely, the model should allow its distribution. In other words, the model has to allow its splitting among the different parts of the system. Further, it can be linked with the representation; one could imagine that there is a model for each Markov Blanket used to represent the system. (2) Distributed Computing Continuum Systems are set within an uncertain environment. On the one side, there is unpredictability in the user's behavior, e.g., in an autonomous vehicle use case, external computations require to follow a car through its trajectory to keep latency at a lower bound [14], but the required hops can not be predicted beforehand, given that the trajectory can change on demand. On the other side, the underlying infrastructure of Distributed Computing Continuum Systems is multi-tenant and multi-proprietary. Hence, another source of uncertainty is from the usage of others tenants of the shared resources.

Further, their complexity and interconnection also generate internal uncertainty, given that the exact knowledge of the system's behavior might not be known; in this regard, there are many sources of internal uncertainties for self-adaptive software systems, as explained in [15], in brief, the model has to handle uncertainty.

To develop the system's model, we look at one concept and one technique. The concept, called **DeepSLOs**, can define and link constraints for a Distributed Computing Continuum System at different abstraction levels. In contrast, the technique, **Causal inference** allows the system to understand its own behavior and to perform a priori analyzed changes on its underlying infrastructure to minimize the effects of uncertainty.

2.2.1. DeepSLOs

DeepSLOs stem from Cloud Service Level Objectives (SLOs). In general, an SLO is a constraint to the underlying infrastructure, simply put, the minimal availability of a service or the maximal CPU usage of a workload. Hence, in the Cloud, when a constraint is violated, an elasticity strategy is triggered, which modifies the system and brings the Service Level Indicator (SLI) within the value specified by the SLO. In contrast, a DeepSLO is a set of hierarchically connected SLOs. This deviates from Cloud SLOs because they are no longer isolated constraints of the system. However, they provide a holistic perspective on the system status given by its constraints. Hence, it is possible to obtain a complete description of the system's performance. Further, DeepSLOs also aim to fully describe the characteristics of the underlying infrastructure, which highlights the infrastructure as a key part of any Distributed Computing Continuum System. Within a single DeepSLO, there are SLOs at different levels of abstraction, which describe the system's performance from different perspectives. Simply put, low-level abstraction SLOs can easily define infrastructure behavior, i.e., the performance of GPUs at the Edge, while higher-level abstraction can deal with application performance, i.e., the accuracy of the inference job at the Edge. Both given examples can be expressed as constraints within a single DeepSLO, each as an SLO but connected through the system dependencies. Still, the first brings information on the hardware performance for the inference job, and the second has a holistic performance on the tasks. In this way, different elastic strategies can be used efficiently depending on the cause of the system's performance degradation.

2.2.2. Causal Inference

Causal inference [16] is a mathematical framework able to discover causal relations between system variables and predict their behavior. Hence, it provides a better understanding of why events occur, solving conflicts when several adaptation strategies are needed. Further, it can predict the outcome of interventions in the system. Simply put, it predicts how the system will behave after a new configuration. Finally, causal inference unfolds the ability to develop counterfactuals to better understand the system's true behavior, providing a key element to the learning capability. On the one side, studying interventions, i.e., applying a specific condition to the variables, can ease the selection of the adaptation strategies; further, using counterfactuals, hypothetical situations can be described to extract learnings and improve the system's resistance against uncertainty. These methods are very relevant for uncertainty management, given that it is possible to achieve knowledge of the system's behavior under conditions not yet met in real operations. Causal inference is usually applied over a directed acyclic graph to perform its analyses and explain the behavior of the system; interestingly, this graph can be provided by the Markov Blanket representation of the system.

One can imagine an application with a peak of demand in some region. Hence, the Distributed Computing Continuum System is required to perform an adaptive action to properly maintain its expected quality standards. However, there can be several ways to tackle that situation, scaling up Cloud or Edge components, using new resources, etc. In such situations, causal inference provides an understanding of how the system will react to these changes, e.g., how costs can be increased, how other services can be affected, how sustainable is the solution, etc. Further, if quality and cost (see Section 2.4) are not fixed, and there are some assumable margins, the manifold of options and their consequences grows. This complexity, also explained through an illustrative use case in [17], can be tackled by a better understanding of the internal system relations provided by causal inference.

2.3. Lifelong Learning Framework

Under the assumption that Distributed Computing Continuum Systems are complex and inhabit a dynamic environment, providing the system with the learning capacity is of foremost importance. Hence, lifelong learning enables improving the models that govern the system, and, due to the nature of these systems, it also needs to be related to composition, i.e., which are the best components of the underlying infrastructure to use in an application. Systems need to be able to change their configuration using other components that are not initially part of the system; in other words, the use of adaptation strategies requires learning, given the huge space of possible configurations. Consequently, they need to make decisions dependent on the system's current state, which cannot be foreseen at design time. Hence, Distributed Computing Continuum Systems need techniques that provide this capacity for continuous improvement in a dynamic setting. In this regard, the Free Energy Principle (FEP) explains the behavior of systems to adapt to their environment, initially developed as a hypothesis on how the brain works [18]. To do so, the FEP aims to maximize the knowledge of the system about its environment, which is needed to continuously improve operation.

Free Energy Principle and Active Inference

The FEP was first defined by K. Friston to describe the behavior of the brain as a system that adapts to its dynamic environment. In brief, the FEP shows that adaptive systems have an internal model of their environment and that this allows adaptive systems to persist in it. Further, the FEP claims that this adaptivity is achieved by minimizing the difference between the internal model that the system possesses of its environment concerning the real environment behavior. Another important observation of the FEP for Distributed Computing Continuum Systems is that regardless of the scale, adaptive systems behave similarly [19].

The active inference is a corollary of the FEP, providing a methodology to develop agents (systems) that can learn from actions following the FEP [20]. Hence, by adapting Distributed Computing Continuum Systems to the active inference methodology, it will be possible to develop systems that progressively learn to adapt within a dynamic environment.

2.4. Business Model

We have learned from Cloud computing that a key aspect of the success of Distributed Computing Continuum Systems is considering and easing their business logic. These systems are multi-tenant and multi-proprietary; hence, we need concepts and methods that allow several stakeholders to collaborate or share part of their infrastructure. We are not considering here which has to be the business model, but two ideas are needed for its emergence. Our intuition is that to enable this, communication and understanding among them need to be the cornerstone. Further, the fact that systems, components, and other involved stakeholders understand different abstractions is a great challenge that needs to be addressed. From this perspective, the use of **Resources**, **Quality**, and **Cost** as the highest-level abstraction state variables enables homogenization among components and systems [21]. Further, they are understandable to all stakeholders given their higher level of abstraction [22].

Finally, security is key to engaging stakeholders. Therefore, **Zero Trust** [23] concepts promise control network flows between all assets, advanced resource protection, fast detection of malicious activities, improved system performance, and secure communication between components. Features that are vital to make Distributed Computing Continuum a reality.

2.4.1. Resources, Quality, and Cost

In previous work on Cloud computing, Resources, Quality, and Cost have been defined as elasticity dimensions [24], which can define the system's overall state. For Distributed Computing Continuum Systems, we are convinced that the same abstractions are needed; however, they are not elasticity dimensions but the highest-level abstractions of the system state, which will ease the system's management decisions. In other words, Resources, Quality, and Cost are required to understand the system's current situation at the highest level of abstraction and how it can deviate toward other possible states. It is important to remark that Resources, Quality, and Cost are chosen because any system stakeholder can interpret them, and agreements at the highest level can be reached, which then can be specified into lower-level agreements tailored to the specific characteristic of each stakeholder.

2.4.2. Security through Zero Trust

Trustworthiness is key among stakeholders, and secure systems are needed to enforce trust. In this regard, Zero Trust [23,25] is an emerging security paradigm where trust among system components needs to be achieved at every step. In other words, it is not enough to be part of the network's system, but the components' behavior is also continuously verified.

3. Discussion

This section first presents some alternative directions to the ones discussed and shows our holistic vision of the ideas and techniques presented. The alternative ideas presented are not exhaustive as this work presents a vision. Hence, this section aims just at providing a broader context to the current research around Distributed Computing Continuum Systems.

3.1. Techniques Discussion

In terms of representation techniques for Distributed Computing Continuum Systems, some works focus on graph-based representations; in [26], the authors specify the task type and system with graphs and leverage such a representation to solve the task offloading problem. Another perspective is given by [27], where their graph-based representation of Distributed Computing Continuum Systems is the input for graph-based neural networks to solve the distributed scheduling problem. Conversely, our representation aims at being general for whichever problem needs to be solved in the Distributed Computing Continuum paradigm. Further, the filtering and nesting capacity of the Markov Blanket, which is also graph-based, is of utmost importance for Distributed Computing Continuum Systems. Take, for instance, the work in [13], which uses propositional logic to represent the Distributed Computing Continuum System; they show how the complexity of solving the task assignment problem is exponential. Hence, a representation able to filter out the useless aspects for the specific case can make it feasible.

There are several approaches that propose a model for the entire Distributed Computing Continuum that can be leveraged to describe and forecast its expected behavior. Most of the time, the description is left to deep neural networks [28–30], which, in general, provide good results, but it is hard to assess its generalization capabilities. There are also approaches that provide a function-based description that allows for solving the problem, e.g., through game theory [31]. This type of model usually requires many parameters and assumptions on these parameters. A similar situation is found when modeling through queue theory [32,33]. From our perspective, queue theory embraces the randomness found in these systems. However, it requires precise modeling for the probability distributions governing the system behavior. Another interesting approach follows the osmotic computing concept [34,35], which uses the analogy of physical pressure to manage Distributed Computing Continuum Systems. Finally, there are also constraint-based approaches for system modeling, such as in [36], where they are able to verify the system's performance. We are proposing a constraint-based perspective from the use of DeepSLOs where we can relate the underlying infrastructure with the application to define the entire Distributed

Computing Continuum System. Still, we understand that the complex and stochastic behavior requires another modeling layer. There, we see fit for causal inference to leverage a probabilistic model with causal relationships.

We relate lifelong learning with the capacity of a system to continue learning and improving its behavior during its life-cyle when the requirements and tasks can change over time. This problem is usually tackled through machine learning, specifically with reinforcement learning in robotics [37,38]. In this regard, an interesting framework for self-adaptive systems is provided by [10]. A new and interesting approach is on graph continuous learning; a survey is presented in [39]. Our intuition to propose active inference from the FEP relates to its ultimate goal of modeling the dynamic environment to optimize the system's adaptive capacity. Further, its mathematical formulation fits well with the use of Markov Blankets, allowing us to easily combine both techniques. We are aware that FEP has been questioned in terms of applicability and generality [40,41]. Nevertheless, its fitness to our scope and the results showed in other articles [42] are promising.

Regarding the business model, there is a vast amount of work on security aspects, where Zero Trust is one of the most prominent perspectives. In terms of high-level abstractions to set common ground on business objectives, the literature is scarce. We propose Cost, Quality, and Resources that stem from previous work and have worked as expected in Cloud systems. In general, objectives can be QoS parameters, but there is no consensus on how to abstract or normalize such metrics.

3.2. Holistic Perspective

The four key aspects discussed in this article are presented in an integrated view in Figure 1.

The Markov Blanket concept provides a nested representation of the system, which allows us to represent the highest-level state of the system (Resources, Quality, and Cost) as a wrapping blanket over the lower-level states. Interestingly, maintaining the Markov Blanket **representation** across abstraction layers enables the usage of the same techniques, regardless of the abstraction layer. Further, SLOs can leverage this structure of blankets to add constraints on the infrastructure at different layers, covering the entire Distributed Computing Continuum System. Hence, they can be understood as the hooks to the infrastructure, allowing the building of a modular and adaptive system's **model**. Combining these SLOs enables the building of the DeepSLO construct. Hence, by means of causal inference, we can define and predict how the relations within a DeepSLO will behave. It will be possible to tailor the adaptive measures to the predictions through the study of interventional situations. Further, the counterfactual capacity of causal inference can generate hypothetical situations that can be input for a lifelong learning framework. Within this framework, **lifelong learning** can be achieved by using methodologies such as active inference from the FEP that will be biased toward actions that can improve the system's knowledge of its environment. Finally, understanding and trustworthiness are required to build a **business model** for Distributed Computing Continuum Systems; hence, both state variables, such as Resources, Quality, and Cost, as well as a secure environment, are key to that development.

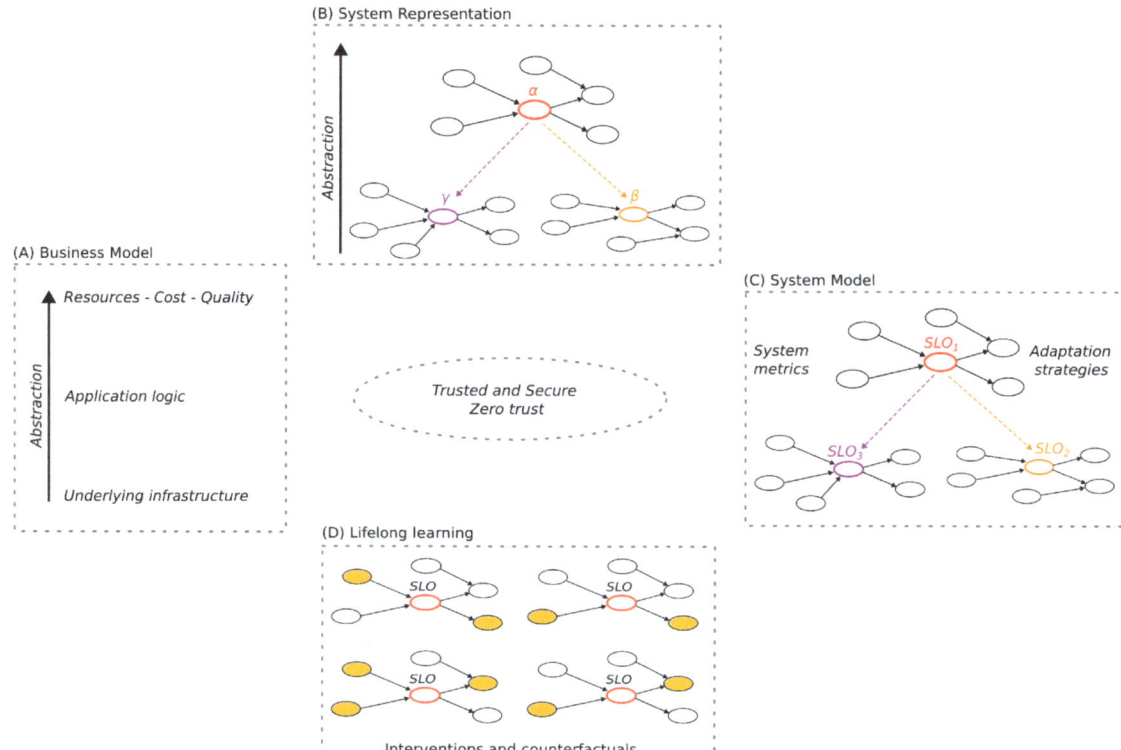

Figure 1. Starting at the center of the figure, we see that all developments require a trusted and secure environment, where Zero Trust techniques will have great relevance due to the heterogeneity and distribution characteristics of the system. On the left (**A**), three different abstractions for a Distributed Computing Continuum System are depicted (the granularity shown can be refined by showing network characteristics or unfolding aspects from the application logic). Each abstraction can easily engage a kind of stakeholder. However, the highest level: Resources-Cost-Quality, aims at being understood by all of them. On top (**B**), there is the representation of the system through Markov Blankets, considering α as a high-level abstraction variable that we can assess by observing the surrounding node, i.e., those factors that affect its value, remarkably thanks to the Markov Blanket approach only those that are relevant are there. Further, α can be decomposed on other lower-level variables (β and γ), providing a nested structure that can cover the entire system. On the right (**C**), we take advantage of this representation to embed SLOs, more specifically, a DeepSLO, in order to model the behavior of the system. As an example, SLO_1 can be related to the quality of a machine-learning inference component, which has its own metrics and adaptation strategies. This is linked to two other lower-level SLOs; in this example, SLO_2 can control the input data by making sure that it has the expected resolution, while SLO_3 controls the required time to perform the inference. Having this decomposition enables a fine-grained capacity for adaptation; other lower-level or higher-level SLOs could be developed if needed to have a broader view of a larger component or a more fine-grained control over a smaller one. On the bottom (**D**), several cases are shown with forced values for metrics and/or elastic strategies. This way, causal inference can improve the systems model to deal with future situations that are not yet known, providing a basis to deal with unexpected or unforeseen scenarios.

4. Conclusions

This article has presented a novel discussion of fundamental aspects that require agreement among the community to enable the development of Distributed Computing

Continuum Systems. This work aims at deviating from the mainstream topics on Distributed Computing Continuum, which usually tackles one of these topics by solving a specific issue, by focusing on the system's representation, its model, its lifelong learning framework, and finally, its business model. This research presents some techniques and concepts that can help address the issues of each of the discussed characteristics, and we compare them with some of the current trends for Distributed Computing Continuum Systems. Further, we show some relations among the presented techniques that can use their synergies to enable future Distributed Computing Continuum Systems.

Author Contributions: All authors contributed equally to conceptualization, investigation and providing resources; writing—original draft preparation and visualization, V.C.P.; writing—review and editing, all authors. All authors have read and agreed to the published version of the manuscript.

Funding: This research received no external funding.

Data Availability Statement: No data was generated in the context of this article.

Conflicts of Interest: The authors declare no conflict of interest.

References

1. Beckman, P.; Dongarra, J.; Ferrier, N.; Fox, G.; Moore, T.; Reed, D.; Beck, M. Harnessing the computing continuum for programming our world. In *Fog Computing*; Zomaya, A., Abbas, A., Khan, S., Eds.; John Wiley & Sons, Ltd.: Hoboken, NJ, USA, 2020; pp. 215–230. [CrossRef]
2. Morichetta, A.; Casamayor Pujol, V.; Dustdar, S. A roadmap on learning and reasoning for distributed computing continuum ecosystems. In Proceedings of the IEEE International Conference on Edge Computing (EDGE), Chicago, IL, USA, 5–10 September 2021; Institute of Electrical and Electronics Engineers: New York, NY, USA, 2021; pp. 25–31. [CrossRef]
3. Costa, B.; Bachiega, J., Jr.; de Carvalho, L.R.; Araujo, A.P. Orchestration in Fog Computing: A Comprehensive Survey. *ACM Comput. Surv. (CSUR)* **2022**, *55*, 1–34. [CrossRef]
4. Dustdar, S.; Casamayor Pujol, V.; Donta, P.K. On distributed computing continuum systems. *IEEE Trans. Knowl. Data Eng.* **2023**, *35*, 4092–4105. [CrossRef]
5. Yu, S.; Chen, X.; Zhou, Z.; Gong, X.; Wu, D. When Deep Reinforcement Learning Meets Federated Learning: Intelligent Multitimescale Resource Management for Multiaccess Edge Computing in 5G Ultradense Network. *IEEE Internet Things J.* **2021**, *8*, 2238–2251. [CrossRef]
6. Xia, X.; Chen, F.; He, Q.; Grundy, J.C.; Abdelrazek, M.; Jin, H. Cost-Effective App Data Distribution in Edge Computing. *IEEE Trans. Parallel Distrib. Syst.* **2021**, *32*, 31–44. [CrossRef]
7. Ullah, A.; Dagdeviren, H.; Ariyattu, R.C.; DesLauriers, J.; Kiss, T.; Bowden, J. MiCADO-Edge: Towards an Application-level Orchestrator for the Cloud-to-Edge Computing Continuum. *J. Grid Comput.* **2021**, *19*, 47. [CrossRef]
8. Hastbacka, D.; Halme, J.; Barna, L.; Hoikka, H.; Pettinen, H.; Larranaga, M.; Bjorkbom, M.; Mesia, H.; Jaatinen, A.; Elo, M. Dynamic Edge and Cloud Service Integration for Industrial IoT and Production Monitoring Applications of Industrial Cyber-Physical Systems. *IEEE Trans. Ind. Inform.* **2022**, *18*, 498–508. [CrossRef]
9. Pusztai, T.; Nastic, S.; Morichetta, A.; Casamayor Pujol, V.; Dustdar, S.; Ding, X.; Vij, D.; Xiong, Y. A Novel Middleware for Efficiently Implementing Complex Cloud-Native SLOs. In Proceedings of the 2021 IEEE 14th International Conference on Cloud Computing (CLOUD), Chicago, IL, USA, 5–10 September 2021.
10. Gheibi, O.; Weyns, D. Lifelong self-adaptation: Self-adaptation meets lifelong machine learning. In Proceedings of the SEAMS'22—17th Symposium on Software Engineering for Adaptive and Self-Managing Systems, Pittsburgh, PA, USA, 22–24 May 2022; Association for Computing Machinery: New York, NY, USA, 2022; pp. 1–12. [CrossRef]
11. Kirchhoff, M.; Parr, T.; Palacios, E.; Friston, K.; Kiverstein, J. The Markov blankets of life: Autonomy, active inference and the free energy principle. *J. R. Soc. Interface* **2018**, *15*, 20170792. [CrossRef]
12. Pearl, J. *Probabilistic Reasoning in Intelligent Systems: Networks of Plausible Inference*; Morgan Kaufmann Publishers Inc.: San Francisco, CA, USA, 1988.
13. Forti, S.; Bisicchia, G.; Brogi, A. Declarative continuous reasoning in the cloud-IoT continuum. *J. Log. Comput.* **2022**, *32*, 206–232. [CrossRef]
14. Rihan, M.; Elwekeil, M.; Yang, Y.; Huang, L.; Xu, C.; Selim, M.M. Deep-VFog: When Artificial Intelligence Meets Fog Computing in V2X. *IEEE Syst. J.* **2021**, *15*, 3492–3505. [CrossRef]
15. Esfahani, N.; Malek, S. Uncertainty in self-adaptive software systems. In *Lecture Notes in Computer Science (including subseries Lecture Notes in Artificial Intelligence and Lecture Notes in Bioinformatics)*; Springer: Berlin/Heidelberg, Germany, 2013; Volume 7475 LNCS, pp. 214–238. ISBN: 9783642358128. [CrossRef]
16. Pearl, J.; Mackenzie, D. *The Book of Why: The New Science of Cause and Effect*, 1st ed.; Basic Books, Inc.: New York, NY, USA, 2018.

17. Casamayor Pujol, V.; Raith, P.; Dustdar, S. Towards a new paradigm for managing computing continuum applications. In Proceedings of the IEEE 3rd International Conference on Cognitive Machine Intelligence, CogMI 2021, Virtual, 13–15 December 2021; Institute of Electrical and Electronics Engineers Inc.: New York, NY, USA, 2021; pp. 180–188. [CrossRef]
18. Friston, K.; Kilner, J.; Harrison, L. A free energy principle for the brain. *J. Physiol. Paris* **2006**, *100*, 70–87. [CrossRef]
19. Palacios, E.R.; Razi, A.; Parr, T.; Kirchhoff, M.; Friston, K. On Markov blankets and hierarchical self-organisation. *J. Theor. Biol.* **2020**, *486*, 110089. [CrossRef]
20. Smith, R.; Friston, K.J.; Whyte, C.J. A step-by-step tutorial on active inference and its application to empirical data. *J. Math. Psychol.* **2022**, *107*, 102632. [CrossRef]
21. Dustdar, S.; Guo, Y.; Satzger, B.; Truong, H.L. Principles of elastic processes. *IEEE Internet Comput.* **2011**, *15*, 66–71. [CrossRef]
22. Nastic, S.; Morichetta, A.; Pusztai, T.; Dustdar, S.; Ding, X.; Vij, D.; Xiong, Y. SLOC: Service level objectives for next generation cloud computing. *IEEE Internet Comput.* **2020**, *24*, 39–50. [CrossRef]
23. Stafford, V. Zero trust architecture. *NIST Spec. Publ.* **2020**, *800*, 207.
24. Truong, H.L.; Dustdar, S.; Leymann, F. Towards the Realization of Multi-dimensional Elasticity for Distributed Cloud Systems. *Procedia Comput. Sci.* **2016**, *97*, 14–23. [CrossRef]
25. Rose, S.; Borchert, O.; Mitchell, S.; Connelly, S. *Zero Trust Architecture*; Technical Report; National Institute of Standards and Technology: Gaithersburg, MD, USA, 2020.
26. LiWang, M.; Gao, Z.; Hosseinalipour, S.; Dai, H. Multi-Task Offloading over Vehicular Clouds under Graph-based Representation. In Proceedings of the ICC 2020—2020 IEEE International Conference on Communications (ICC), Virtual, 7–11 June 2020; pp. 1–7. [CrossRef]
27. Zhao, Z.; Verma, G.; Rao, C.; Swami, A.; Segarra, S. Distributed scheduling using graph neural networks. In Proceedings of the ICASSP, IEEE International Conference on Acoustics, Speech and Signal Processing—Proceedings, Toronto, ON, Canada, 6–11 June 2021; Institute of Electrical and Electronics Engineers Inc.: New York, NY, USA, 2021; pp. 4720–4724. [CrossRef]
28. Yu, Z.; Hu, J.; Min, G.; Wang, Z.; Miao, W.; Li, S. Privacy-Preserving Federated Deep Learning for Cooperative Hierarchical Caching in Fog Computing. *IEEE Internet Things J.* **2022**, *9*, 22246–22255. [CrossRef]
29. Zhang, K.; Cao, J.; Zhang, Y. Adaptive Digital Twin and Multi-agent Deep Reinforcement Learning for Vehicular Edge Computing and Networks. *IEEE Trans. Ind. Inform.* **2022**, *18*, 1405–1413. [CrossRef]
30. Sheng, S.; Chen, P.; Chen, Z.; Wu, L.; Yao, Y. Deep Reinforcement Learning-Based Task Scheduling in IoT Edge Computing. *Sensors* **2021**, *21*, 1666. [CrossRef]
31. Xia, X.; Chen, F.; He, Q.; Cui, G.; Grundy, J.C.; Abdelrazek, M.; Xu, X.; Jin, H. Data, User and Power Allocations for Caching in Multi-Access Edge Computing. *IEEE Trans. Parallel Distrib. Syst.* **2022**, *33*, 1144–1155. [CrossRef]
32. Tadakamalla, V.; Menasce, D. Autonomic Elasticity Control for Multi-server Queues under Generic Workload Surges in Cloud Environments. *IEEE Trans. Cloud Comput.* **2020**, *10*, 984–995. [CrossRef]
33. Guo, S.; Wu, D.; Zhang, H.; Yuan, D. Queueing Network Model and Average Delay Analysis for Mobile Edge Computing. In Proceedings of the 2018 International Conference on Computing, Networking and Communications, ICNC 2018, Maui, HI, USA, 5–8 March 2018; pp. 172–176. ISBN: 9781538636527. [CrossRef]
34. Villari, M.; Fazio, M.; Dustdar, S.; Rana, O.; Ranjan, R. Osmotic Computing: A New Paradigm for Edge/Cloud Integration. *IEEE Cloud Comput.* **2016**, *3*, 76–83. [CrossRef]
35. Gamal, I.; Abdel-Galil, H.; Ghalwash, A. Osmotic Message-Oriented Middleware for Internet of Things. *Computers* **2022**, *11*, 56. [CrossRef]
36. Camara, J.; Muccini, H.; Vaidhyanathan, K. Quantitative verification-aided machine learning: A tandem approach for architecting self-adaptive IoT systems. In Proceedings of the IEEE 17th International Conference on Software Architecture, ICSA 2020, Salvador, Brazil, 16–20 March 2020; pp. 11–22. ISBN: 9781728146591. [CrossRef]
37. Thrun, S. Lifelong Learning Algorithms. In *Learning to Learn*; Thrun, S., Pratt, L., Eds.; Springer US: Boston, MA, USA, 1998; pp. 181–209. [CrossRef]
38. Yang, F.; Yang, C.; Liu, H.; Sun, F. Evaluations of the Gap between Supervised and Reinforcement Lifelong Learning on Robotic Manipulation Tasks. In Proceedings of the 5th Conference on Robot Learning. PMLR, London, UK, 8–11 November 2022; pp. 547–556.
39. Febrinanto, F.G.; Xia, F.; Moore, K.; Thapa, C.; Aggarwal, C. Graph Lifelong Learning: A Survey. *IEEE Comput. Intell. Mag.* **2023**, *18*, 32–51. [CrossRef]
40. Aguilera, M.; Millidge, B.; Tschantz, A.; Buckley, C.L. How particular is the physics of the free energy principle? *Phys. Life Rev.* **2021**, *40*, 24–50. [CrossRef]
41. Raja, V.; Valluri, D.; Baggs, E.; Chemero, A.; Anderson, M.L. The Markov blanket trick: On the scope of the free energy principle and active inference. *Phys. Life Rev.* **2021**, *39*, 49–72. [CrossRef] [PubMed]
42. Da Costa, L.; Parr, T.; Sajid, N.; Veselic, S.; Neacsu, V.; Friston, K. Active inference on discrete state-spaces: A synthesis. *J. Math. Psychol.* **2020**, *99*, 102447. [CrossRef]

Disclaimer/Publisher's Note: The statements, opinions and data contained in all publications are solely those of the individual author(s) and contributor(s) and not of MDPI and/or the editor(s). MDPI and/or the editor(s) disclaim responsibility for any injury to people or property resulting from any ideas, methods, instructions or products referred to in the content.

Article

Comparative Analysis of Membership Inference Attacks in Federated and Centralized Learning [†]

Ali Abbasi Tadi [1,*], Saroj Dayal [1], Dima Alhadidi [1] and Noman Mohammed [2]

1. School of Computer Science, University of Windsor, Windsor, ON N9B 3P4, Canada; sdayal@uwindsor.ca (S.D.); dima.alhadidi@uwindsor.ca (D.A.)
2. Department of Computer Science, University of Manitoba, Winnipeg, MB R3T 2N2, Canada; noman.mohammed@umanitoba.ca
* Correspondence: abbasit@uwindsor.ca
† This paper is an extended version of our paper published in International Database Engineered Applications Symposium Conference, Heraklion, Crete, Greece, 5–7 May 2023. Entitled 'Comparative Analysis of Membership Inference Attacks in Federated Learning'.

Abstract: The vulnerability of machine learning models to membership inference attacks, which aim to determine whether a specific record belongs to the training dataset, is explored in this paper. Federated learning allows multiple parties to independently train a model without sharing or centralizing their data, offering privacy advantages. However, when private datasets are used in federated learning and model access is granted, the risk of membership inference attacks emerges, potentially compromising sensitive data. To address this, effective defenses in a federated learning environment must be developed without compromising the utility of the target model. This study empirically investigates and compares membership inference attack methodologies in both federated and centralized learning environments, utilizing diverse optimizers and assessing attacks with and without defenses on image and tabular datasets. The findings demonstrate that a combination of knowledge distillation and conventional mitigation techniques (such as Gaussian dropout, Gaussian noise, and activity regularization) significantly mitigates the risk of information leakage in both federated and centralized settings.

Keywords: federated learning; membership inference attack; privacy; machine learning

Citation: Abbasi Tadi, A.; Dayal, S.; Alhadidi, D.; Mohammed, N. Comparative Analysis of Membership Inference Attacks in Federated and Centralized Learning. *Information* 2023, 14, 620. https://doi.org/10.3390/info14110620

Academic Editor: Peter Revesz

Received: 30 September 2023
Revised: 17 November 2023
Accepted: 18 November 2023
Published: 19 November 2023

Copyright: © 2023 by the authors. Licensee MDPI, Basel, Switzerland. This article is an open access article distributed under the terms and conditions of the Creative Commons Attribution (CC BY) license (https://creativecommons.org/licenses/by/4.0/).

1. Introduction

Machine learning (ML) is gaining popularity thanks to the increasing availability of extensive datasets and technological advancements [1,2]. Centralized learning (CL) techniques become impractical in the context of abundant private data as they mandate transmitting and processing data through a central server. Google's federated learning (FL) has emerged as a distributed machine learning paradigm since its inception in 2017 [3]. In FL, a central server supports participants in the training model by exchanging trained models or gradients of training data without revealing raw or sensitive information either to the central server or other participants. The application of FL is crucial, particularly in processing sensitive and personal data, such as in healthcare, where ML is increasingly prevalent, especially in compliance with GDPR [4] and HIPAA [5] regulations. Despite its advancements, FL is susceptible to membership inference attacks (MIA), a method employed to gain insights into training data. Although FL primarily aims for privacy protection, attackers can infer specific data by intercepting FL updates transmitted between training parties and the central server [6,7]. For instance, if an attacker is aware that patient data are part of the model's training set, they could deduce the patient's current health status [8]. Prior research has explored membership inference attacks (MIA) in a centralized environment where data are owned by a single data owner. It is imperative to extend this investigation to MIA in federated learning (FL). This article undertakes an analysis of various MIA techniques initially proposed in the centralized learning (CL) environment [9–11]. The examination encompasses their applicability in the FL environment and evaluates the effectiveness of countermeasures to mitigate these attacks in both FL and CL environments. An earlier version of this work has already been published [12], focusing solely on MIA in the FL environment. In that iteration, we scrutinized nine mitigation techniques [9,10,13–19] against MIA attacks and showed that knowledge distillation [19] performs better in reducing the attack recall while keeping accuracy as high as possible. We also conducted some experiments to observe the effects of three various optimizers, Stochastic Gradient Descent (SGD) [20], Root Mean Squared Propagation (RMSProp) [21], and Adaptive Gradient (Adagrad) [22], in deep learning on MIA recall and FL model accuracy. We found no difference between these optimizers on MIA recall. In this paper, we investigated two more optimizers and three more countermeasures in both CL and FL environments, and we compared the results. To the best of our knowledge, this study is the first comprehensive study that investigates the MIA in both CL and FL environments and applies twelve mitigation techniques against MIA with five various optimizers for the target model. Our contributions in this paper are summarized below.

- We conducted a comprehensive study of the effectiveness of the membership inference attack in the FL and CL environments considering different attack techniques, optimizers, datasets, and countermeasures. Existing related work focuses on the CL environment and the effectiveness of one single countermeasure. In this paper, we investigated the FL environment, compared it with the CL environment, and studied the effectiveness of combining two mitigation techniques together.
- We compared the effectiveness of four well-known membership inference attacks [9–11] in the CL and FL environments considering different mitigation techniques: dropout [16], Monte Carlo dropout [13], batch normalization [14], Gaussian noise [23], Gaussian dropout [16], activity regularization [24], masking [17], and knowledge distillation [19].
- We compared the accuracy of models in the CL and the FL environments using five optimizers: SGD, RMSProp, Adagrad, incorporation of Nesterov momentum into Adam (Nadam) [25], and Adaptive Learning Rate method (Adadelta) [26] using four real datasets, MNIST [27], Fashion-MNIST (FMNIST) [28], CIFAR-10 [29], and Purchase [30]. We found that using the Adadelta optimizer alone, for image datasets, can mitigate the MIA significantly while preserving the accuracy of the model.
- We established a trade-off relationship between model accuracy and attack recall. Our investigation revealed that employing knowledge distillation in conjunction with either Gaussian noise, Gaussian dropout, or activity regularization yields the most favorable balance between model accuracy and attack recall across both image and tabular datasets.

The remainder of this article is organized as follows. In the Section 2, we presented the related work. In Section 3, we explained the different attacks on a model for membership inference. Countermeasures are detailed in Section 4. The setup and the results of the experiments are described and analyzed in Section 5. Finally, we conclude our article in Section 6.

2. Related Work

This section summarizes the related work focusing on the MIA in CL and FL (Table 1).

Table 1. Related work summary.

Authors	CL or FL	Attack	Defense
Shokri et al. [9]	CL	✓	✓
Salem et al. [10]	CL	✓	✓
Nasr et al. [31]	CL, FL	✓	×
Liu et al. [11]	CL	✓	×
Carlini et al. [2]	CL	✓	×
Conti et al. [32]	CL	✓	✓
Zheng et al. [33]	CL	×	✓
Shejwalkar et al. [34]	CL	×	✓
Lee et al. [35]	FL	×	✓
Su et al. [36]	FL	×	✓
Xie et al. [37]	FL	×	✓

2.1. MIA against CL

Shokri et al. [9] performed the first MIA on ML models to identify the presence of a data sample in the training set of the ML model with black-box access. Shokri et al. [9] created a target model, shadow models, and attack models, and they made two main assumptions. First, the attacker must create multiple shadow models, each with the same structure as the target model. Second, the dataset used to train shadow models comes from the same distribution as the target model's training data. Subsequently, Salem et al. [10] widened the scope of the MIA of Shokri et al. [9]. They showed that the MIA is possible without having any prior assumption of the target model dataset or having multiple shadow models. Nasr et al. [31] showed that more reasonable attack scenarios are possible in both FL and CL environments. They designed a white-box attack on the target model in FL and CL by assuming different adversary prior knowledge. Lan Liu et al. [11] studied perturbations in feature space and found that the sensitivity of trained data to a fully trained machine learning model is lower than that of untrained data. Lan Liu et al. [11] calculated sensitivity by comparing the sensitivity values of different data samples using a Jacobian matrix, which measures the relationship between the target's predictions and the feature value of the target sample.

Numerous attacks in the existing literature draw inspiration from Shokri's research [9]. Carlini et al. [2] introduced a novel attack called the Likelihood Ratio Attack (LiRA), which amalgamates concepts from various research papers. They advocate for a shift in the evaluation metric for MIA by recommending the use of the true positive rate (recall) while maintaining a very low false alarm rate. Their findings reveal that, when measured by recall, many attacks prove to be less effective than previously believed. In our study, we adopt the use of recall, rather than accuracy, as the measure of MIA attack effectiveness.

2.2. MIA against FL

Nasr et al. [31] showed that MIA seriously compromises the privacy of FL participants even when the universal model achieves high prediction accuracy. A common defense against such attacks is the differential privacy (DP) [38] approach, which manipulates each update with some random noise. However, it suffers from a significant loss of FL classification accuracy. Bai et al. [39] proposed a homomorphic-cryptography-based privacy enhancement mechanism impacting MIA. They used homomorphic cryptography to encrypt the collaborators' parameters and added a parameter selection method to the FL system aggregator to select specific participant updates with a given probability. Another FL MIA defense technique is the digestive neural network (DNN) [35], which modifies inputs and skews updates, maximizing FL classification accuracy and minimizing inference attack accuracy. Wang et al. [36] proposed a new privacy mechanism called the Federated Regularization Learning Model to prevent information leakage in FL. Xie et al. [37] proposed an adversarial noise generation method that was added to the attack features of the attack model on MIA against FL.

3. Attack Techniques for Membership Inference

In this section, we summarize the different methods of MIA [9–11] that we applied in this paper. The summary of the considered membership attacks is shown in Table 2. We employed four well-known attacks in this paper, and each of them has its own characteristics.

Table 2. Comparison of the considered attacks.

Attack Type	Shadow Model		Target's Model Training Data Distribution	Prediction Sensitivity
	No. Shadow Models	Target Model Structure		
Attack 1 [9]	10	✓	✓	-
Attack 2 [10]	1	-	✓	-
Attack 3 [10]	1	-	-	-
Attack 4 [11]	-	-	-	✓

3.1. Shokri et al.'s MIA

MIA can be formulated [40] as follows:

$$M_{Attack}(K_{M_{Target}}(x,y)) \rightarrow 0,1 \quad (1)$$

Given a data sample (x, y) and additional knowledge $K_{M_{Target}}$ about the target model M_{Target}, the attacker typically tries to create an attack model M_{Attack} to eventually return either 0 or 1, where 0 indicates the sample is not a member of the training set and 1 indicates the sample is a member of the training set. The additional knowledge can be the distribution of the target data and the type of the target model. Figure 1 summarizes the general idea of the first MIA on ML models proposed by Shokri et al. [9].

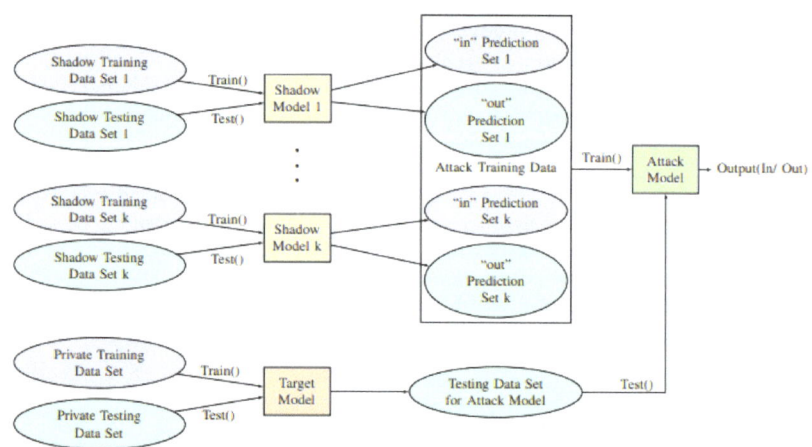

Figure 1. Overview of MIA on ML models [9].

The target model takes a data sample as input and generates the probability prediction vector after training. Suppose D_{Target}^{Train} is the private training dataset of the target model M_{Target}, where (x_i, y_i) are the labeled data records. In this labeled dataset, (x_i) represents the input to the target model, while (y_i) represents the class label of x_i in the set $1, 2, ..., C_{Target}$. The output of the target model M_{Target} is a vector of probabilities of size C_{Target}, where the elements range from 0 to 1 and they sum to 1. Multiple shadow models are created by the attacker to mimic the behavior of the target model and to generate the data needed to train the attack model. The attacker creates several (n) shadow models $M_{Shadow}^i()$, where each shadow model i is trained on the dataset D_{Shadow}^i. The attacker first splits its dataset D_{Shadow}^i into two sets, D_{Shadow}^{iTrain} and D_{Shadow}^{iTest}, such that $D_{Shadow}^{iTrain} \cap D_{Shadow}^{iTrial} = \phi$. Then, the attacker trains each shadow model M_{Shadow}^i with the training set D_{Shadow}^{iTrain} and tests the same model with D_{Shadow}^{iTest} test dataset. The attack model is a collection of models, one for each output class of target data. D_{Attack}^{Train} is the attack model's training dataset, which contains labeled data records (x_i, y_i) and the probability vector generated by the shadow model for each data sample x_i. The label for x_i in the attack model is either "in" if x_i is used to train the shadow model or "out" if x_i is used to test the shadow model. This attack is named Attack 1 in our experiments.

3.2. Salem et al.'s MIA

Early demonstrations by Shokri et al. [9] on the feasibility of MIA are based on many assumptions, e.g. the use of multiple shadow models, knowledge of the structure of the target model, and the availability of a dataset from the same distribution as the training data of the target model. Salem et al [10] diminished all these key assumptions, showing that the MIA is generally applicable at low cost and carries greater risk than previously thought [10]. They provided two MIA attacks: I) with the knowledge of dataset distribution, model architecture, and only one shadow model, and II) with no knowledge about dataset distribution and model architecture. The former attack is named Attack 2 and the latter one is named Attack 3 in Table 2.

3.3. Prediction Sensitivity MIA

The idea behind this attack is that training data from a fully trained ML model generally have lower prediction sensitivity than untrained data (i.e., test data). The overview of this attack [11] is shown in Figure 2. The only allowed interaction between the attacker and the target model M is to query M with a sample x and then obtain the prediction result. The target model M maps the n-dimensional vector $x \in \mathbb{R}^n$ to the output m-dimensional $y \in \mathbb{R}^m$. The Jacobian matrix of M is a matrix $m \times n$ whose element in the ith row and jth column is $J_{ij} = \frac{\partial y_i}{\partial x_j} (i \in [1, 2, \ldots, n]$ and $j \in [1, 2, \ldots, m])$:

$$J(x; M) = \begin{bmatrix} \frac{\partial M(x)}{\partial x_1} & \cdots & \frac{\partial M(x)}{\partial x_n} \end{bmatrix} = \begin{bmatrix} \frac{\partial y_1}{\partial x_1} & \cdots & \frac{\partial y_1}{\partial x_n} \\ \vdots & \ddots & \vdots \\ \frac{\partial y_m}{\partial x_1} & \cdots & \frac{\partial y_m}{\partial x_n} \end{bmatrix} \quad (2)$$

where $y = M(x)$. The input sample is $x = [x_1, x_2, \ldots, x_n]$, and the corresponding prediction is $y = [y_1, y_2, \ldots, y_m]$. $\frac{\partial y_i}{\partial x_j}$ is the relationship between the change in the input record's i-th feature value and the change in the prediction probability that this sample belongs to j-th class.

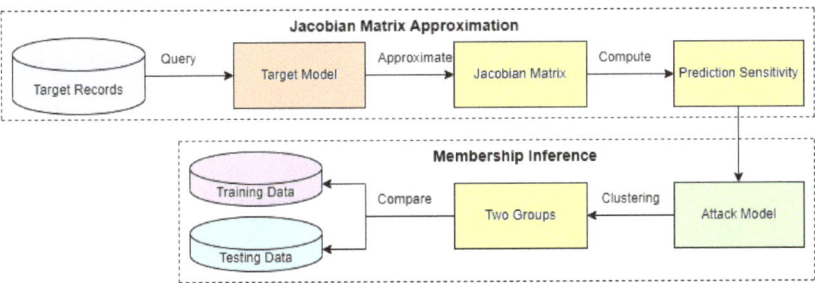

Figure 2. Overview of MIA using Jacobian matrix and prediction sensitivity [11].

The Jacobian matrix comprises a series of first-order partial derivatives. The derivatives can be approximated by calculating the numerical differentiation with the following equation:

$$\frac{\partial y_j}{\partial x_i} \approx \frac{M(x + \epsilon) - M(x - \epsilon)}{2\epsilon}, \quad (3)$$

where ϵ is a small value added to the input sample's i-th feature value. Add ϵ to the i-th feature value of the target sample x_t, whose membership property to know provides two modified samples to query the target model and derive the partial derivatives of the i-th feature for the target model: $\frac{\partial M(x)}{\partial x_i} = \left[\frac{\partial y_1}{\partial x_i}, \frac{\partial y_2}{\partial x_i} \ldots, \frac{\partial y_m}{\partial x_i}\right]$. Then, for each feature in x, this process is repeated to combine the partial derivatives into the Jacobian matrix. For simplicity, the approximation of the Jacobian matrix is defined as $J(x; M)$. The L-2 norm of $J(x; M)$ represents the prediction sensitivity for the target sample, as described by Novak et al. [41]. For a $m \times n$ matrix A, the L-2 norm of A can be computed as follows:

$$||A||_2 = (\sum_{i=1}^{m} \sum_{j=1}^{n} |a_{ij}|^2)^{\frac{1}{2}} \qquad (4)$$

where i and j are the row and column number of the matrix element a_{ij}, respectively. There is a difference in prediction sensitivity between samples from the training set and samples from the testing set. Once prediction sensitivity is calculated, an unsupervised clustering method (like k-means) partitions a set of target records (prediction sensitivity values) into two subsets. The cluster with the lowest mean sensitivity compared to the members of the M's training set is chosen. Then, during the inference stage, the samples are clustered into three or more groups and ordered by an average norm. Finally, the groups with lower average norms are predicted from the target model's training set, whereas others are not.

4. Defense Mechanisms

Attackers take advantage of the fact that ML models behave differently during the prediction with new data than with training data to differentiate members from nonmembers. This property is associated with the degree of overfitting, which is measured by the generalization gap. The generalization gap is the difference between the accuracy of the model between training and testing time. When overfitting is high, the model is more vulnerable to MIA. Therefore, whatever method is used to reduce overfitting is also profitable for MIA reduction. We applied the following methods to see how they mitigate the MIA.

- **Dropout (D):** It prevents overfitting by randomly deleting units in the neural network and allows for an approximately efficient combination of many different neural network architectures [16]. This was suggested by Salem et al. [10] and implemented as an MIA mitigation technique in ML models in a centralized framework.
- **Monte Carlo Dropout (MCD):** It is proposed by Gal et al. [13]. It captures the uncertainty of the model. Various networks (where several neurons have been randomly disabled) can be visualized as Monte Carlo samples from the space of all available models. This provides a mathematical basis for the model to infer its uncertainty, often improving its performance. This work allows dropout to be applied to the neural network during model inference [42]. Therefore, instead of making one prediction, multiple predictions are made, one for each model (already prepared with random disabled neurons), and their distributions are averaged. Then, the average is considered as the final prediction.
- **Batch Normalization (BN):** This is a technique that improves accuracy by normalizing activations in the middle layers of deep neural networks [14]. Normalization is used as a defense in label-only MIA, and the results show that both regularization and normalization can slightly decrease the average accuracy of the attack [32].
- **Gaussian Noise (GN):** This is the most practical perturbation-based model for describing the nonlinear effects caused by additive Gaussian noise [23]. GN is used to ignore adversarial attacks [15].
- **Gaussian Dropout (GD):** It is the integration of Gaussian noise with the random probability of nodes. Unlike standard dropout, nodes are not entirely deleted. Instead of ignoring neurons, they are subject to Gaussian noise. From Srivatsava's experiments [16], it appears that using the Gaussian dropout reduced computation time

because the weights did not have to be scaled each time to match the skipped nodes, as in the standard dropout.
- **Activity Regularization (AR)**: It is a technique used to encourage the model to have specific properties regarding the activations (outputs) of neurons in the network during training. The purpose of activity regularization is to prevent overfitting and encourage certain desirable characteristics in the network's behavior. The L1 regularizer and the L2 regularizer are two regularization techniques [24]. L1 regularization penalizes the sum of the absolute values of the weights, while L2 regularization penalizes the sum of the squares of the weights. Shokri et al. [9] used a conventional L2 regularizer as a defense technique to overcome MIA in ML neural network models.
- **Masking (M)**: It tells the sequence processing layers that some steps are missing from the input and should be ignored during data processing [17]. If all input tensor values in that timestep are equal to the mask value, the timestep is masked (ignored) in all subsequent layers of that timestep.
- **Differential Privacy (DP)**: Differentially Private Stochastic Gradient Descent (DPSGD) is a differentially private version of the Stochastic Gradient Descent (SGD) algorithm that happens during model training [18] and incorporates gradient updates with some additive Gaussian noise to provide differential privacy. DP [43–45] is a solid standard to ensure the privacy of distributed datasets.
- **Knowledge Distillation (KD)**: It distills and transfers knowledge from one deep neural network (DNN) to another DNN [19,46]. According to many MIA mitigation articles, KD outperforms the cutting edge approaches [33,34] in terms of MIA mitigation, while other FL articles support that it facilitates effective communication [47–49] to maintain the heterogeneity of the collaborating parties.
- **Combination of KD with AR (AR–KD)**: In our early experiments [12], we noticed that, in most test cases, KD lowers the recall while preserving the model accuracy. In this work, we are combining AR as a mitigation technique with KD. To the best of our knowledge, this is the first work that combines AR and KD and evaluates its results both in CL and FL.
- **Combination of KD with GN (GN–KD)**: Like AR, we are also combining GN and KD to see how they affect the attack recall and model accuracy. This paper is also the first paper that combines GN and KD and evaluates the performance of this combination in CL and FL environments.
- **Combination of KD with GD (GD–KD)**: We also combine KD and GD to see their effects on attack recall and model accuracy using five various optimizers on image datasets. To our knowledge, there is no work that combines these two methods to evaluate how they behave against MIA. Therefore, this is the first paper that combines these methods and analyses them in both CL and FL environments.

5. Performance Analysis

In this section, a summary of the experimental setup and results is provided. We performed our experiments on a 2.30 GHz 12th Gen Intel(R) Core(TM) i7-12700H processor with 16.00 GB RAM on the x64-based Windows 11 OS. We used open-source frameworks and standard libraries, such as Keras and Tensorflow in Python. The code of this work is available at [50].

5.1. Experimental Setup

In the following, we detail the experimental setup.

5.1.1. Datasets

The datasets of our experiments are CIFAR-10 [29], MNIST [27], FMNIST [28], and Purchase [30]. These datasets are the benchmark to validate the MIA, and they are the same as those used in recent related work [51]. CIFAR-10, MNIST, and FMNIST are image datasets in which, by normalizing, we fit the image pixel data in the range [0,1], which helps to train the model more accurately. Purchase is a tabular dataset that has 600 dimensions and 100 labels. We used one-hot encoding of this dataset to be able to feed it into the neural network [51]. Each dataset is split into 30,000 for training and 10,000 for testing. For training in the FL environment, the training dataset is uniformly divided between three FL participants to train the local models based on the FedAvg [3] algorithm separately and update the central server to reach a global optimal model.

5.1.2. Model Architecture

The models are based on the Keras sequential function and a linear stack of neural network layers. In these models, we first defined the flattened input layer, followed by three dense layers. The MNIST and FMNIST input sizes are 28×28, while the CIFAR-10 input sizes are 32×32. The Purchase dataset input size is considered 600 since it has 600 features. We added all countermeasure layers in between the dense layers. As knowledge distillation is an architectural mitigation technique, we ran a separate experiment to see its performance. We specified an output size of 10 as the labels for each class in the MNIST, FMNIST, and CIFAR-10 datasets range between 0 and 9. Also, we set the output size of 100 for the Purchase dataset as the labels for this dataset range between 0 and 99. In addition, we set the activation function for the output layer to softmax to make the outputs sum to 1.

5.1.3. Training Setup

For training, we used SGD, RMSProp, Adagrad, Nadam, and Adadelta optimizers, with a learning rate equal to 0.01. The loss function for all the optimizers is the categorical cross-entropy. We have a batch size of 32 and epochs of 10 for each participant during training. We reproduced the FL process, including local participant training and FedAvg aggregation. The scheme of data flow is illustrated in Figure 3.

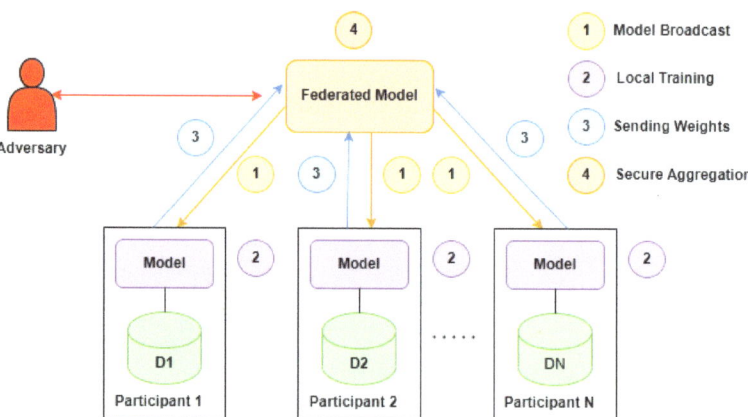

Figure 3. Overview of the FL system.

5.1.4. Evaluation Metrics

We focus on test accuracy as an evaluation metric for the FL model and recall as an evaluation metric for successful attacks in the FL setting. The recall (true positive rate) represents the fraction of the members of the training dataset that are correctly inferred as members by the attacker.

5.1.5. Comparison Methods

We investigated the performance of four attacks, as mentioned in Table 2. Attack 1 employs multiple shadow models mimicking both the structure and the data distribution of the target model. Attack 2 applies a single shadow model. The structure of the model is different. However, the training data distribution imitates the target model. Unlike Attack 1 and Attack 2, in Attack 3, both the structure of the model and the training data distribution differ from the target model. Finally, Attack 4 applies the Jacobian matrix paradigm, which brings us an entirely different membership inference attack using the target model.

5.2. Experimental Results

In this section, we compared FL and CL. We also experimentally analyzed the effect of the MIA and the effect of the mitigation techniques in both environments, considering image and tabular datasets.

5.2.1. CL vs. FL

Many studies thoroughly compared the CL and FL approaches [52,53]. FL is concluded as a network-efficient alternative to CL [54]. In our comparison of the two approaches, as shown in Figures 4–7, CL outperformed FL regarding accuracy in most cases, which is expected. In Figure 7, the accuracy in CL is considerably lower than the accuracy in FL for GN, GD, and AR. This is justified by the nature of the tabular dataset, which seems to be overfitted using Adadelta and Nadam optimizers in the CL environment, and overfitting is removed when we apply these optimizers in the FL environment. The accuracy values are also tabulated in Tables 3 and 4 for CL and FL environments, respectively. In all figures and tables, the WC is the value for the model accuracy (or attack recall) without having any countermeasure included in the model. Figures 8–11 illustrate attack recall in our experiments. An interesting aspect to note is related to the Adadelta optimizer in image datasets. If we examine Adadelta's performance in image datasets in Figures 4–6, we can observe that there is minimal loss in accuracy when using this optimizer. However, our experiments depicted in Figures 8–10 indicate that, even when we do not implement any countermeasure (WC) to mitigate membership inference attacks (MIA), Adadelta is capable of functioning as a countermeasure without significantly compromising utility. It is evident that utilizing Adadelta alone results in a substantial reduction in the recall of the MIA attack. However, for tabular datasets, Adadelta is not performing significantly differently from other optimizers, as shown in Figure 11. In all the tables in this paper, the value in parentheses shows the difference between that countermeasure and its corresponding value in the without countermeasure (WC) column. WC shows the values when we do not use any countermeasure.

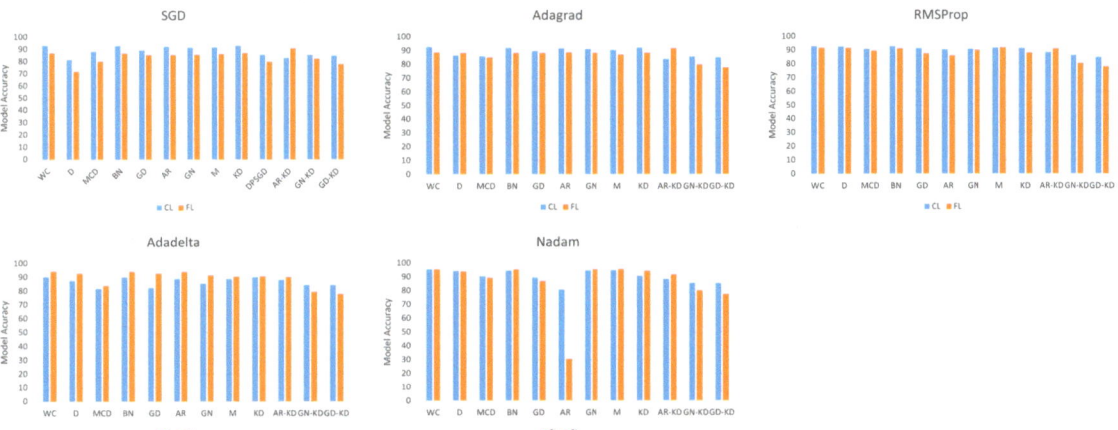

Figure 4. Comparison of model accuracy of CL and FL using various optimizers and countermeasures—MNIST dataset.

Table 3. CL model accuracy.

Datasets	Optimizers	WC	D	MCD	BN	GD	AR	GN	M	KD	DP	AR-KD	GN-KD	GD-KD
MNIST	SGD	93.1	81.6(−11.5)	88.3(−4.8)	92.8(−0.3)	89.1(−4)	92.1(−1)	91.4(−1.7)	91.6(−1.5)	**92.8(−0.3)**	85.5(−7.6)	82.8(−10.3)	85.2(−7.9)	84.8(−8.3)
	Adagrad	92.6	86.5(−6.1)	85.7(−6.9)	**92.1(−0.5)**	89.6(−3)	91.6(−1)	91.1(−1.5)	90.3(−2.3)	92(−0.6)	-	83.8(−8.8)	85.6(−7)	84.9(−7.7)
	RMSProp	92.8	92.5(−0.3)	90.9(−1.9)	**92.7(−0.1)**	91.3(−1.5)	90.4(−2.4)	90.7(−2.1)	91.5(−1.3)	91.2(−1.6)	-	88.3(−4.5)	86.1(−6.7)	84.8(−8)
	Nadam	**95.5**	94.1(−1.4)	90.3(−5.2)	94.4(−1.1)	89.4(−6.1)	80.7(−14.8)	94.5(−1)	**94.7(−0.8)**	90.6(−4.9)	-	88.3(−7.2)	85.4(−10.1)	85.2(−10.3)
	Adadelta	90.2	87.5(−2.7)	81.6(−8.6)	**90.1(−0.1)**	82.3(−7.9)	88.8(−1.4)	85.4(−4.8)	88.8(−1.4)	90.1(−0.1)	-	88(−2.2)	84.3(−5.9)	84.3(−5.9)
FMNIST	SGD	**88.6**	84.7(−3.9)	84.1(−4.5)	86.4(−2.2)	85.2(−3.3)	87.3(−1.3)	85.8(−2.8)	**88.1(−0.5)**	86.9(−1.7)	84.2(−4.4)	82.2(−6.4)	74.7(−13.9)	73.2(−15.4)
	Adagrad	85.8	78.6(−7.2)	75.9(−9.9)	84.6(−1.2)	81.9(−3.9)	83.8(−2)	79.1(−6.7)	**88.7(+2.9)**	82.3(−3.5)	-	83.1(−2.7)	75.3(−10.5)	73.8(−12)
	RMSProp	83.6	79.5(−4.1)	78.5(−5.1)	82.9(−0.7)	78.7(−4.9)	82.6(−1)	81.7(−1.9)	**83.1(−0.5)**	81.9(−1.7)	-	81.5(−2.1)	75.5(−8.1)	73.2(−10.4)
	Nadam	85.9	82.5(−3.4)	78.8(−7.1)	**83.4(−2.5)**	74.4(−11.5)	66.2(−19.7)	83(−2.9)	83.2(−2.7)	82.5(−3.4)	-	82.4(−3.5)	75.6(−10.3)	72.6(−13.3)
	Adadelta	83.8	77.6(−6.2)	75.1(−8.7)	**82.7(−1.1)**	73(−10.8)	77.4(−6.4)	77.8(−6)	81.8(−2)	82.3(−1.5)	-	82(−1.8)	74.2(−9.6)	71.3(−12.5)
CIFAR-10	SGD	85.7	74.2(−11.5)	74.9(−10.8)	82.8(−2.9)	83.1(−2.6)	**84.3(−1.4)**	78.6(−7.1)	82.5(−3.2)	83.9(−1.8)	74.6(−11.1)	82.6(−3.1)	75.1(−10.6)	74.2(−11.5)
	Adagrad	**88.4**	75.7(−12.7)	72.6(−15.8)	85.8(−2.6)	83.6(−4.8)	**87.2(−1.2)**	81.9(−6.5)	82.5(−5.9)	85.3(−3.1)	-	83.9(−4.5)	76.3(−12.1)	75.1(−13.3)
	RMSProp	86.3	81.6(−4.7)	76.4(−9.9)	**85.7(−0.6)**	84.9(−1.4)	84.2(−2.1)	82.1(−4.2)	83.6(−2.7)	81.5(−4.8)	-	80.5(−5.8)	77(−9.3)	75.9(−10.4)
	Nadam	75.3	73.4(−1.9)	73.6(−1.7)	**81.6(6.3)**	80.6(5.3)	72.3(7)	78.2(2.9)	80.8(5.5)	81.1(5.8)	-	79.8(4.5)	76.2(0.9)	75.6(0.3)
	Adadelta	78.2	71.2(−7)	70.5(−7.7)	83.1(4.9)	82.1(3.9)	75.5(7.3)	79.4(1.2)	81.3(3.1)	**84.5(6.3)**	-	82.1(3.9)	75.9(−2.3)	73.8(−4.4)
Purchase	SGD	79.3	72(−7.3)	79.2(−0.1)	70.5(−8.8)	57(−22.3)	3.8(−75.5)	76.8(−2.5)	79.8(0.5)	79.2(−0.1)	70.3(−9)	**82.6(3.3)**	75.1(−4.2)	74.2(−5.1)
	Adagrad	82.6	76.2(−6.4)	83.1(0.5)	69.2(−13.4)	64.7(−17.9)	4.4(−78.2)	80.7(−1.9)	82.9(0.3)	78.8(−3.8)	-	**83.9(1.3)**	76.3(−6.3)	75.1(−7.5)
	RMSProp	56.4	24.1(−32.3)	51.5(−4.9)	67.4(11)	8.5(−47.9)	5.2(−51.2)	51.7(−4.7)	52.6(−3.8)	77.8(21.4)	-	**80.5(24.1)**	77(20.6)	75.9(19.5)
	Nadam	65.1	41.4(−23.7)	66.6(1.5)	68.9(3.8)	13.2(−51.9)	8.2(−56.9)	60.9(−4.2)	67.5(2.4)	79.4(14.3)	-	79.8(14.7)	76.2(11.1)	75.6(10.5)
	Adadelta	28.1	15.5(−12.6)	29.1(1)	24.6(−3.5)	3.1(−25)	2.5(−25.6)	14.3(−13.8)	29.1(1)	80.3(52.2)	-	**82.1(54)**	75.9(47.8)	73.8(45.7)

Table 4. FL model accuracy.

Datasets	Optimizers	WC	D	MCD	BN	GD	AR	GN	M	KD	DP	AR-KD	GN-KD	GD-KD
MNIST	SGD	87	72(−1.5)	80(−7)	86.7(−0.3)	85.5(−1.5)	85.4(−1.6)	85.6(−1.4)	86.2(−0.8)	86.9(−0.1)	79.9(−7.1)	90.9(3.9)	82.3(−4.7)	77.9(−9.1)
	Adagrad	88.7	88.2(−0.5)	84.9(−3.8)	88.3(−0.4)	88.1(−0.6)	88.5(−0.2)	88.3(−0.4)	87(−1.7)	88.2(−0.5)	-	91.6(2.9)	79.7(−9)	77.5(−11.2)
	RMSProp	91.7	91.6(−0.1)	89.5(−2.2)	91.1(−0.6)	87.4(−4.3)	86(−5.7)	90(−1.7)	91.7(0)	87.9(−3.8)	-	90.9(−0.8)	80.4(−11.3)	78(−13.7)
	Nadam	95.5	93.9(−1.6)	89.3(−6.2)	95.3(−0.2)	86.7(−8.8)	30.4(−65.1)	95.3(−0.2)	95.4(−0.1)	94.1(−1.4)	-	91.5(−4)	80.1(−15.4)	77.4(−18.1)
	Adadelta	94.3	92.8(−1.5)	83.9(−10.4)	94.1(−0.2)	92.8(−1.5)	94(−0.3)	91.6(−2.7)	90.5(−3.8)	90.7(−3.6)	-	90.2(−4.1)	79.3(−15)	77.9(−16.4)
FMNIST	SGD	81.3	79.8(−1.5)	76.4(−4.9)	81(−0.3)	75.7(−5.6)	80.5(−0.8)	74.9(−6.4)	80.3(−1)	80.9(−0.4)	77.6(−3.7)	83.5(2.2)	73.3(−8)	69.4(−11.9)
	Adagrad	82.6	82(−0.6)	78.9(−3.7)	81.6(−1)	79.9(−2.7)	82.2(−0.4)	78.4(−4.2)	80.6(−2)	80.7(−1.9)	-	83.5(0.9)	73.2(−9.4)	69.6(−13)
	RMSProp	91.7	76.8(−14.9)	74.4(−17.3)	71(−20.7)	72.3(−19.4)	68.2(−23.5)	55(−36.7)	76.1(−15.6)	75.8(−15.9)	-	83.1(−8.6)	73.3(−18.4)	69.6(−22.1)
	Nadam	84.6	82.2(−2.4)	78.6(−6)	83(−1.6)	69.3(−15.3)	52.5(−32.1)	73.8(−10.8)	83.2(−1.4)	83.9(−0.7)	-	83.3(−1.3)	73.3(−11.3)	69(−15.6)
	Adadelta	84.1	81.1(−3)	75.8(−8.3)	81.1(−3)	72(−12.1)	81.6(−2.5)	75.4(−8.7)	82.8(−1.3)	83(−1.1)	-	83(−1.1)	72.8(−11.3)	67.8(−16.3)
CIFAR-10	SGD	79.5	73(−6.5)	65.9(−13.6)	79.3(−0.2)	73.2(−6.3)	74.6(−4.9)	74.9(−4.6)	73.1(−6.4)	79.3(−0.2)	75.7(−3.8)	82.8(3.3)	75.3(−4.2)	68.5(−11)
	Adagrad	76.3	67(−9.3)	54(−22.3)	76.2(−0.1)	72.9(−3.4)	71.4(−4.9)	71.1(−5.2)	69.9(−6.4)	75.7(−0.6)	-	79.7(3.4)	70.2(−6.1)	66.7(−9.6)
	RMSProp	72.8	61.2(−11.6)	53.6(−19.2)	72.4(−0.4)	70.9(−1.9)	72.2(−0.6)	71(−1.8)	68.6(−4.2)	72.1(−0.7)	-	71.3(−1.5)	64.8(−8)	61.1(−11.7)
	Nadam	80.3	75.6(−4.7)	70.9(−9.4)	79.8(−0.5)	72.3(−8)	74.6(−5.7)	75.2(−5.1)	76.8(−3.5)	77.6(−2.7)	-	79.4(−0.9)	70.8(−9.5)	70.3(−10)
	Adadelta	78.6	77.8(−4.8)	72.6(−10)	78.2(−4.4)	70.1(−12.5)	71.3(−11.3)	77.8(−4.8)	75.9(−6.7)	76.3(−6.3)	-	81.5(−1.1)	65.8(−16.8)	66.8(−15.8)
Purchase	SGD	78.9	77.8(−1.1)	78.5(−0.4)	78.3(−0.6)	79(0.1)	79.6(0.7)	78.3(−0.6)	79.3(0.4)	78.8(−0.1)	76.5(−2.4)	78(−0.9)	75.5(−3.4)	43.9(−35)
	Adagrad	81.3	80.2(−1.1)	80.4(−0.9)	80.4(−0.9)	80(−1.3)	81.4(0.1)	80.2(−1.1)	80.1(−1.2)	77.6(−3.7)	-	79(−2.3)	77.4(−3.9)	43.8(−37.5)
	RMSProp	21.6	20.6(−1)	19.4(−2.2)	22.7(1.1)	23.8(2.2)	21.8(0.2)	23(1.4)	23.9(2.3)	76.6(55)	-	78.8(57.2)	76.9(55.3)	46.8(25.2)
	Nadam	30.1	28.7(−1.4)	32(1.9)	29.9(−0.2)	29.3(−0.8)	28.6(−1.5)	27.9(−2.2)	25.5(−4.6)	79.3(49.2)	-	78.8(48.7)	74.1(44)	45.3(15.2)
	Adadelta	30.8	29.6(−1.2)	32.8(2)	34.4(3.6)	33.8(3)	32.2(1.4)	32.6(1.8)	31.4(0.6)	78.6(47.8)	-	77.8(47)	77(46.2)	42.9(12.1)

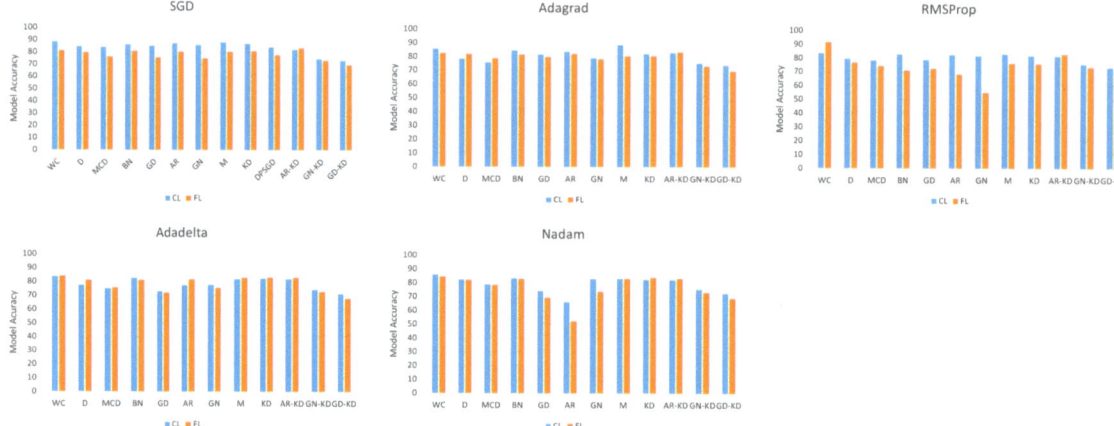

Figure 5. Comparison of model accuracy of CL and FL using various optimizers and countermeasures—FMNIST dataset.

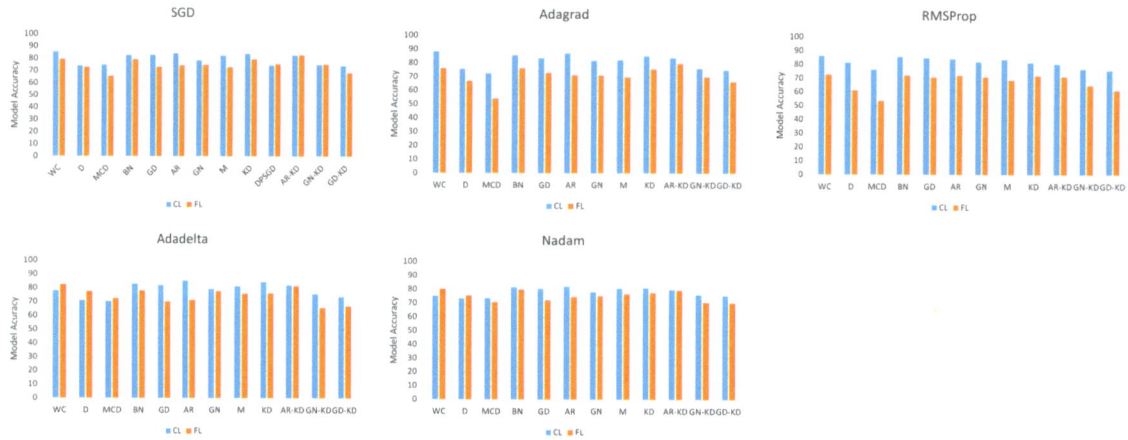

Figure 6. Comparison of model accuracy of CL and FL using various optimizers and countermeasures—CIFAR-10 dataset.

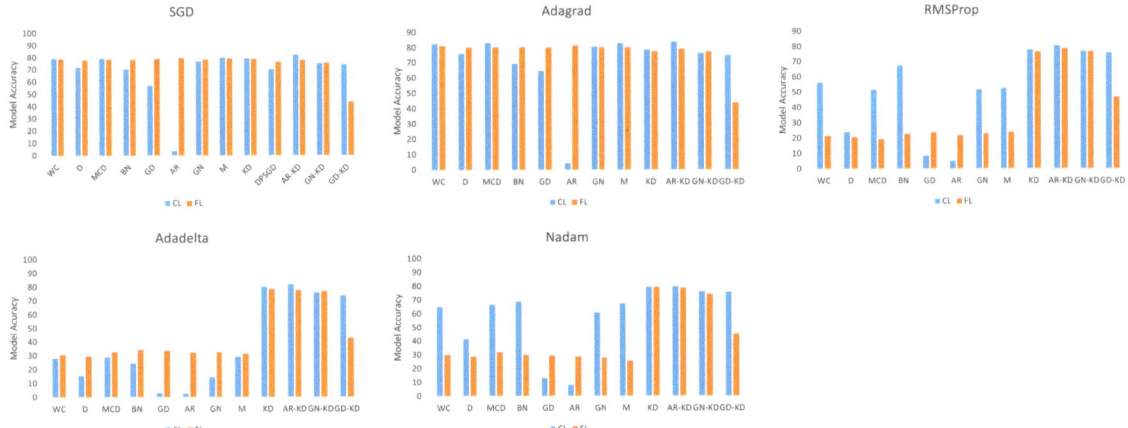

Figure 7. Comparison of model accuracy of CL and FL using various optimizers and countermeasures—Purchase dataset.

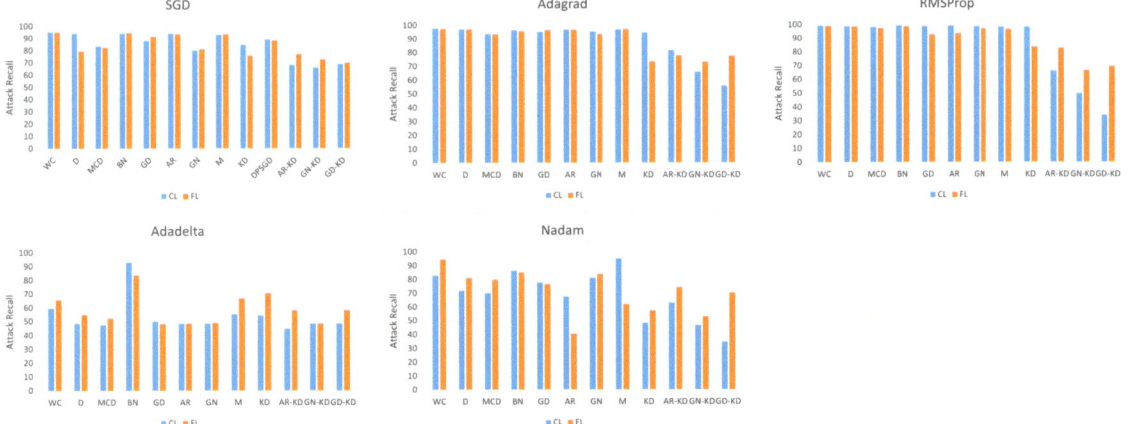

Figure 8. Comparison of Attack 1 recall on CL and FL using various optimizers and countermeasures—MNIST dataset.

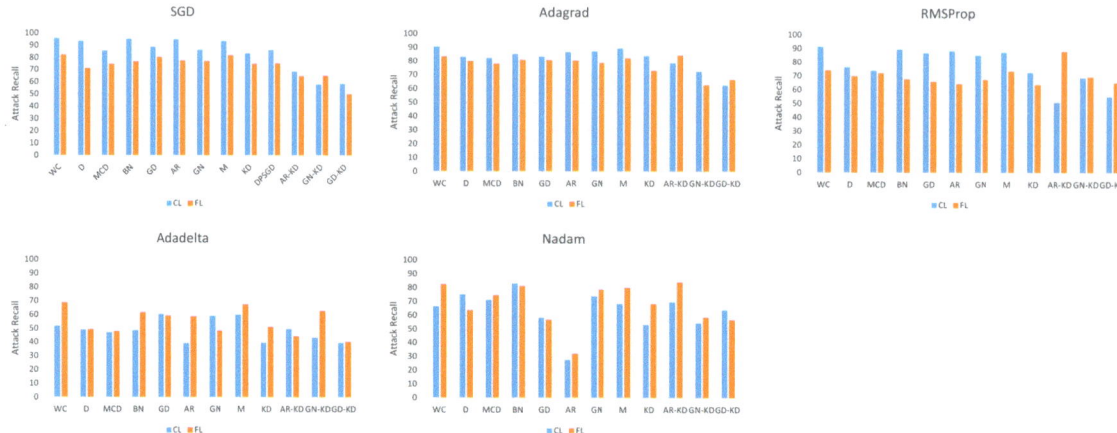

Figure 9. Comparison of Attack 1 Recall on CL and FL using various optimizers and countermeasures—FMNIST dataset.

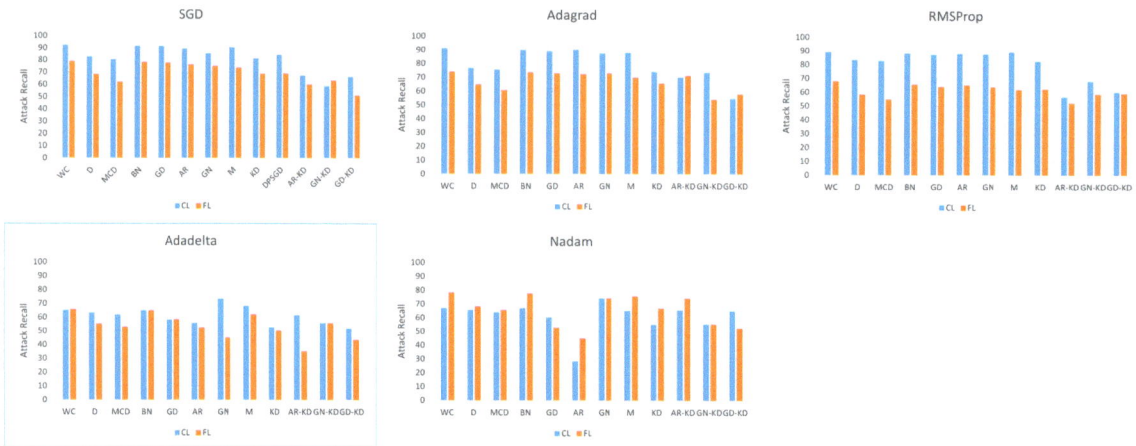

Figure 10. Comparison of Attack 1 recall on CL and FL using various optimizers and countermeasures—CIFAR-10 dataset.

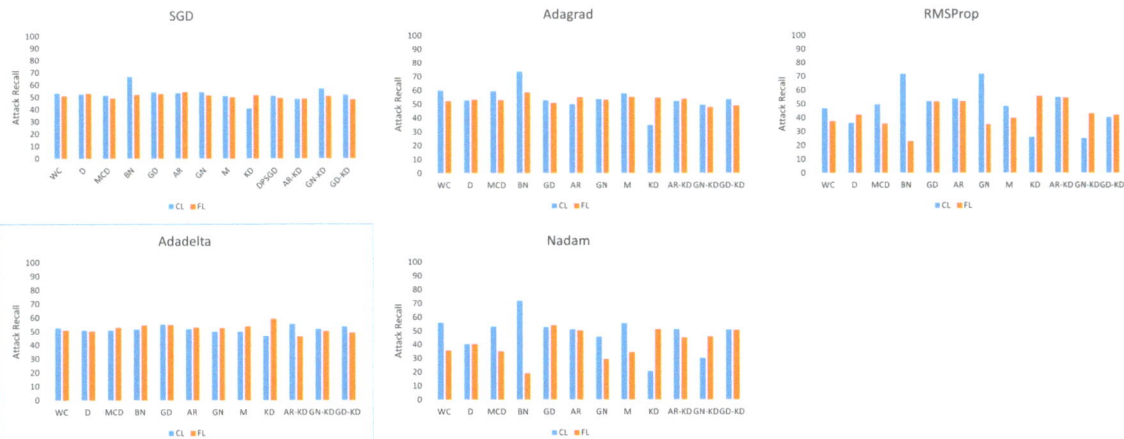

Figure 11. Comparison of Attack 1 recall on CL and FL using various optimizers and countermeasures—Purchase dataset.

Generally, the recall of Attack 1 is almost the same, if not less, in FL compared to the recall in CL considering different mitigation techniques. Figure 12 illustrates five various optimizers' effects as well as various countermeasures' effects on the FL model accuracy, where the y-axis provides the test accuracy of the FL model. As DP-SGD is specialized for SGD optimizer, we applied DP only on SGD optimizer and not with other optimizers. The first group in all the plots is WC, which represents the baseline without countermeasures. We have provided the full details of our experiments in CL and FL environments in Tables 3 and 4, respectively.

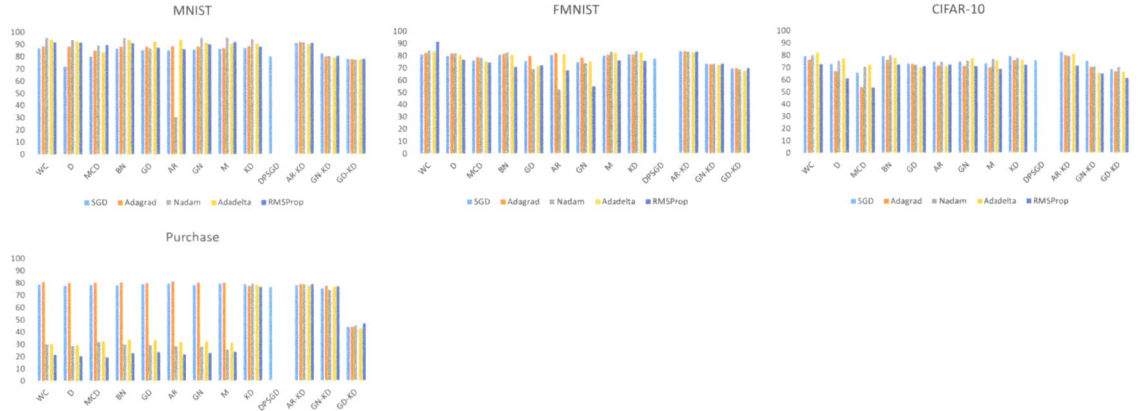

Figure 12. A comparison of FL model accuracy with five various optimizers, with and without countermeasures—MNIST, FMNIST, CIFAR-10, and Purchase datasets.

- **CL model accuracy without countermeasure:** As per Table 3, the highest CL model accuracy results for Nadam, SGD, Adagrad, and Adagrad on the MNIST, FMNIST, CIFAR-10, and Purchase datasets, respectively. In contrast, Nadam on the CIFAR-10, Adadelta on MNIST, FMNIST, and Purchase yield the lowest accuracy. Generally speaking, depending on the dataset, the optimizer, and the batch size used in each round of training, the values for the model accuracy change.

- **CL model accuracy with countermeasures**: As per Table 4, the combination that yields the highest CL model accuracy for MNIST after applying countermeasures belongs to Nadam with M. When we apply M as the countermeasure and Nadam as the optimizer, the accuracy of the model slightly decreases compared to the case when we use no countermeasure (WC). Subsequently, this is followed by an increase in the attack recall when using Nadam with M, as per Table 5. In general, Nadam with M slightly decreases model accuracy and significantly increases attack recall for the MNIST dataset, while Adadelta with MCD provides the lowest model accuracy. For the FMNIST dataset, when we use Adagrad with M, we have even higher accuracy than no countermeasure. However, the attack recall is subsequently high, as shown in Table 5. In CIFAR-10, AR and BN hold the highest accuracy, while MCD has the lowest accuracy. In the Purchase dataset, AR–KD yields the highest accuracy for all optimizers, even better than without countermeasures. This happens while attack recall in the Purchase dataset, as per Table 5, is reduced for SGD and Adagrad.
- **FL model accuracy without countermeasure**: As shown in Table 4, the highest FL model accuracy belongs to Nadam, RMSProp, Adadelta, and Adagrad on the MNIST, FMNIST, CIFAR-10, and Purchase datasets, respectively, whereas RMSProp on the CIFAR-10 and Purchase datasets as well as SGD on MNIST and FMNIST yield the lowest accuracy. In general, FL model accuracy is the lowest for Purchase and the highest for MNIST. This is justified by the nature of the datasets and the distribution of their features, which make each data record more distinguishable from the others. The reason why some optimizers are performing very well for specific datasets in the CL environment and not performing well for the same dataset in the FL environment is that these optimizers are sensitive to the FedAvg algorithm, where we average the total weights that are computed locally by the clients to generate the global model.
- **FL model accuracy with countermeasures**: As per Table 4, BN has no significant effect on the CIFAR-10 model accuracy. For CIFAR-10, the highest accuracy belongs to AR–KD when using the SGD optimizer, and the lowest accuracy belongs to MCD when using the Adagrad optimizer. For MNIST and FMNIST, the countermeasure that maintains the maximum accuracy varies between different optimizers. For instance, in FMNIST, the mitigation technique that keeps the model accuracy at its maximum value is AR–KD for four optimizers: SGD, Adagrad, RMSProp, and Adadelta. Also, for Nadam, KD yields the highest accuracy in the FL environment. For FMNIST, the lowest accuracy belongs to AR when using the Nadam optimizer. For MNIST, the best accuracy goes for AR–KD when using SGD and Adagrad, whereas M provides the highest accuracy in RMSProp and Nadam. Also, BN provides the highest accuracy when using Adadelta. The lowest accuracy for MNIST belongs to GD–KD when using the Nadam optimizer. The highest accuracy for the Purchase dataset belongs to AR when using SGD and Adagrad, as well as KD when using Nadam and Adadelta.

Table 5. CL attack recall.

Datasets	Optimizers	Attacks	WC	D	MCD	BN	GD	AR	GN	M	KD	DP	AR-KD	GN-KD	GD-KD
MNIST	SGD	Attack-1	95.40	94.2(−1.2)	83.4(−12)	94.4(−1)	87.9(−7.5)	94(−1.4)	79.8(−15.6)	93.1(−2.3)	84.6(−10.8)	89.2(−6.2)	68.1(−27.3)	65.8(−29.6)	68.6(−26.8)
		Attack-2	94.90	93.7(−1.2)	83.2(−11.7)	94.8(−0.1)	86.8(−8.1)	93.6(−1.3)	83.2(−1.7)	93.9(−1)	82.4(−12.5)	88.3(−6.6)	65.4(−29.5)	63.2(−31.7)	65.2(−29.7)
		Attack-3	90.70	85.3(−5.4)	74.5(−16.2)	88.7(−2)	82.9(−7.8)	83.2(−7.5)	88.6(−2.1)	89.3(−1.4)	80.5(−10.2)	82.4(−8.3)	63.7(−27)	60.4(−30.3)	63.4(−27.3)
		Attack-4	87.10	32(−55.1)	34.6(−52.5)	28.7(−58.4)	24.5(−62.2)	35.2(−51.8)	37.4(−49.7)	24.6(−62.5)	22.6(−64.5)	26.8(−60.3)	24.7(−62.4)	32.5(−54.6)	28(−59.1)
	Adagrad	Attack-1	97.80	97.2(−0.6)	93.8(−4)	96.6(−1.2)	95.2(−2.6)	96.9(−0.9)	95.4(−2.4)	96.9(−0.9)	94.6(−3.2)		81.6(−16.2)	65.9(−31.9)	55.9(−41.9)
		Attack-2	97.70	92.4(−5.3)	93.5(−4.2)	95.7(−2)	95.1(−2.6)	96.4(−1.3)	93.6(−4.1)	95.9(−1.8)	76.1(−21.6)		66.2(−31.5)	64.1(−33.6)	55.3(−42.4)
		Attack-3	91.30	87.4(−3.9)	76.6(−14.7)	89.5(−1.8)	86.2(−5.1)	86.3(−5)	81.3(−10)	85.4(−5.9)	74.9(−16.4)		63.4(−27.9)	58.9(−32.4)	52.1(−39.2)
		Attack-4	86.90	33.6(−53.3)	32.1(−54.8)	35.2(−51.7)	34.1(−52.8)	39.7(−47.2)	31.8(−55.1)	35.3(−51.6)	31.4(−55.5)		20(−66.9)	96(9.1)	84(−2.9)
	RMSProp	Attack-1	99.40	98.9(−0.5)	98.3(−1.1)	99.3(−0.1)	98.8(−0.6)	99.1(−0.3)	98.6(−0.8)	98.4(−1)	98.2(−1.2)		66.2(−33.2)	49.7(−49.7)	34.2(−65.2)
		Attack-2	99.20	98.6(−0.6)	97.3(−1.9)	98.7(−0.5)	97.2(−2)	98.3(−0.9)	97.7(−1.5)	97(−2.2)	96.4(−2.8)		64.3(−34.9)	45.5(−53.7)	33.1(−66.1)
		Attack-3	98.60	96.3(−2.3)	94.1(−4.5)	98.2(−0.4)	93.1(−5.5)	95.6(−3)	94.3(−4.3)	94.8(−3.8)	92.5(−6.1)		58.9(−39.7)	43.8(−54.8)	32.4(−66.2)
		Attack-4	89.90	23(−66.9)	38.7(−51.2)	39.5(−50.4)	37.3(−52.6)	39.4(−50.4)	34.6(−55.3)	32.8(−57.1)	31.3(−58.6)		24(−65.9)	16(−73.9)	20(−69.9)
	Nadam	Attack-1	82.90	71.9(−11)	70.1(−12.8)	86.3(3.4)	77.8(−5.1)	67.7(−15.2)	81.1(−1.8)	95(12.1)	48.3(−34.6)		63.1(−19.8)	46.5(−36.4)	34.6(−48.3)
		Attack-2	80.70	70.3(−10.4)	68.5(−12.2)	85.6(4.9)	76.4(−4.3)	66.4(−14.3)	79.5(−1.2)	92.9(12.2)	45.5(−35.2)		61.2(−19.5)	44.6(−36.1)	32.1(−48.6)
		Attack-3	78.50	71.5(−7)	69.4(−9.1)	77.5(−1)	65.1(−13.4)	58.9(−19.6)	73.6(−4.9)	78.2(−0.3)	44.7(−33.8)		59.8(−18.7)	39.5(−39)	29.8(−48.7)
		Attack-4	80.30	69(−11.3)	67.9(−12.4)	78(−2.3)	32(−48.3)	28(−52.3)	84.4(4.1)	43.3(−37)	52(−28.3)		30(−50.3)	24.2(−56.1)	12(−68.3)
	Adadelta	Attack-1	59.70	48.4(−11.3)	47.3(−12.4)	93(33.3)	49.8(−9.9)	48.3(−11.4)	48.3(−11.4)	55.3(−4.4)	54.4(−5.3)		44.6(−15.1)	48.3(−11.4)	48.3(−11.4)
		Attack-2	58.80	47.3(−11.5)	46.7(−12.1)	91.8(33)	45.5(−13.3)	47.8(−11)	44.2(−14.6)	52.8(−6)	53.1(−5.7)		43.1(−15.7)	45.6(−13.2)	46.9(−11.9)
		Attack-3	56.50	43.6(−12.9)	46.8(−9.7)	92.2(35.7)	49.3(−7.2)	45.6(−10.9)	41.3(−15.2)	51.9(−4.6)	52.2(−4.3)		42.6(−13.9)	43.5(−13)	45.8(−10.7)
		Attack-4	44.00	36(−8)	34.8(−9.2)	76.2(32.2)	81(−36)	20(−24)	20(−24)	12(−32)	32.8(−11.2)		16(−28)	24(−20)	20(−24)
FMNIST	SGD	Attack-1	95.8	93.7(−2.1)	85.6(−10.2)	95.3(−0.5)	88.9(−6.9)	94.7(−1.1)	86.5(−9.3)	93.4(−2.4)	83.5(−12.3)	86.3(−9.5)	68.4(−27.4)	57.9(−37.9)	58.4(−37.4)
		Attack-2	93.6	86.4(−7.2)	83.2(−10.4)	93.1(−0.5)	85.9(−7.7)	92.8(−0.8)	86.1(−7.5)	92.5(−1.1)	82.9(−10.7)	85.7(−7.9)	65.2(−28.4)	55.6(−38)	56.2(−37.4)
		Attack-3	90.2	82.2(−8)	81.6(−8.6)	89.6(−0.6)	84.9(−5.3)	89.1(−1.1)	83.7(−6.5)	88.5(−1.7)	81.9(−8.3)	83.1(−7.1)	61.3(−28.9)	51.2(−39)	54.2(−36)
		Attack-4	82.1	27(−55.1)	25.8(−56.3)	33.6(−48.5)	38.4(−43.7)	42.8(−39.3)	29.5(−52.6)	37.5(−44.6)	21.7(−60.4)	27.9(−54.2)	35.9(−46.2)	22.8(−59.3)	8(−74.1)
	Adagrad	Attack-1	90.6	83.2(−7.4)	82.4(−8.2)	85.4(−5.2)	83.6(−7)	86.9(−3.7)	87.6(−3)	89.5(−1.1)	84.2(−6.2)		78.8(−11.8)	72.6(−18)	62.7(−27.9)
		Attack-2	87.2	82.6(−4.6)	81.6(−5.6)	84.3(−2.9)	81.9(−5.3)	85.3(−1.9)	86.8(−0.4)	85.7(−1.5)	81.8(−5.4)		76.5(−10.7)	68.5(−18.7)	58.6(−28.6)
		Attack-3	85.7	82.4(−3.3)	81.1(−4.6)	82.8(−2.9)	78.4(−7.3)	82.9(−2.8)	82.7(−3)	83.6(−2.1)	79.3(−6.4)		75.1(−10.6)	65.3(−20.4)	55.9(−29.8)
		Attack-4	80.9	46.2(−34.7)	26.9(−54)	36.7(−44.2)	26.4(−54.5)	35.4(−45.5)	32(−48.9)	43.2(−37.7)	31.2(−49.7)		88(7.1)	68(−12.9)	24(−56.9)
	RMSProp	Attack-1	91.5	76.4(−15.1)	73.8(−17.7)	89.4(−2.1)	86.9(−4.6)	88.4(−3.1)	85.2(−6.1)	87.5(−4)	72.6(−18.9)		51(−40.5)	68.7(−22.8)	55.1(−36.4)
		Attack-2	89.2	73.6(−15.6)	72.7(−16.5)	86.3(−2.9)	82.8(−6.4)	87.2(−2)	83.1(−6.1)	85.3(−3.9)	71.8(−17.4)		47.3(−41.9)	65.3(−23.9)	52.6(−36.6)
		Attack-3	85.3	71.9(−13.4)	70.6(−14.7)	84.3(−1)	81.4(−3.9)	82.8(−2.5)	80.1(−5.2)	81.6(−3.7)	69.3(−16)		45.9(−39.4)	58.8(−26.5)	50.8(−34.5)
		Attack-4	70.8	20.1(−50.7)	20.5(−50.3)	35.9(−34.9)	26.2(−44.6)	34.8(−36)	28.6(−42.2)	29.6(−41.2)	26.1(−44.7)		24(−46.8)	92(21.2)	28(−42.8)
	Nadam	Attack-1	66.5	75.2(8.7)	71.2(4.7)	83(16.5)	58.4(−8.1)	27.5(−39)	74.1(7.6)	68.4(1.9)	53.3(−13.2)		69.6(3.1)	54.5(−12)	64.1(−2.4)
		Attack-2	65.7	73.5(7.8)	70.4(4.7)	81.3(15.6)	55.3(−10.4)	25.6(−40.1)	73.1(7.4)	65.2(−0.5)	51.6(−14.1)		65.8(0.1)	52.6(−13.1)	63.2(−2.5)
		Attack-3	60.3	69.8(9.5)	65.9(5.6)	79.7(19.4)	51.9(−8.4)	26.6(−33.7)	70.4(10.1)	62.8(2.5)	49.8(−10.5)		64.2(3.9)	50.8(−9.5)	61.5(1.2)
		Attack-4	28	66(38)	58.7(30.7)	14(−14)	88(60)	12(−16)	92(64)	4(−24)	44(16)		76.2(48.2)	32(4)	84(56)
	Adadelta	Attack-1	51.8	49.1(−2.7)	47.2(−4.6)	48.8(−3)	60.5(8.7)	39.2(−12.6)	59.2(7.4)	60(8.2)	39.4(−12.4)		49.7(−2.1)	43.3(−8.5)	39.5(−12.3)
		Attack-2	50.7	47.1(−3.6)	45.3(−5.4)	50.3(−0.4)	59.6(8.9)	37.4(−13.3)	58.1(7.4)	58.1(7.4)	37.6(−13.1)		45.6(−5.1)	42.1(−8.6)	35.8(−14.9)
		Attack-3	49.6	46.8(−2.8)	44.4(−5.2)	49.5(−0.1)	55.3(5.7)	35.5(−14.1)	57.6(8)	57.3(7.7)	35.9(−13.7)		43.8(−5.8)	39.6(−10)	34.5(−15.1)
		Attack-4	92	88(−4)	78.3(−13.7)	92(0)	80(−12)	16(−76)	84(−8)	12(−80)	20(−72)		7(−85)	66.3(−25.7)	39.9(−52.1)

Table 5. Cont.

Datasets	Optimizers	Attacks	WC	D	MCD	BN	GD	AR	GN	M	KD	DP	AR-KD	GN-KD	GD-KD
CIFAR-10	SGD	Attack-1	92.6	82.8(−9.8)	80.5(−12.1)	91.7(−0.9)	91.5(−1.1)	89.4(−3.2)	85.3(−7.3)	90.4(−2.2)	81.3(−11.3)	84.2(−8.4)	67.3(−25.3)	58.6(−34)	66.3(−26.3)
		Attack-2	90.4	80.2(−10.2)	79.9(−10.5)	89.2(−1.2)	89.6(−0.8)	84.2(−6.2)	83.7(−6.7)	86.9(−3.5)	79.4(−11)	81.8(−8.6)	63.2(−27.2)	55.4(−35)	64.8(−25.6)
		Attack-3	84.7	77.9(−6.8)	75.1(−9.6)	82.8(−1.9)	82.6(−2.1)	81.7(−3)	82.3(−2.4)	82.8(−1.9)	75.2(−9.5)	76.4(−8.3)	62.1(−22.6)	53.7(−31)	60.1(−24.6)
		Attack-4	78.4	36.8(−41.6)	35.6(−42.8)	42.7(−35.7)	40.9(−37.5)	40.5(−37.9)	39.1(−39.3)	40.6(−37.8)	33.9(−44.5)	35.2(−43.2)	25.6(−52.8)	38.4(−40)	26(−52.4)
	Adagrad	Attack-1	91.3	76.8(−14.5)	75.7(−15.6)	90.2(−1.1)	89.4(−1.9)	90.4(−0.9)	87.6(−3.7)	88.1(−3.3)	74.2(−17.1)	-	70.2(−21.1)	73.8	54.8(−36.5)
		Attack-2	89.4	73.6(−15.8)	72.4(−17)	88.3(−1.1)	87.9(−1.5)	85.3(−4.1)	84.1(−5.3)	83.8(−5.6)	72.8(−16.6)	-	68.3(−21.1)	69.7(−19.7)	51.9(−37.5)
		Attack-3	83.6	68.9(−14.7)	65.7(−17.9)	83.1(−0.5)	80.7(−2.9)	81.6(−2.1)	81.6(−2)	82.4(−1.2)	66.9(−16.7)	-	65.4(−18.2)	68.3(−15.3)	50.1(−33.5)
		Attack-4	72.5	28.6(−43.9)	25.3(−47.2)	36.4(−36.1)	34.9(−37.6)	37.7(−34.8)	30.3(−42.2)	33.4(−39.1)	25.9(−46.6)	-	71.3(−1.2)	53.9(−18.6)	24(−48.5)
	RMSProp	Attack-1	89.3	83.6(−5.7)	82.9(−6.4)	88.4(−0.9)	87.3(−2)	88(−1.3)	87.6(−1.7)	89.1(−0.2)	82.5(−6.8)	-	56.8(−32.5)	68.2(−21.1)	60.3(−29)
		Attack-2	84.2	78.4(−5.8)	75.9(−8.3)	83.9(−0.3)	82.4(−1.8)	81.9(−2.3)	82.8(−1.4)	82.7(−1.5)	76.8(−7.4)	-	60.9(−23.3)	65.8(−18.4)	58.9(−25.3)
		Attack-3	81.6	74.9(−6.7)	72.9(−8.7)	80.6(−1)	78.5(−3.1)	77.3(−4.3)	78.2(−3.4)	80.1(−1.5)	71.4(−10.2)	-	61.2(−20.4)	67.3(−14.3)	54.6(−27)
		Attack-4	68.5	24.7(−43.8)	22.6(−45.9)	32.9(−35.6)	30.5(−38)	31.8(−36.7)	29.2(−39.3)	31.3(−37.2)	23.4(−45.1)	-	48.8(−19.7)	60.8(−7.7)	28.8(−39.7)
	Nadam	Attack-1	67.2	65.8(−1.4)	64.2(−3)	67.1(−0.1)	60.6(−6.6)	28.6(−38.6)	74.3(7.1)	65.3(−1.9)	55.3(−11.9)	-	65.8(−1.4)	55.6(−11.6)	65.1(−2.1)
		Attack-2	65.8	63.3(−2.5)	62.1(−3.7)	64.2(−1.6)	58.3(−7.5)	25.7(−40.1)	73.6(7.8)	61.4(−4.4)	51.2(−14.6)	-	62.3(−3.5)	54.3(−11.5)	63.8(−2)
		Attack-3	63.2	61.1(−2.1)	60.7(−2.5)	62.1(−1.1)	57.8(−5.4)	24.5(−38.7)	74.8(11.6)	63.1(−0.1)	50.8(−12.4)	-	60.9(−2.3)	52.8(−10.4)	60.9(−2.3)
		Attack-4	68.3	48.3(−20)	45.6(−22.7)	67.5(−0.8)	61.2(−7.1)	29.6(−38.7)	88.5(20.2)	72.1(3.8)	45.3(−23)	-	44.2(−24.1)	58.9(−9.4)	18(−50.3)
	Adadelta	Attack-1	65.3	63.5(−1.8)	61.9(−3.4)	65.1(−0.2)	58.3(−7)	55.8(−9.5)	73.5(8.2)	68.3(3)	52.8(−12.5)	-	61.7(−3.6)	55.8(−9.5)	51.8(−13.5)
		Attack-2	64.2	61.3(−2.9)	60.2(−4)	64.1(−0.1)	57.6(−6.6)	53.9(−10.3)	73.6(9.4)	65.1(0.9)	53.4(−10.8)	-	60.5(−3.7)	54.9(−9.3)	49.3(−14.9)
		Attack-3	63.1	59.8(−3.3)	57.8(−5.3)	62.3(−0.8)	55.8(−7.3)	52.1(−11)	74.5(11.4)	66.3(3.2)	51.1(−12)	-	58.8(−4.3)	53.1(−10)	45.2(−17.9)
		Attack-4	58	53(−5)	51.4(−6.6)	38(−20)	35(−23)	48.6(−9.4)	70.8(12.8)	28.8(−29.2)	24(−34)	-	48(−10)	32(−26)	40.8(−17.2)
Purchase	SGD	Attack-1	53.3	52.5(−0.8)	51.6(−1.7)	66.9(13.6)	54.2(0.9)	53.5(0.2)	54.2(0.9)	51.1(−2.2)	41(−12.3)	51.2(−2.1)	48.7(−4.6)	57(3.7)	52(−1.3)
		Attack-2	52.2	51.2(−1)	50.6(−1.6)	66.1(13.9)	53.8(1.6)	52.5(0.3)	53.3(1.1)	50.1(−2.1)	40.5(−11.7)	50.1(−2.1)	47.5(−4.7)	56.5(4.3)	51.8(−0.4)
		Attack-3	51.8	50.1(−1.7)	50.1(−1.7)	65.8(14)	52.2(0.4)	51.2(−0.6)	52.1(0.3)	49.8(−2)	39.2(−12.6)	49.9(−1.9)	46.9(−4.9)	55.3(3.5)	50.9(−0.9)
		Attack-4	88	12(−76)	66.8(−21.2)	70.3(−17.7)	32.6(−55.4)	54.3(−33.7)	16(−72)	12(−76)	20(−68)	66(−22)	26(−62)	28(−60)	24(−64)
	Adagrad	Attack-1	59.9	52.9(−7)	59.4(−0.5)	73.7(13.8)	53.1(−6.8)	50.1(−9.8)	53.9(−6)	57.9(−2)	35(−24.9)	-	52.3(−7.6)	49.6(−10.3)	53.7(−6.2)
		Attack-2	58.4	51.3(−7.1)	58.8(0.4)	73.1(14.7)	52.8(−5.6)	49.4(−9)	53.1(−5.3)	56.7(−1.7)	32.5(−25.9)	-	51.9(−6.5)	48.5(−9.9)	52.9(−5.5)
		Attack-3	57.2	51.3(−5.9)	58.1(0.9)	72.8(15.6)	52(−5.2)	48.9(−8.3)	52.8(−4.4)	55.9(−1.3)	31.8(−25.4)	-	51.1(−6.1)	48.9(−8.3)	52.1(−5.1)
		Attack-4	16	16(0)	24(8)	16(0)	28(12)	14(−2)	12(−4)	24(8)	26(10)	-	20(4)	8(−8)	25(9)
	RMSProp	Attack-1	46.9	36.4(−10.5)	49.8(2.9)	71.8(24.9)	52.1(5.2)	53.9(7)	71.8(24.9)	48.4(1.5)	26(−20.9)	-	54.9(8)	25.2(−21.7)	40.5(−6.4)
		Attack-2	45.3	36.1(−9.2)	48.9(3.6)	70.5(25.2)	51.5(5.9)	52.9(7.6)	70.2(24.9)	47.9(2.6)	25.4(−19.9)	-	53.8(8.5)	24.7(−20.6)	39.7(−5.6)
		Attack-3	44.3	35.2(−9.1)	48.1(3.8)	69.9(25.6)	51.1(6.8)	52.17(7.8)	68.9(24.6)	47.1(2.8)	24.9(−19.4)	-	53.1(8.8)	23.9(−20.4)	39.1(−5.2)
		Attack-4	92	24(−68)	32(−60)	12(−80)	16(−76)	92(0)	12(−80)	24(−68)	76(−16)	-	20(−72)	8(−84)	36(−56)
	Nadam	Attack-1	56.2	40.4(−15.8)	53.4(−2.8)	72.1(15.9)	53.1(−3.1)	51.3(−4.9)	46(−10.2)	55.8(−0.4)	20.8(−35.4)	-	51.3(−4.9)	30.2(−26)	51(−5.2)
		Attack-2	55.3	39.8(−15.5)	52.2(−3.1)	71.8(16.5)	52.3(−3)	50.8(−4.5)	45.7(−9.6)	55.1(−0.2)	19.9(−35.4)	-	50.9(−4.4)	29.2(−26.1)	50.8(−4.5)
		Attack-3	55.1	39.2(−15.9)	51.3(−3.8)	70.9(15.8)	51.3(−3.8)	49.2(−5.9)	44.3(−10.8)	55.8(0.7)	18.4(−36.7)	-	50.1(−5)	28.7(−26.4)	49.9(−5.2)
		Attack-4	84	92(8)	24(−60)	16(−68)	24(−60)	76(−8)	11(−73)	8(−76)	28(−56)	-	8(−76)	18(−66)	12(−72)
	Adadelta	Attack-1	52.7	51.1(−1.6)	51.1(−1.6)	51.8(−0.9)	55.3(2.6)	51.9(−0.8)	50(−2.7)	50(−2.7)	46.8(−5.9)	-	55.6(2.9)	52(−0.7)	53.7(1)
		Attack-2	51.3	50.2(−1.1)	49.8(−1.5)	50.3(−1)	52.7(1.4)	49.8(−1.5)	48.7(−2.6)	49.2(−2.1)	45.2(−6.1)	-	53.2(1.9)	51.2(−0.1)	51.3(0)
		Attack-3	49.8	48.3(−1.5)	49.5(−0.3)	48.2(−1.6)	50.8(1)	49.2(−0.6)	47.6(−2.2)	48.6(−1.2)	43.2(−6.6)	-	50.1(0.3)	49.8(0)	50.1(0.3)
		Attack-4	84	4(−80)	72(−12)	8(−76)	24(−60)	64(−20)	28(−56)	62(−22)	8(−76)	-	18(−66)	16(−68)	16(−68)

5.2.2. Attack Recall

Reducing the attacks' recall is the best sign that implies MIA mitigation. Figure 13 illustrates the results of the four aforementioned attacks applying five optimizers, with(out) countermeasures on four datasets, MNIST, FMNIST, CIFAR-10, and Purchase, respectively. The y-axis represents the recall of the attack. The attack recall in CL is tabulated in Table 5, whereas the attack recall in FL is tabulated in Table 6 on various datasets and optimizers.

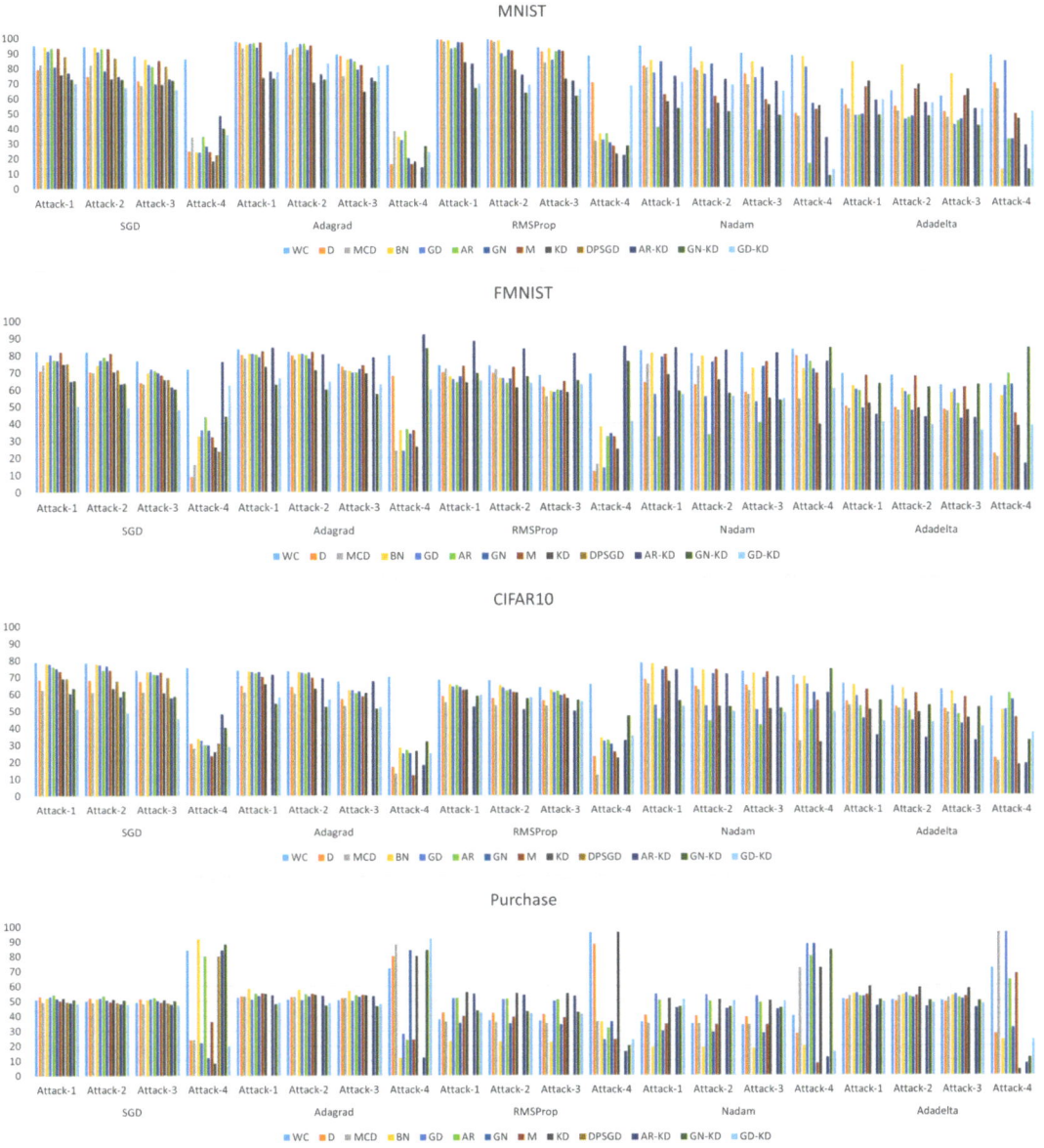

Figure 13. A comparison of the four attacks on the FL environment using five optimizers with and without countermeasures.

Table 6. FL attack recall.

Datasets	Optimizers	Attacks	WC	D	MCD	BN	GD	AR	GN	M	KD	DP	AR-KD	GN-KD	GD-KD
MNIST	SGD	Attack-1	95.2	79.3(−15.9)	82.4(−12.8)	94.5(−0.7)	91.6(−3.6)	93.6(−1.6)	81(−14.2)	93.4(−1.8)	75.7(−19.5)	88(−7.2)	76.9(−18.3)	72.6(−22.6)	69.8(−25.4)
		Attack-2	94.5	74.6(−19.9)	82(−12.5)	94.3(−0.2)	91(−3.5)	93.1(−1.4)	78.2(−16.3)	93(−1.5)	72.8(−21.7)	86.7(−7.8)	74.5(−20)	72.5(−22)	67.2(−27.3)
		Attack-3	88.2	71.7(−16.5)	68.4(−19.8)	86.2(−2)	82.5(−5.7)	81.1(−7.1)	69.4(−18.8)	85(−3.2)	68.9(−19.3)	81.2(−7)	72.6(−15.6)	71.8(−16.4)	65.8(−22.4)
		Attack-4	86	24.8(−61.2)	34(−52)	24.4(−61.6)	24.1(−61.9)	34.5(−51.5)	28(−58)	24.3(−61.7)	18(−68)	22.1(−63.9)	48.4(−37.6)	40(−46)	35.5(−50.5)
	Adagrad	Attack-1	97.6	97(−0.6)	93.4(−4.2)	95.8(−1.8)	96.4(−1.2)	96.7(−0.9)	93.5(−4.1)	97(−0.6)	73.4(−24.2)	-	77.6(−20)	73(−24.6)	77.4(−20.2)
		Attack-2	97.5	89(−8.5)	93.1(−4.4)	94(−3.5)	96.1(−1.4)	96.3(−1.2)	92.1(−5.4)	95(−2.5)	70.3(−27.2)	-	75.8(−21.7)	72.3(−25.2)	83.2(−14.3)
		Attack-3	89.2	88.1(−1.1)	74.7(−14.5)	86(−3.2)	86.2(−3)	84.5(−4.7)	79(−10.2)	81.9(−7.3)	64.2(−25)	-	73.6(−15.6)	71.2(−18)	81.6(−7.6)
		Attack-4	82	16(−66)	38(−44)	34.3(47.7)	32(−50)	38.4(−43.6)	20(−62)	16(−66)	17.8(−64.2)	-	13.9(−68.1)	28(−54)	24(−58)
	RMSProp	Attack-1	99	98.7(−0.3)	97.4(−1.6)	98.6(−0.4)	92.8(−6.2)	93.5(−5.5)	97(−2)	96.6(−2.4)	83.6(−15.4)	-	82.7(−16.3)	66.5(−32.5)	69.4(−29.6)
		Attack-2	98.9	98.2(−0.7)	97(−1.9)	98.3(−0.6)	89.4(−9.5)	87.7(−11.3)	91.9(−7)	91.3(−7.6)	78.6(−20.4)	-	75.2(−23.7)	63.2(−35.7)	68.7(−30.2)
		Attack-3	93.5	90.8(−2.7)	83.4(−10.1)	92.9(−0.6)	85.3(−8.2)	91(−2.5)	91.7(−1.8)	91.05(−2.4)	72.5(−21)	-	71.1(−22.4)	61.1(−32.4)	65.9(−27.6)
		Attack-4	88	70(−18)	31(−57)	36(−52)	32(−58)	36.2(51.8)	30(−58)	28(−60)	22.6(−65.4)	-	21.6(−66.4)	28(−60)	68(−20)
	Nadam	Attack-1	94.5	81.1(−13.4)	79.8(−14.7)	85.1(−9.4)	76.5(−18)	40.5(−54)	83.9(−10.6)	62.1(−32.4)	57.4(−37.1)	-	74.2(−20.3)	52.9(−41.6)	70.2(−24.3)
		Attack-2	93.8	79.5(−14.3)	78.2(−15.6)	84.2(−9.6)	75.6(−18.2)	39.5(−54.3)	82.3(−11.5)	60.9(−32.9)	56.2(−37.6)	-	72.1(−21.7)	50.7(−43.1)	68.5(−25.3)
		Attack-3	89.5	75.8(−13.7)	68.5(−21)	83.9(−5.6)	73.2(−16.3)	38.7(−50.8)	79.8(−9.7)	58.6(−30.9)	55.3(−34.2)	-	70.6(−18.9)	48.2(−41.3)	64.3(−25.2)
		Attack-4	88	49.3(−38.7)	47.5(−40.5)	87.2(−0.8)	80(−8)	16(−72)	56(−32)	52(−36)	54.6(−33.4)	-	33(−55)	8(−80)	12(−76)
	Adadelta	Attack-1	65.7	55.1(−10.6)	52.3(−13.4)	83.7(18)	48.1(−17.6)	48.3(−17.4)	48.8(−16.9)	66.8(1.1)	70.7(5)	-	58.1(−7.6)	48.3(−17.4)	58.4(−7.3)
		Attack-2	64.3	53.9(−10.4)	50.8(−13.5)	81.8(17.5)	45.4(−18.9)	46.7(−17.6)	47.5(−16.8)	65.2(0.9)	68.5(4.2)	-	56.3(−8)	47.6(−16.7)	56.3(−8)
		Attack-3	60.9	50.2(−10.7)	46.7(−14.2)	75.6(14.7)	41.8(−19.1)	44.3(−16.6)	45.4(−15.5)	60.8(−0.1)	65.3(4.4)	-	52.3(−8.6)	41.1(−19.8)	52.1(−8.8)
		Attack-4	88	69.3(−18.7)	65.4(−22.6)	121(−76)	84(−4)	32(−56)	32(−56)	49(−39)	45.6(−42.4)	-	28(−60)	12(−76)	50.6(−37.4)
FMNIST	SGD	Attack-1	82.4	71(−11.4)	74.6(−7.4)	76.7(−5.7)	80.5(−1.9)	77.5(−4.9)	77(−5.4)	81.9(−0.5)	74.8(−7.6)	75.1(−7.3)	64.7(−17.7)	65.2(−17.2)	50.1(−32.3)
		Attack-2	82.1	70.3(−11.8)	69.8(−12.3)	74.4(−7.7)	77.2(−4.9)	79.1(−3)	76.9(−5.2)	81.1(−1)	70.2(−11.9)	71.6(−10.5)	63.2(−18.9)	63.5(−18.6)	49.2(−32.9)
		Attack-3	76.8	64.1(−12.7)	63.2(−13.6)	69.8(−7)	71.9(−4.9)	71.1(−5.7)	69.7(−7.1)	68.3(−8.5)	65.6(−11.2)	65.8(−11)	61.3(−15.5)	60.2(−16.6)	47.9(−28.9)
		Attack-4	72	9(−63)	16(−56)	32.8(−39.2)	36.2(−35.8)	44(−28)	36(−36)	32(−40)	26.1(−45.9)	23.8(−48.2)	76.1(4.1)	44(−28)	62.6(−9.4)
	Adagrad	Attack-1	83.6	80.2(−3.4)	78.2(−5.4)	81.1(−2.5)	81(−2.6)	80.6(−2)	78.9(−4.7)	82.5(−0.2)	73.2(−10.4)	-	84.5(0.9)	62.9(−20.7)	66.8(−16.8)
		Attack-2	82.2	80(−2.2)	77.5(−4.7)	80.9(−1.3)	81(−1.2)	80.2(−2)	78.1(−4.1)	82(−0.2)	71.2(−11)	-	80.6(−1.6)	59.8(−22.4)	64.8(−17.4)
		Attack-3	75.1	73.4(−1.7)	71.3(−3.8)	71.1(−4)	69.9(−5.2)	70(−5.1)	72.1(−3)	74.2(−0.9)	69.4(−5.7)	-	78.8(3.7)	57.3(−17.8)	63.1(−12)
		Attack-4	80	68(−12)	24(−56)	36.4(−43.6)	24(−56)	36.8(−43.2)	34(−46)	36(−44)	26.3(−53.7)	-	92(12)	84(4)	59.8(−20.2)
	RMSProp	Attack-1	74.2	69.9(−4.3)	72.3(−1.9)	67.8(−6.4)	66(−8.2)	64.3(−9.9)	67.5(−6.7)	73.6(−0.6)	63.9(−10.3)	-	88.1(13.9)	69.4(−4.8)	65.3(−8.9)
		Attack-2	73.9	69.4(−4.5)	71.7(−2.2)	66.7(−7.2)	66.5(−7.4)	63.8(−10.1)	65(−7.8)	73(−0.9)	60.8(−13.1)	-	83.7(9.8)	67.5(−6.4)	63.7(−10.2)
		Attack-3	68.2	61.3(−6.9)	55.8(−12.4)	59(−9.2)	58.3(−9.9)	59.6(−8.6)	59.3(−8.9)	64.6(−3.6)	58.1(−10.1)	-	80.9(12.7)	65.1(−3.1)	62.9(−5.3)
		Attack-4	69	12(−57)	16(−53)	38(−31)	14(−55)	32.3(−36.7)	34(−35)	32(−37)	24.7(−44.3)	-	85(16)	76.1(7.1)	41(−28)
	Nadam	Attack-1	82.6	63.9(−18.7)	74.7(−7.9)	81.25(−1.3)	56.9(−25.7)	32.1(−50.5)	78.8(−3.8)	80.2(−2.4)	68.4(−14.2)	-	84.2(1.6)	58.8(−23.8)	56.9(−25.7)
		Attack-2	80.7	62.5(−18.2)	73.2(−7.5)	79.4(−1.3)	55.5(−25.2)	33.2(−47.5)	75.6(−5.1)	78.6(−2.1)	65.2(−15.5)	-	82.6(1.9)	57.4(−23.3)	55.8(−24.9)
		Attack-3	81.5	58.1(−23.4)	56.8(−24.7)	72.3(−9.2)	52.3(−29.2)	40.2(−41.3)	73.1(−8.4)	75.9(−5.6)	54.5(−27)	-	80.9(−0.6)	53.4(−28.1)	54.3(−27.2)
		Attack-4	83.2	79.2(−4)	54(−29.2)	72(−11.2)	80(−3.2)	76(−7.2)	71.6(−11.6)	69.1(−14.1)	39(−44.2)	-	76(−7.2)	84(0.8)	60(−23.2)
	Adadelta	Attack-1	68.9	49.4(−19.5)	48.2(−20.7)	61.8(−7.1)	59.4(−9.5)	58.9(−10)	48.5(−20.4)	67.6(−1.3)	51.2(−17.7)	-	44.6(−24.3)	62.9(−6)	40.3(−28.6)
		Attack-2	67.8	48.7(−19.1)	47.3(−20.5)	60.2(−7.6)	58.2(−9.6)	56.5(−11.3)	46.8(−21)	67.2(−0.6)	48.5(−19.3)	-	43.1(−24.7)	60.8(−7)	38.6(−29.2)
		Attack-3	62.2	47.6(−14.6)	46.5(−15.7)	57.8(−4.4)	59.5(−2.7)	51.2(−11)	42.3(−19.9)	60.8(−1.4)	47.3(−14.9)	-	42.6(−19.6)	62.4(0.2)	35.4(−26.8)
		Attack-4	62.8	22(−40.8)	20(−42.8)	55.9(−6.9)	61.6(−1.2)	68.8(6)	62.4(−0.4)	45.3(−17.5)	37.9(−24.9)	-	16(−46.8)	84(21.2)	38.2(−24.6)

Table 6. Cont.

Datasets	Optimizers	Attacks	WC	D	MCD	BN	GD	AR	GN	M	KD	DP	AR-KD	GN-KD	GD-KD
CIFAR-10	SGD	Attack-1	79	68.5(−10.5)	62.2(−16.8)	78.3(−0.7)	77.8(−1.2)	76.3(−2.7)	75.2(−3.8)	73.6(−5.4)	68.9(−10.1)	69.1(−9.9)	60.2(−18.8)	63.5(−15.7)	51.2(−27.8)
		Attack-2	78.6	68.2(−10.4)	61.1(−17.5)	78.1(−0.5)	77.4(−1.2)	74.3(−4.3)	76.7(−1.9)	74.2(−4.4)	63.1(−15.5)	67.6(−11)	58.3(−20.3)	61.7(−16.9)	48.9(−29.7)
		Attack-3	74.3	67.4(−6.9)	60.9(−13.4)	73.4(−0.9)	73.2(−1.1)	71.5(−2.8)	71.2(−3.1)	72.8(−1.5)	60.4(−13.9)	69.6(−4.7)	57.4(−16.9)	58.4(−15.9)	45.3(−29)
		Attack-4	75.6	31(−44.6)	28(−47.6)	33.9(−41.7)	32.6(−43)	30(−45.6)	29.8(−45.8)	23.4(−52.2)	25.8(−49.8)	30.9(−44.7)	48(−27.6)	40(−35.6)	29(−46.6)
	Adagrad	Attack-1	74.2	65(−9.2)	61(−13.2)	73.8(−0.4)	73.1(−1.1)	72.4(−1.8)	73(−1.2)	70(−4.2)	65.8(−8.4)		71.4(−2.8)	54.2(−20)	58.1(−16.1)
		Attack-2	73.7	64.3(−9.4)	60(−13.7)	73.1(−0.6)	72.6(−1.1)	72.2(−1.5)	72.8(−0.9)	69.5(−4.2)	63.1(−10.6)		69.2(−4.5)	52.4(−21.3)	56.8(−16.9)
		Attack-3	67.4	56.9(−10.5)	53(−14.4)	62.4(−5)	62.2(−5.2)	60.4(−7)	61.5(−5.9)	58.4(−9)	60.4(−7)		67.4(0)	51.1(−16.3)	52.4(−15)
		Attack-4	70.1	17(−53.1)	13(−57.1)	28.4(−41.7)	25(−45.1)	27.1(−43)	25.2(−44.9)	12(−58.1)	26.2(−43.9)		18(−52.1)	32(−38.1)	25(−45.1)
	RMSProp	Attack-1	68.2	58.6(−9.6)	55(−13.2)	65.8(−2.4)	64.2(−4)	65.3(−2.9)	64(−4.2)	62.1(−6.1)	62.3(−5.9)		52.4(−15.8)	58.7(−9.5)	59.3(−8.9)
		Attack-2	67.9	57.3(−10.6)	53.2(−14.7)	65.1(−2.8)	63.6(−5.3)	61.9(−6)	62.4(−5.5)	60.8(−7.1)	60.6(−7.3)		50.6(−17.3)	57.1(−10.8)	57.9(−10)
		Attack-3	63.7	56(−7.7)	52.9(−10.8)	62.3(−1.4)	60.8(−2.9)	61.7(−2)	59(−4.7)	59.4(−4.3)	57.2(−6.5)		49.7(−14)	56.2(−7.5)	55.4(−8.3)
		Attack-4	65.6	23(−42.6)	12(−53.6)	34(−31.6)	32.1(−33.5)	32.7(−32.9)	30.3(−35.3)	25.6(−40)	21.9(−43.7)		32.4(−33.2)	46.8(−18.8)	35(−30.6)
	Nadam	Attack-1	78.4	68.5(−9.9)	65.8(−12.6)	77.8(−0.6)	53.1(−25.3)	45.2(−33.2)	74.3(−4.1)	75.9(−2.5)	67.1(−11.3)		74.3(−4.1)	55.6(−22.8)	52.6(−25.8)
		Attack-2	75.2	64.2(−11)	62.3(−12.9)	74.2(−1)	52.8(−22.4)	43.8(−31.4)	71.6(−3.6)	74.2(−1)	52.3(−22.9)		71.4(−3.8)	52.1(−23.1)	49.4(−25.8)
		Attack-3	73.1	64.8(−8.3)	61.7(−11.4)	72.1(−1)	50.3(−22.8)	41.6(−31.5)	69.2(−3.9)	72.6(−0.5)	50.9(−22.2)		69.8(−3.3)	51.3(−21.8)	48.7(−24.4)
		Attack-4	70.6	65.2(−5.4)	32(−38.6)	70.1(−0.5)	65.4(−5.2)	50.3(−20.3)	60.5(−10.1)	55.6(−15)	31.2(−39.4)		60.3(−10.3)	74.6(4)	49.1(−21.5)
	Adadelta	Attack-1	65.9	55.3(−10.6)	53.1(−12.8)	65.1(−0.8)	58.5(−7.4)	52.6(−13.3)	45.2(−20.7)	62.1(−3.8)	50.3(−15.6)		35.3(−30.6)	55.8(−10.1)	43.7(−22.2)
		Attack-2	64.5	52.1(−12.4)	51.2(−13.3)	63.1(−1.4)	56.2(−8.3)	49.5(−15)	43.8(−20.7)	59.8(−4.7)	48.7(−15.8)		33.7(−30.8)	52.9(−11.6)	42.9(−21.6)
		Attack-3	62.3	50.6(−11.7)	48.5(−13.8)	61.1(−1.2)	53.1(−9.2)	47.6(−14.7)	41.9(−20.4)	57.6(−4.7)	45.4(−16.9)		32.2(−30.1)	51.6(−10.7)	40.3(−22)
		Attack-4	58	22(−36)	20(−38)	50(−8)	50.3(−7.7)	60(2)	56.2(−1.8)	45.7(−12.3)	18(−40)		18.6(−39.4)	32.4(−25.6)	36.9(−21.1)
Purchase	SGD	Attack-1	52.2	53.1(1.9)	49.3(−1.9)	52.2(1)	52.9(1.7)	54.5(3.3)	51.6(0.4)	50.1(−1.1)	51.9(0.7)	49.5(−2.3)	48.9(−2.3)	50.9(−0.3)	48.4(−2.8)
		Attack-2	50.2	52.2(2)	48.9(−1.3)	51.8(1.6)	52.1(1.9)	53.7(3.5)	50.8(0.6)	49.5(−0.7)	51.1(0.9)	49.1(−1.1)	48.1(−2.1)	50.6(0.4)	47.9(−2.3)
		Attack-3	49.3	51.6(2.3)	48.1(−1.2)	51.1(1.8)	51.5(2.2)	52.4(3.1)	50.1(0.8)	49(−0.3)	50.8(1.5)	48.7(−0.6)	47.8(−1.5)	49.9(0.6)	47.1(−2.2)
		Attack-4	84	24(−60)	24(−60)	92(8)	22(−62)	80(−4)	12(−72)	36(−48)	8(−76)	80(−4)	84(0)	88(4)	20(−64)
	Adagrad	Attack-1	52.4	53.3(0.9)	53.1(0.7)	58.7(6.3)	51.1(−1.3)	55.2(2.8)	53.4(1)	55.3(2.9)	54.9(2.5)		54(1.6)	47.9(−4.5)	49.1(−3.3)
		Attack-2	51.2	52.8(1.6)	52.8(1.6)	57.8(6.6)	50.8(−0.4)	54.9(3.7)	53.1(1.9)	54.8(3.6)	54.2(3)		53.5(2.3)	47(−4.2)	48.8(−2.4)
		Attack-3	50.8	52.1(1.3)	52.1(1.3)	57(6.2)	50.2(−0.6)	54.2(3.4)	52.8(2)	54.1(3.3)	53.7(2.9)		53.1(2.3)	46.5(−4.3)	48.1(−2.7)
		Attack-4	72	80(8)	88(16)	12(−60)	28(−44)	24(−48)	84(12)	24(−48)	80(8)		12(−60)	84(12)	92(20)
	RMSProp	Attack-1	37.8	42.3(4.5)	36(−1.8)	23.1(−14.7)	51.8(14)	52(14.2)	35.2(−2.6)	39.8(2)	55.8(18)		54.7(16.9)	43.4(5.6)	42.2(4.4)
		Attack-2	37.1	41.9(4.8)	35.7(−1.4)	22.8(−14.3)	51.2(14.1)	51.6(14.5)	34.8(−2.3)	39.1(2)	55.1(18)		53.9(16.8)	42.8(5.7)	41.8(4.7)
		Attack-3	36.8	41(4.2)	35(−1.8)	22.5(−14.3)	49.7(12.9)	51(14.2)	34.2(−2.6)	38.7(1.9)	54.9(18.1)		53.1(16.3)	42.2(5.4)	41.2(4.4)
		Attack-4	96	88(−8)	36(−60)	36(−60)	24(−72)	32(−64)	36(−60)	24(−72)	96(0)		16(−80)	20(−76)	24(−72)
	Nadam	Attack-1	35.8	40.4(4.6)	35.1(−0.7)	54.7(18.6)	54.4(18.6)	50.5(14.7)	29.6(−6.2)	34.4(−1.4)	51.4(15.6)		45.3(9.5)	46.1(10.3)	50.9(15.1)
		Attack-2	34.7	39.8(5.1)	34.7(0)	54.1(18.7)	53.9(19.2)	49.8(15.1)	28.9(−5.8)	34(−0.7)	50.7(16)		44.8(10.1)	45.8(11.1)	50.1(15.4)
		Attack-3	34.1	39(4.9)	34(−0.1)	53.3(3.3)	53.1(19)	49.1(15)	28.3(−5.8)	33.8(−0.3)	50(15.9)		44.3(10.2)	45.3(11.2)	49.8(15.7)
		Attack-4	40	28(−12)	72(32)	20(−20)	88(48)	80(40)	88(48)	8(−32)	72(32)		12(−28)	84(44)	16(−24)
	Adadelta	Attack-1	51.1	50.5(−0.6)	53.1(2)	54.7(3.6)	55.1(4)	53.1(2)	52.8(1.7)	53.9(2.8)	59.4(8.3)		46.4(−4.7)	50.5(−0.6)	50.3(−1.8)
		Attack-2	50.4	50.1(−0.3)	52.9(2.5)	54.1(3.7)	54.8(4.4)	52.8(2.4)	51.9(1.5)	53.2(2.8)	58.7(8.3)		45.9(−4.5)	50(−0.4)	49.8(−0.2)
		Attack-3	50	49.1(−0.9)	52.1(2.1)	53.3(3.3)	54.3(4.3)	52.1(2.1)	51.1(1.1)	52.8(2.8)	58(8)		45.3(−4.7)	49.8(−0.2)	48.1(−1.9)
		Attack-4	72	28(−44)	96(24)	24(−48)	96(24)	64(−8)	32(−40)	68(−4)	4(−68)		8(−64)	12(−60)	24(−48)

- **CL attacks recall without countermeasure:** As shown in Table 5, for the MNIST dataset, the strongest attack is Attack 1 when we apply RMSProp. The recall value of this attack without any countermeasure is 99.4%, which is the highest among other attacks. For Attack 1, only changing the optimizer to Adadelta drops this value to 59.7% without using any countermeasure. Also, the weakest attack goes for Attack 4 whenr using Adadelta optimization. The recall value of this attack is 44%. For the FMNIST dataset, the strongest attack is Attack 1 with the SGD optimizer and the weakest attack is Attack 4 with the Nadam optimizer. For CIFAR-10, the strongest attack is Attack 1 with SGD optimizer and the weakest is Attack 4 with Adadelta optimizer. For the Purchase dataset, the best attack is Attack 4 with RMSProp optimizer and the worst attack is Attack 4 with Adagrad optimizer.
- **CL attacks recall with countermeasures:** As per Table 5, different mitigation techniques provide various recall values in every attack. We observe that the strongest attack in the case of MNIST, which is Attack 1 with RMSProp, is defended by GD–KD by a reduction of 65.2% of recall value, which is impressive. Using GD–KD is only reducing the model accuracy by 8% according to Table 3. We can conclude that, in the CL environment, GD–KD provides the strongest defense with the lowest model accuracy degradation. This is very important in developing future ML models. For the FMNIST dataset, the strongest attack belongs to Attack 1 when using SGD. This attack in the case of FMNIST is also defended by GD–KD by a reduction of 37.4% in recall value, although the strongest defense for this particular attack and dataset is GN–KD with a 37.9% recall reduction. It is noteworthy that GD–KD and GN–KD drop model accuracy by 15.4% and 13.9%, respectively, as shown in Table 3. The same thing holds true for the CIFAR-10 dataset. The strongest attack is Attack 1 with SGD optimizer for this dataset, and GN–KD is capable of defending this attack by a reduction in attack recall by 34%. Also, in the Purchase dataset, the strongest attack, which is Attack 4 with RMSProp optimizer, is defended by GN–KD and resulted in recall value reduction by 84%. In general, we observe that, in the CL environment, in most of the experiments, the combinations of KD and another countermeasure provides lower attack recall values than other mitigation techniques. This means that these combinations are the best to defend MIA against ML in the CL environment.
- **FL attacks recall without countermeasure:** As shown in Figure 13 and Table 6, for the MNIST dataset, we observe that the highest attack recall (99%) belongs to Attack 1 with RMSProp. This value is significantly reduced to 65.7% by only changing the optimizer to Adadelta. It is impressive to see that changing the optimizer to Adadelta will not drop model accuracy significantly. According to Table 4, using Adadelta reduces FL model accuracy by approximately 1% compared to Nadam. For the FMNIST dataset, Attack 1 with Adagrad provides the highest attack recall value (83.6%). When we change the optimizer to Adadelta, we witness a drop in attack recall to 68.9% without any mitigation technique. The same as Adadelta in MNIST, we are seeing a slight drop in accuracy from 91.7% to 84.1% according to Table 4. For the CIFAR-10 dataset, the highest attack recall is 79% for Attack 1 with SGD optimizer. This value is dropped to 65.9% by only changing the optimizer to Adadelta. Similar to MNIST and FMNIST, this change has not had a significant impact on the accuracy of the FL model. As shown in Table 4, the accuracy of CIFAR-10, when using Adadelta as an optimizer, only drops by roughly 2%. For the Purchase dataset, the best attack is Attack 4 with the RMSProp optimizer with 96% recall value. Also, without applying any countermeasure, the lowest recall value for this dataset belongs to Attack 3 with the Nadam optimizer.
- **FL attacks recall with countermeasures:** As shown in Table 6, it is evident that the various mitigation techniques exhibit varying performance. However, in general, the combinations of KD with either GD, GN, or AR consistently offer improved protection while preserving the model's utility. For MNIST with RMSProp, GN–KD effectively reduces the recall of Attack 1 by 32.5%, which is the most potent attack in our FL MNIST experiments. Remarkably, this reduction is achieved with only an 11%

decrease in FL model accuracy, as indicated in Table 4. In the case of FMNIST, Table 6 reveals that Attack 1 with Adagrad exhibits a high recall value of 83.6%. However, this attack can be mitigated by GN–KD, resulting in a 20.7% reduction in recall. It is worth noting that this defense strategy incurs a modest accuracy drop of 9.7%, as reflected in Table 4. In CIFAR-10, the strongest attack is Attack 1 with SGD, boasting a recall value of 92.6%. GN–KD is capable of reducing this recall to 58.6% while causing a minimal 4.2% drop in FL accuracy, as detailed in Table 4. In the Purchase dataset, the most potent attack, Attack 4, using the RMSProp optimizer, experiences an 80% reduction in effectiveness with a recall value of 96% when AR–KD is applied. Notably, AR–KD not only avoids a decline in accuracy for the Purchase dataset with the RMSProp optimizer but also substantially boosts accuracy by 52%. This improvement is attributed to the capacity of AR–KD to modify the model's architecture, thereby averting overfitting.

5.2.3. Accuracy–Recall Trade-Off

To obtain a clear comparison between the efficiency of the countermeasures, we calculated the ratio of accuracy over recall. The higher the ratio is, the better the trade-off we are achieving. Figure 14 illustrates the accuracy–recall ratio of each countermeasure. As shown in Figure 14, for almost all optimizers, the highest trade-off belongs to one of the combinatory approaches (either AR–KD, GN–KD, or GD–KD). This figure proves that the combinational approaches that we tested provide a better trade-off between the accuracy of the target model and MIA attack recall. The higher value of this trade-off conveys the message that the mitigation technique keeps the accuracy of the target model high and reduces the attack recall as much as possible.

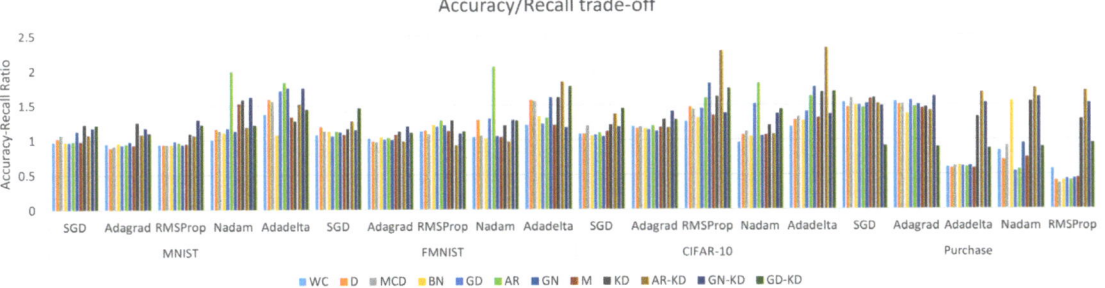

Figure 14. The ratio of the accuracy of the model over recall of the attack model in FL environment.

5.2.4. Privacy and Utility

Concluding from Tables 3–6, it is noted that combination of KD with either AR, GN, or GD has significant advantages over using each one of them separately as well as over other conventional countermeasures. Experiments are showing that the new combinations of countermeasures successfully handle the trade-off between privacy and utility. Generally speaking, in all datasets and almost all optimizers (AR, GD, and GN), KD is capable of reducing the attack recall while preserving the accuracy of the model at a high level. Not only do they preserve the utility of the model at a high level but also, due to the nature of KD, in some cases, they increase model accuracy as well.

6. Conclusions

This research paper presents a thorough examination of the accuracy of centralized and federated learning models, as well as the recall rates associated with different membership inference attacks. Additionally, it evaluates the effectiveness of various defense mechanisms within both centralized and federated learning environments. Our experimental findings reveal that Attack 1 [9] yields the highest advantage for potential attackers, while Attack 4 [11] is the least favorable for malicious actors. Among the defense strategies examined, the combination of knowledge distillation (KD) with activity regularization (AR), Gaussian dropout (GD), or Gaussian noise (GN) emerges as the most effective in the context of centralized and federated learning. Notably, these three combinations stand out for their ability to effectively balance the trade-off between preserving privacy and maintaining utility. This comparative analysis holds significant importance for guiding future advancements in model development.

Author Contributions: Conceptualization, A.A.T. and D.A.; methodology, A.A.T., S.D. and D.A.; software, A.A.T., S.D. and N.M.; validation, A.A.T., D.A. and N.M.; formal analysis, A.A.T.; investigation, A.A.T. and N.M.; resources, D.A.; data curation, A.A.T.; writing—original draft preparation, A.A.T., S.D. and D.A.; writing—review and editing, A.A.T. and D.A.; visualization, A.A.T.; supervision, D.A.; project administration, D.A.; funding acquisition, D.A. All authors have read and agreed to the published version of the manuscript.

Funding: This research is supported by the Natural Sciences and Engineering Research Council of Canada (NSERC) Discovery Grant (RGPIN-2019-05689).

Data Availability Statement: The codes and data are available at https://github.com/University-of-Windsor/ComparitiveAnalysis.

Conflicts of Interest: The authors declare no conflict of interest.

References

1. Niknam, S.; Dhillon, H.S.; Reed, J.H. Federated learning for wireless communications: Motivation, opportunities, and challenges. *IEEE Commun. Mag.* **2020**, *58*, 46–51. [CrossRef]
2. Carlini, N.; Chien, S.; Nasr, M.; Song, S.; Terzis, A.; Tramer, F. Membership inference attacks from first principles. In Proceedings of the 2022 IEEE Symposium on Security and Privacy (SP), San Francisco, CA, USA, 22–26 May 2022; pp. 1897–1914.
3. McMahan, B.; Moore, E.; Ramage, D.; Hampson, S.; y Arcas, B.A. Communication-efficient learning of deep networks from decentralized data. In Proceedings of the Artificial Intelligence and Statistics, Fort Lauderdale, FL, USA, 20–22 April 2017; pp. 1273–1282.
4. Regulation, P. General data protection regulation. *Intouch* **2018**, *25*, 1–5.
5. Act, A. Health insurance portability and accountability act of 1996. *Public Law* **1996**, *104*, 191.
6. Carlini, N.; Liu, C.; Erlingsson, Ú.; Kos, J.; Song, D. The Secret Sharer: Evaluating and Testing Unintended Memorization in Neural Networks. In Proceedings of the USENIX Security Symposium, Santa Clara, CA, USA, 14–16 August 2019; Volume 267.
7. Melis, L.; Song, C.; De Cristofaro, E.; Shmatikov, V. Exploiting unintended feature leakage in collaborative learning. In Proceedings of the 2019 IEEE Symposium on Security and Privacy (SP), San Francisco, CA, USA, 19–23 May 2019; pp. 691–706.
8. Backes, M.; Berrang, P.; Humbert, M.; Manoharan, P. Membership privacy in MicroRNA-based studies. In Proceedings of the 2016 ACM SIGSAC Conference on Computer and Communications Security, Vienna, Austria, 24–28 October 2016; pp. 319–330.
9. Shokri, R.; Stronati, M.; Song, C.; Shmatikov, V. Membership inference attacks against machine learning models. In Proceedings of the 2017 IEEE Symposium on Security and Privacy (SP), San Jose, CA, USA, 22–24 May 2017; pp. 3–18.
10. Salem, A.; Zhang, Y.; Humbert, M.; Berrang, P.; Fritz, M.; Backes, M. Ml-leaks: Model and data independent membership inference attacks and defenses on machine learning models. *arXiv* **2018**, arXiv:1806.01246.
11. Liu, L.; Wang, Y.; Liu, G.; Peng, K.; Wang, C. Membership Inference Attacks Against Machine Learning Models via Prediction Sensitivity. *IEEE Trans. Dependable Secur. Comput.* **2022**, *20*, 2341–2347. [CrossRef]
12. Dayal, S.; Alhadidi, D.; Abbasi Tadi, A.; Mohammed, N. Comparative Analysis of Membership Inference Attacks in Federated Learning. In Proceedings of the 27th International Database Engineered Applications Symposium, Heraklion, Greece, 5–7 May 2023; pp. 185–192.
13. Gal, Y.; Ghahramani, Z. Dropout as a bayesian approximation: Representing model uncertainty in deep learning. In Proceedings of the International Conference on Machine Learning, New York, NY, USA, 20–22 June 2016; pp. 1050–1059.
14. Bjorck, N.; Gomes, C.P.; Selman, B.; Weinberger, K.Q. Understanding batch normalization. In Proceedings of the Advances in Neural Information Processing Systems 31 (NeurIPS 2018), Montreal, QC, Canada, 8 December 2018; pp. 31–40.
15. Xiao, Y.; Yan, C.; Lyu, S.; Pei, Q.; Liu, X.; Zhang, N.; Dong, M. Defed: An Edge Feature Enhanced Image Denoised Networks Against Adversarial Attacks for Secure Internet-of-Things. *IEEE Internet Things J.* **2022**, *10*, 6836–6848. [CrossRef]
16. Srivastava, N.; Hinton, G.; Krizhevsky, A.; Sutskever, I.; Salakhutdinov, R. Dropout: A simple way to prevent neural networks from overfitting. *J. Mach. Learn. Res.* **2014**, *15*, 1929–1958.
17. Keras Documentation: Masking Layer. Available online: https://keras.io/api/layers/core_layers/masking/ (accessed on 29 September 2023).

18. Abadi, M.; Chu, A.; Goodfellow, I.; McMahan, H.B.; Mironov, I.; Talwar, K.; Zhang, L. Deep learning with differential privacy. In Proceedings of the 2016 ACM SIGSAC Conference on Computer and Communications Security, Vienna, Austria, 24–28 October 2016; pp. 308–318.
19. Yim, J.; Joo, D.; Bae, J.; Kim, J. A gift from knowledge distillation: Fast optimization, network minimization and transfer learning. In Proceedings of the IEEE Conference on Computer Vision and Pattern Recognition, Honolulu, HI, USA, 21–26 July 2017; pp. 4133–4141.
20. Bottou, L. Large-scale machine learning with stochastic gradient descent. In Proceedings of the COMPSTAT'2010: 19th International Conference on Computational Statistics, Paris, France, 22–27 August 2010; pp. 177–186.
21. Tieleman, T.; Hinton, G. Lecture 6.5-rmsprop: Divide the gradient by a running average of its recent magnitude. *COURSERA Neural Netw. Mach. Learn.* **2012**, *4*, 26–31.
22. McMahan, H.B.; Streeter, M. Adaptive bound optimization for online convex optimization. *arXiv* **2010**, arXiv:1002.4908.
23. Poggiolini, P. The GN model of non-linear propagation in uncompensated coherent optical systems. *J. Light. Technol.* **2012**, *30*, 3857–3879. [CrossRef]
24. Keras Documentation: Activityregularization Layer. Available online: https://keras.io/api/layers/regularization_layers/activity_regularization/ (accessed on 29 September 2023).
25. Dozat, T. Incorporating Nesterov Momentum into Adam. Available online: https://openreview.net/forum?id=OM0jvwB8jIp57ZJjtNEZ (accessed on 29 September 2023).
26. Zeiler, M.D. Adadelta: An adaptive learning rate method. *arXiv* **2012**, arXiv:1212.5701.
27. Deng, L. The mnist database of handwritten digit images for machine learning research [best of the web]. *IEEE Signal Process. Mag.* **2012**, *29*, 141–142. [CrossRef]
28. Xiao, H.; Rasul, K.; Vollgraf, R. Fashion-mnist: A novel image dataset for benchmarking machine learning algorithms. *arXiv* **2017**, arXiv:1708.07747.
29. Krizhevsky, A.; Hinton, G. *Learning Multiple Layers of Features from Tiny Images*; University of Toronto: Toronto, ON, Canada, 2009.
30. Datasets. Available online: https://www.comp.nus.edu.sg/~reza/files/datasets.html (accessed on 29 September 2023).
31. Nasr, M.; Shokri, R.; Houmansadr, A. Comprehensive privacy analysis of deep learning: Passive and active white-box inference attacks against centralized and federated learning. In Proceedings of the 2019 IEEE Symposium on Security and Privacy (SP), San Francisco, CA, USA, 19–23 May 2019; pp. 739–753.
32. Conti, M.; Li, J.; Picek, S.; Xu, J. Label-Only Membership Inference Attack against Node-Level Graph Neural Networks. In Proceedings of the 15th ACM Workshop on Artificial Intelligence and Security, Los Angeles, CA, USA, 11 November 2022; pp. 1–12.
33. Zheng, J.; Cao, Y.; Wang, H. Resisting membership inference attacks through knowledge distillation. *Neurocomputing* **2021**, *452*, 114–126. [CrossRef]
34. Shejwalkar, V.; Houmansadr, A. Membership privacy for machine learning models through knowledge transfer. In Proceedings of the AAAI Conference on Artificial Intelligence, Virtually, 2–9 February 2021; Volume 35, pp. 9549–9557.
35. Lee, H.; Kim, J.; Ahn, S.; Hussain, R.; Cho, S.; Son, J. Digestive neural networks: A novel defense strategy against inference attacks in federated learning. *Comput. Secur.* **2021**, *109*, 102378. [CrossRef]
36. Su, T.; Wang, M.; Wang, Z. Federated Regularization Learning: An Accurate and Safe Method for Federated Learning. In Proceedings of the 2021 IEEE 3rd International Conference on Artificial Intelligence Circuits and Systems (AICAS), Washington, DC, USA, 6–9 June 2021; pp. 1–4.
37. Xie, Y.; Chen, B.; Zhang, J.; Wu, D. Defending against Membership Inference Attacks in Federated learning via Adversarial Example. In Proceedings of the 2021 17th International Conference on Mobility, Sensing and Networking (MSN), Exeter, UK, 13–15 December 2021; pp. 153–160.
38. Firdaus, M.; Larasati, H.T.; Rhee, K.H. A Secure Federated Learning Framework using Blockchain and Differential Privacy. In Proceedings of the 2022 IEEE 9th International Conference on Cyber Security and Cloud Computing (CSCloud)/2022 IEEE 8th International Conference on Edge Computing and Scalable Cloud (EdgeCom), Xi'an, China, 25–27 June 2022; pp. 18–23.
39. Bai, Y.; Fan, M. A method to improve the privacy and security for federated learning. In Proceedings of the 2021 IEEE 6th International Conference on Computer and Communication Systems (ICCCS), Las Vegas, CA, USA, 4–6 October 2021; pp. 704–708.
40. Chen, H.; Li, H.; Dong, G.; Hao, M.; Xu, G.; Huang, X.; Liu, Z. Practical membership inference attack against collaborative inference in industrial IoT. *IEEE Trans. Ind. Infor.* **2020**, *18*, 477–487. [CrossRef]
41. Novak, R.; Bahri, Y.; Abolafia, D.A.; Pennington, J.; Sohl-Dickstein, J. Sensitivity and generalization in neural networks: An empirical study. *arXiv* **2018**, arXiv:1802.08760.
42. Milanés-Hermosilla, D.; Trujillo Codorniú, R.; López-Baracaldo, R.; Sagaró-Zamora, R.; Delisle-Rodriguez, D.; Villarejo-Mayor, J.J.; Núñez-Álvarez, J.R. Monte Carlo Dropout for Uncertainty Estimation and Motor Imagery Classification. *Sensors* **2021**, *21*, 7241. [CrossRef]
43. Dwork, C.; McSherry, F.; Nissim, K.; Smith, A. Calibrating noise to sensitivity in private data analysis. In Proceedings of the Theory of Cryptography Conference, New York, NY, USA, 4–7 March 2006; pp. 265–284.
44. Dwork, C. A firm foundation for private data analysis. *Commun. ACM* **2011**, *54*, 86–95. [CrossRef]
45. Dwork, C.; Roth, A. The algorithmic foundations of differential privacy. *Found. Trends Theor. Comput. Sci.* **2014**, *9*, 211–407. [CrossRef]

46. Hinton, G.; Vinyals, O.; Dean, J. Distilling the knowledge in a neural network. *arXiv* **2015**, arXiv:1503.02531.
47. Wu, C.; Wu, F.; Lyu, L.; Huang, Y.; Xie, X. Communication-efficient federated learning via knowledge distillation. *Nat. Commun.* **2022**, *13*, 2032. [CrossRef]
48. Jiang, D.; Shan, C.; Zhang, Z. Federated learning algorithm based on knowledge distillation. In Proceedings of the 2020 International Conference on Artificial Intelligence and Computer Engineering (ICAICE), Beijing, China, 23–25 October 2020; pp. 163–167.
49. Li, X.; Chen, B.; Lu, W. FedDKD: Federated learning with decentralized knowledge distillation. *Appl. Intell.* **2023**, *53*, 18547–18563. [CrossRef]
50. Available online: https://github.com/University-of-Windsor/ComparitiveAnalysis (accessed on 29 September 2023).
51. Yuan, X.; Zhang, L. Membership Inference Attacks and Defenses in Neural Network Pruning. In Proceedings of the 31st USENIX Security Symposium (USENIX Security 22), Boston, MA, USA, 10–12 August 2022.
52. Asad, M.; Moustafa, A.; Ito, T. Federated learning versus classical machine learning: A convergence comparison. *arXiv* **2021**, arXiv:2107.10976.
53. Peng, S.; Yang, Y.; Mao, M.; Park, D.S. Centralized Machine Learning Versus Federated Averaging: A Comparison using MNIST Dataset. *KSII Trans. Internet Inf. Syst. (TIIS)* **2022**, *16*, 742–756.
54. Drainakis, G.; Katsaros, K.V.; Pantazopoulos, P.; Sourlas, V.; Amditis, A. Federated vs. centralized machine learning under privacy-elastic users: A comparative analysis. In Proceedings of the 2020 IEEE 19th International Symposium on Network Computing and Applications (NCA), Cambridge, MA, USA, 24–27 November 2020; pp. 1–8.

Disclaimer/Publisher's Note: The statements, opinions and data contained in all publications are solely those of the individual author(s) and contributor(s) and not of MDPI and/or the editor(s). MDPI and/or the editor(s) disclaim responsibility for any injury to people or property resulting from any ideas, methods, instructions or products referred to in the content.

Review

Exploring Federated Learning Tendencies Using a Semantic Keyword Clustering Approach

Francisco Enguix [1,*], Carlos Carrascosa [1] and Jaime Rincon [2]

1 Valencian Research Institute for Artificial Intelligence (VRAIN), Universitat Politècnica de València (UPV), 46022 Valencia, Spain; carrasco@dsic.upv.es
2 Departamento de Digitalización, Escuela Politécnica Superior, Universidad de Burgos, 09006 Miranda de Ebro, Spain; jarincon@ubu.es
* Correspondence: fraenan@upv.es

Abstract: This paper presents a novel approach to analyzing trends in federated learning (FL) using automatic semantic keyword clustering. The authors collected a dataset of FL research papers from the Scopus database and extracted keywords to form a collection representing the FL research landscape. They employed natural language processing (NLP) techniques, specifically a pre-trained transformer model, to convert keywords into vector embeddings. Agglomerative clustering was then used to identify major thematic trends and sub-areas within FL. The study provides a granular view of the thematic landscape and captures the broader dynamics of research activity in FL. The key focus areas are divided into theoretical areas and practical applications of FL. The authors make their FL paper dataset and keyword clustering results publicly available. This data-driven approach moves beyond manual literature reviews and offers a comprehensive overview of the current evolution of FL.

Keywords: federated learning; analysis; review; multi-agent system (MAS)

1. Introduction

Federated learning (FL) has emerged as a revolutionary paradigm in collaborative machine learning [1]. It empowers multiple devices or institutions to train a model while collectively safeguarding data privacy. This decentralized approach contrasts traditional methods where data are centralized for model training, potentially compromising user privacy and data ownership. FL accomplishes this collaborative learning by keeping raw data distributed on individual devices, and instead of sharing the raw data, participants exchange the model updates.

The field of FL is experiencing explosive growth, leading to a vast and ever-expanding body of research literature. This presents a significant challenge to researchers attempting to identify current trends and emerging sub-areas within FL. Traditional manual literature reviews with a global approach, while valuable, become increasingly impractical for analyzing field trends as the number of publications and the intricate interplay of FL concepts continue to grow exponentially, as depicted in Figure 1. To address this challenge, this paper proposes the use of an automated semantic keyword clustering technique as a critical tool for analyzing FL research trends.

Automated semantic keyword clustering leverages advanced natural language processing (NLP) techniques to extract meaningful data from the vast amount of interconnected areas in FL. Using pre-trained transformer models [2], the research article keywords can be transformed into dense vector spaces that capture their semantic relationships. This empowers the creation of clusters based on thematic relevance, revealing the underlying thematic structure of the FL research landscape.

Citation: Enguix, F.; Carrascosa, C.; Rincon, J. Exploring Federated Learning Tendencies Using a Semantic Keyword Clustering Approach. *Information* **2024**, *15*, 379. https://doi.org/10.3390/info15070379

Academic Editor: Peter Z. Revesz

Received: 7 May 2024
Revised: 18 June 2024
Accepted: 26 June 2024
Published: 28 June 2024

Copyright: © 2024 by the authors. Licensee MDPI, Basel, Switzerland. This article is an open access article distributed under the terms and conditions of the Creative Commons Attribution (CC BY) license (https://creativecommons.org/licenses/by/4.0/).

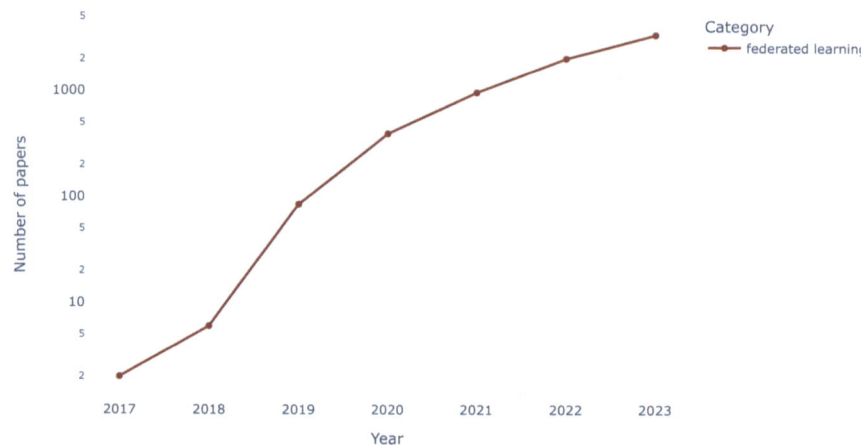

Figure 1. Number of papers containing the keyword "federated learning" across the years, on a logarithmic scale on the y-axis, based on the public dataset presented in Section 2.

This paper presents a semantically-based literature analysis of the 7 953 papers about FL. The primary objective is to uncover and explore the major theoretical categories and practical application areas of FL and examine the current trends of the field, along with the emerging sub-areas that have received less research attention. First, we formulate a series of research questions (RQs) that guide the investigation. These RQs delve into the current trends in FL (RQ1), the tendencies of these trends (RQ2), and the application domains where FL techniques are finding utility (RQ3 and RQ4). Recognizing the potential in under-explored areas, we propose additional research questions (RQ5 and RQ6) that focus on identifying emerging sub-areas within FL that have received limited research focus, and investigating how existing FL techniques can be adapted to address these application domains. The final question (RQ7) looks ahead to predict potential future directions and areas of growth. Formally, we formulated the following research questions:

RQ1: What are the current trends in FL?
RQ2: What are the tendencies of the current trends in FL?
RQ3: What are the application domains where FL techniques are applied?
RQ4: What are the tendencies of the application domains?
RQ5: What are the emerging sub-areas within FL?
RQ6: What are the tendencies of the emerging sub-areas?
RQ7: What are the potential future trends of FL?

A data-mining technique and a transformer-based semantic analysis of the literature's keywords will be employed to address these RQs and uncover the trends and tendencies within this extensive collection. This approach permits automatically grouping keywords into clusters, revealing the thematic relationships and dominant topics within the current body of FL research.

The Structure of the Survey

The structure of this survey is designed to address the research questions and present the findings. Figure 2 provides a classification scheme outlining these categories. Then, we will delve deeper into each category and explore the relevant advancements from the existing literature. Following the introduction, this paper unfolds across several key sections:

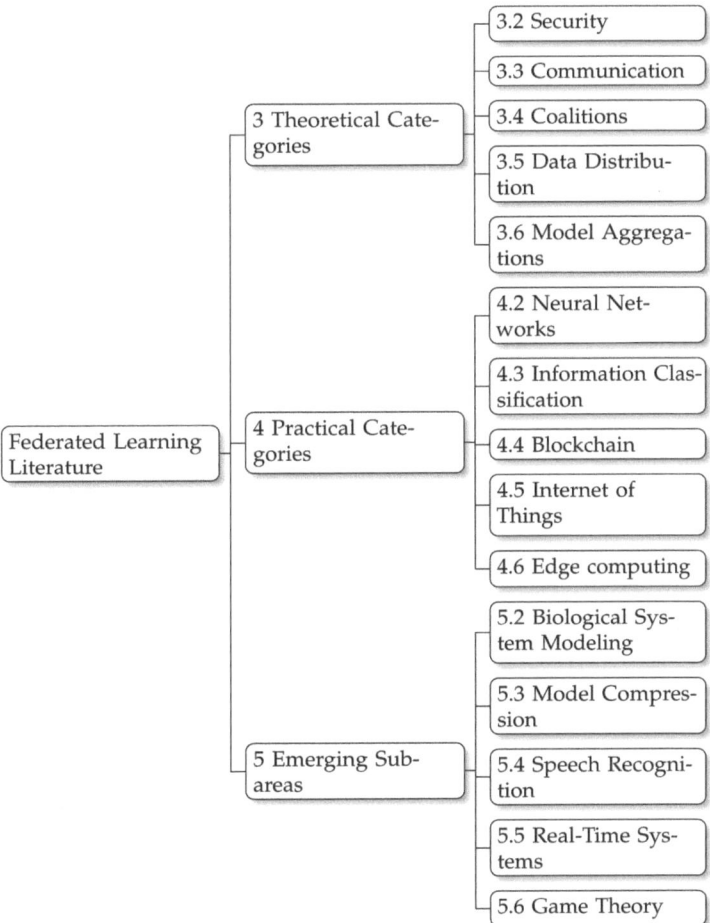

Figure 2. Taxonomy of this paper.

- Research Method (Section 2). This section delves into the approach employed to analyze trends and sub-areas within FL. It details the utilization of keyword extraction and automated clustering techniques to gain insights from the vast FL research landscape.
- Theoretical Categories (Section 3). Here, we present a detailed analysis of the key theoretical areas of FL. This section explores crucial aspects such as security mechanisms, communication protocols, coalition formation, data distribution strategies, and model aggregation techniques.
- Practical Categories (Section 4). Shifting the focus to the practical applications of FL, this section examines its implementation in various domains. We will explore how FL empowers neural networks, facilitates information classification tasks, integrates with blockchain technology, and finds applications in the Internet of Things (IoT) and edge computing environments.
- Emerging Sub-Areas (Section 5). This section explores the sub-areas of FL research that have emerged as a result of previous research directions. Here, we will identify and analyze these emerging trends that hold significant promise for the future development of the field, including biological system modeling, model compression techniques, advancements in speech recognition, the application of FL to real-time systems, and the utilization of game theory for improved performance.

- Conclusion (Section 6). Building upon the foundation in the preceding sections, this section will synthesize the key findings. It will address the research questions and explore potential future research directions of FL.

2. Research Method

This study aims to analyze the current trends and sub-areas within the field of FL while examining the tendencies over the years. We leverage a data-driven approach that utilizes keyword extraction and automated clustering techniques to achieve this.

Our analysis begins by collecting a comprehensive dataset of research papers from the Scopus database. We employ a query to identify relevant publications of FL in computer science that are written in English. The exact query we used in advance Scopus searcher is: TITLE-ABS-KEY ("federated learning") AND (LIMIT-TO (EXACTKEYWORD, "Federated Learning")) AND (LIMIT-TO (SUBAREA, "COMP")) AND (LIMIT-TO (LANGUAGE, "English")).

The result of this query, on 15 April 2024, reveals 7953 results without counting the 11 duplicated papers. Subsequently, we extract the keywords from each paper, forming a collection of 22,841 unique keywords that represent the research landscape in FL. Then, to uncover the underlying thematic structure within this keyword collection, we turn to Natural Language Processing (NLP) techniques. We employ a pre-trained transformer model, specifically the `all-mpnet-base-v2` model, to convert each keyword into a 768-dimensional dense vector space. We used the `all-mpnet-base-v2` transformer because it is trained for a total number of sentence pairs above 1 billion sentences (https://huggingface.co/sentence-transformers/all-mpnet-base-v2, accessed on 26 April 2024) and this corpus includes the Semantic Scholar Open Research Corpus (S2ORC), which is a general-purpose corpus for NLP and text mining research over scientific papers [3]. In addition, the `all-mpnet-base-v2` model has the best average performance between the performance of sentence embeddings and the performance of semantic search, over all the Hugging Face pre-trained sentence transformers models (https://www.sbert.net/docs/pretrained_models.html, accessed on 26 April 2024).

These embeddings capture the semantic relationships between keywords, allowing us to group them based on their semantic meaning. We perform agglomerative clustering on the vector embeddings to identify the major thematic trends and sub-areas. This clustering algorithm starts with each keyword as an individual cluster and iteratively merges the most similar clusters based on a distance metric. In this case, we utilize the Euclidean metric to measure the distance between cluster centroids and Ward's linkage to determine the optimal merging strategy. We used the Euclidean distance because effectively captures the inherent semantic relationships among the keywords, ensuring that the clustering process reflects true semantic groupings [4]. Moreover, the Euclidean distance is computationally efficient, facilitating the iterative process of agglomerative clustering, which involves repeated distance calculations between clusters. The final number of clusters, set at 100, provides a granular view of the thematic landscape while maintaining a manageable number of groups for analysis.

By examining the keywords within each cluster, we can identify the key thematic trends and sub-areas that are currently shaping the field of FL. In Table 1 are shown the number of papers of five keyword groups, over the years, of each category presented on this paper. The number of papers, over all the years, of the keywords groups can be found in Tables A1 and A2. This novel data-driven approach allows us to move beyond manual literature reviews and capture the broader dynamics of research activity within the domain. We can then delve deeper into specific clusters to understand the research questions, methodologies, and potential applications that are driving the current evolution of FL.

We made the FL paper dataset public and the keyword clustering results. You can find those files under the following public GitHub repository: https://github.com/FranEnguix/datasets/tree/main/2024%20FL%20Tendencies (accessed on 26 April 2024).

Table 1. The number of papers over the years of the selected keyword groups.

Category	Total	2017	2018	2019	2020	2021	2022	2023	2024
communication	1110	1	1	20	84	182	300	382	140
security	1076	1	1	7	43	110	255	498	161
coalition	942	1	1	8	77	104	297	355	99
data distribution	671	0	0	6	27	72	170	297	99
model aggregations	574	0	0	5	34	92	139	232	72
neural networks	2592	2	3	28	137	327	657	1097	341
classification (of information)	1292	0	1	10	65	172	321	536	187
blockchain	1281	1	0	21	68	147	340	515	189
Internet of Things	1262	0	1	12	53	116	328	541	211
edge computing	1142	0	0	16	73	158	325	417	153
biological system modeling	288	0	0	0	5	26	59	140	58
model compression	277	0	0	3	18	43	58	109	46
speech recognition	273	0	0	1	26	30	84	99	33
real-time systems	241	0	1	5	18	35	53	94	35
game theory	232	0	0	6	16	23	57	90	40

3. Main Theoretical Categories

This section dissects the research landscape by analyzing the publication trends within the following core theoretical areas: security, communication, coalitions, data distribution, and model aggregation. Our analysis, presented in the following subsections, leverages a data-driven trend analysis approach examining the yearly publication volume across these categories. Subsequently, we will present each category, highlighting the novel advances in each sub-area.

3.1. Data Analysis

While the current main sub-areas of FL started with just a handful of publications in 2017 and 2018, there has been a steady rise across all categories, with a sharp increase from 2019 onward, as depicted in Figure 3. This growth highlights the growing interest in FL as a method to collaboratively train ML models without compromising data privacy. Notably, as Figure 4 exposed, the category of "security" shows the most significant rise, reflecting a growing focus on addressing potential vulnerabilities in FL systems. Interestingly, "communication" research, though increasing, has not grown at the same exponential rate as other categories. This suggests that researchers might be prioritizing core security and privacy challenges over delving deeper into optimizing communication efficiency in FL. Overall, the data indicate a maturing field of FL research with a focus on building robust and secure systems for collaborative ML.

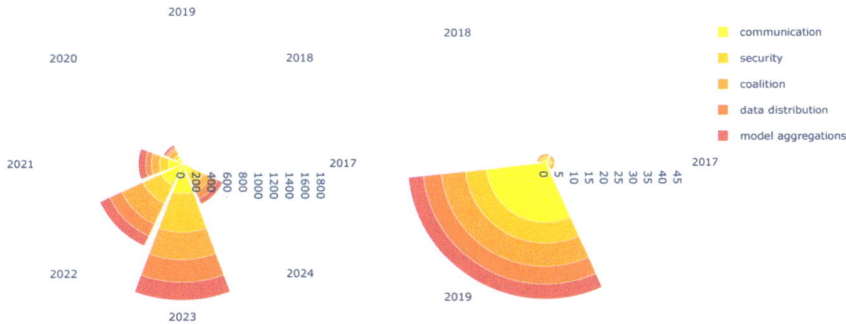

Figure 3. All theoretical keyword category groups over the years.

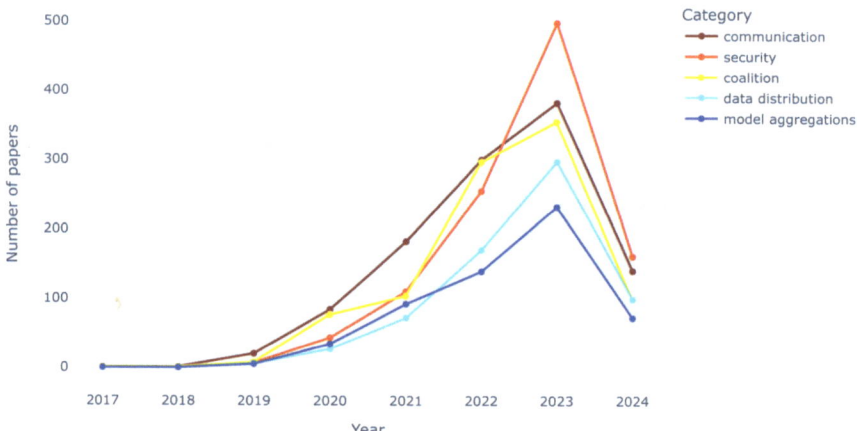

Figure 4. Tendencies of the selected keyword groups over the years.

3.2. Security

FL offers a compelling solution for collaborative machine learning while safeguarding data privacy. However, its core strength—keeping data distributed across devices—also presents a significant security challenge. The FL security research addresses these challenges through a multi-pronged approach, focusing on protecting both model parameters and the underlying data.

3.2.1. Model Inversion Attacks

One major concern is model inversion attacks. In these attacks, malicious participants attempt to reconstruct the training data used to build the model by analyzing the model updates exchanged during the FL process [5–8]. Researchers are developing differential privacy techniques to address this [9,10]. Differential privacy injects controlled noise into model updates, making it statistically impossible to infer any information about individual data points used for training. This technique provides strong data privacy for participants [11].

3.2.2. Poisoning Attacks

Another security threat involves poisoning attacks. Here, malicious actors attempt to manipulate the training process by injecting poisoned data or updates. This can lead to a degraded or biased model [12].

Data Poisoning Attacks

In these attacks, malicious actors inject tampered data points into the training process, aiming to manipulate the FL model to their advantage. These points are designed to mislead the FL model during training, forcing it to learn incorrect patterns or biased outputs that benefit the attacker.

Since FL relies on local participant updates, it can be challenging to detect poisoned data points, especially if they are disguised. Additionally, the distributed nature of FL makes it difficult to pinpoint the source of the attack. Also, there is a novel approach that directly inverts the loss function, generating strong malicious gradients at each training iteration to push the model away from the optimal solution [13].

Techniques like outlier detection algorithms can identify suspicious data points during aggregation [14]. Additionally, robust aggregation methods that down-weight or eliminate extreme updates can further reduce the impact of poisoned data [15].

Model Poisoning Attacks

Unlike data poisoning attacks that focus on corrupting data points, model poisoning attacks target the model updates exchanged during FL. Malicious participants can contribute strategically modified model updates that steer the global model in the desired direction.

Successful model poisoning attacks can cause the global model to learn faulty patterns or biased outputs. This can lead to inaccurate predictions, hindering the functionality of the FL system and potentially causing harm depending on the application. A novel technique named the model shuffle attack (MSA) introduces a unique method of shuffling and scaling model parameters. While the attacker's model appears accurate during testing, it secretly disrupts the training of the global model [16]. This sabotage can significantly slow convergence or even prevent the global model from learning effectively.

Several approaches can help mitigate model poisoning attacks. Cryptographic techniques like SMPC combined with blockchain [17] can be employed to prevent participants from directly observing the model updates, making it harder to inject malicious modifications. Additionally, federated Byzantine fault tolerance (Byzantine-FL) protocols can identify and exclude unreliable or malicious participants from the training process [15], safeguarding the integrity of the federated model.

3.2.3. Membership Inference Attacks

These attacks attempt to determine whether a specific data point belongs to a particular participant's dataset that contributes to the FL training process.

Attackers can potentially infer membership by analyzing the model's predictions on strategically crafted data points. If the model's behavior deviates significantly for a certain input compared to the general prediction pattern, it might indicate the presence of that data point in a participant's training dataset. The PAPI attack is a novel poisoning-assisted property inference attack that targets properties of the training data that are not directly relevant to the model's purpose [18]. By strategically manipulating data labels, a malicious participant can leverage updates to the central model to infer these sensitive properties, even from benign participants.

Existing works have proposed homomorphic encryption and secure multiparty computation (SMC) to address this issue, but these approaches do not apply to large-scale systems with limited computation resources. Differential privacy methods inject noise into the model updates during training, making it statistically harder to link specific data points to participants. Still, it brings a substantial trade-off between privacy budget and model performance. A novel FL framework, based on the computational Diffie–Hellman (CDH) problem to encrypt local models, safeguards against inference attacks [19]. The framework achieves this with minimal impact on model accuracy and computational/communication costs and eliminates the need for secure pairwise communication channels.

3.2.4. Backdoor Attacks

A backdoor attack is a malicious attempt to manipulate a model during training. This is achieved by introducing triggers embedded in the training data. When a sample containing this trigger is fed to the model, it will be misled into producing a specific, attacker-defined output, while functioning normally for all other data. These attacks can be untargeted, aiming to simply degrade the model's overall performance, or targeted, aiming to force the model to misclassify specific triggered samples into a particular category [20].

The attacker achieves this by poisoning the training data. Pairs of data points are created: one being the original training sample and its correct label, and another being the same sample altered with the backdoor trigger and a desired, potentially incorrect, label. The attacker can manipulate the model's learning process by strategically including these poisoned pairs in a small portion of the training data without raising major red flags. This way, the backdoor becomes embedded in the final model, causing it to malfunction when encountering the specific trigger.

The attacks occur during the training phase and rely on a universal trigger that can be added to any sample to activate the backdoor functionality. Backdoor attacks can be particularly concerning as they can bypass standard privacy-preserving techniques in FL. An attacker might steer the model's predictions toward a specific outcome, as they are designed to be subtle and difficult to detect.

An example of a backdoor attack is Cerberus Poisoning (CerP), a new distributed backdoor attack against FL systems [21]. CerP works by having multiple attackers collaborate to fine-tune a backdoor trigger for each of their devices. This makes the poisoned models from the attackers appear more similar to the unpoisoned models from honest users, allowing CerP to bypass existing defenses and successfully embed a backdoor in the final FL model.

While some defenses against label-flipping attacks exist, backdoor attacks are a significant threat. The defense mechanism defending poisoning attacks in FL (DPA-FL) tackles this issue in two phases [22]. First, it compares model weights from participants to identify significant differences, potentially indicating a malicious actor. Second, it tests the aggregated model's accuracy on a dataset, potentially revealing attackers through low performance.

3.3. Communication

Initially, a common depiction featured a central server orchestrating model aggregation, while clients performed local training. This configuration, known as centralized FL, typically employs a star topology. In contrast, decentralized FL, adopting a mesh topology, has gained prominence. In decentralized FL, no central server exists. Instead, clients use peer-to-peer (P2P) communication, exchanging local models directly. This decentralized approach enhances privacy and mitigates reliance on potentially untrusted central servers.

3.3.1. Centralized FL (CFL)

CFL takes a coordinated approach to training a model while keeping data private. Unlike traditional centralized learning—where all data go to one place—CFL leverages a central server to manage the process without ever directly accessing the raw data residing on participants' devices or institutions. This server acts as a conductor, first distributing a starting global model to all participants.

Participants train this model locally on their own datasets, tailoring it to their specific data. Afterward, only the updated model weights, representing the learning from the local training, are uploaded back to the central server. This server then plays a crucial role by aggregating these updates from multiple participants. Combining the knowledge embedded in each update, the central server refines the global model, effectively incorporating the insights from all the distributed datasets. This iterative process of distributing, training locally, and aggregating updates continues until the desired level of model performance is achieved.

3.3.2. Decentralized FL (DFL)

DFL presents an alternative approach that tackles limitations inherent to the central server in CFL. Unlike CFL, DFL dismantles the single point of control, fostering a collaborative learning environment that is both more distributed and potentially more privacy-preserving. This paradigm thrives on direct communication between participating devices or institutions, eliminating the need for a central server altogether. This P2P approach offers potential benefits in reducing communication overhead compared to CFL, as updates can be exchanged directly between participants.

However, removing the central server also complicates the training process. DFL relies on techniques like consensus algorithms [23] to ensure all participants agree on the current state of the global model, a task that becomes more intricate without a central authority. Additionally, ensuring robust security measures remains an active area of research in DFL [24]. DFL offers advantages in privacy and potentially reduces the communication burden compared with the CFL architecture.

3.4. Coalitions

The traditional FL framework treats all participants as equals, raising challenges in efficiency and communication overhead. This section explores the concept of coalitions in FL, a method for grouping agents based on specific criteria. These groupings, known as coalitions, can be formed based on the semantic similarity of the data participants manage or can be formed based on the geographic location and communication radius of participants. Here, we explore these two key approaches to coalition formation:

3.4.1. Semantic-Based Formation

In semantic-based formation, agents are grouped based on the similarity of their data. This ensures that participants within a coalition contribute data that share similar meanings and underlying patterns. This approach can be further classified into:

Static Formation

Here, coalitions are formed based on pre-defined semantic criteria. This could involve analyzing the metadata associated with the data held by each agent and initially classifying the agents into clusters. With static coalitions, once agents are grouped together, these coalitions remain fixed throughout the training process.

Dynamic Formation

Coalitions are formed or reformed continuously based on the semantic similarity of the data itself. ML techniques like automatic semantic clustering, topic modeling, or content analysis can be employed to dynamically assess data similarity and adjust coalition membership accordingly.

3.4.2. Positional-Based Formation

Positional-based formation relies on the geographical proximity of agents and their communication range. This approach is particularly relevant for scenarios where the agents are in different locations and when agents are moving.

Static Formation

Agents within a specific geographical region with a fixed communication range or that are neighbors in the communication graph are grouped into a coalition. In static coalitions, after the initial formation of groups, the group memberships do not change over time.

Dynamic Formation

In dynamic formation, agents can form or leave coalitions based on real-time location updates or changes to their communication range. This could be beneficial in scenarios where data collection is ongoing and the spatial distribution of the agents is constantly changing. Wireless ad hoc networks (WANETs) are examples of this scenario, where agents join or leave groups based on their availability within the wireless range [25].

3.5. Data Distribution

FL deals with training a model collaboratively across multiple participants, each holding their own private data. However, the data distribution across participants can be imbalanced, leading to challenges.

One of FL's major challenges lies in handling statistical heterogeneity within the data. In this context, statistical heterogeneity refers to the non-IID nature of FL data, which deviates from the assumption of identical data distributions across clients. Unlike traditional centralized machine learning, where data are typically drawn from a single source, FL data originates from diverse clients, each with its own unique data distribution. These variations can impact the quality of local models and subsequently affect the performance of the aggregated global model.

3.5.1. Label Distribution Skew

Label distribution skew refers to the unequal distribution of class labels within the training data held by different clients. Some clients may possess a surplus of data belonging to specific classes, while others may have a scarcity for the same classes. This imbalance can significantly impact the performance of the model. Imagine that participant A primarily has data for the class "cat" and very little for "dog", while participant B has the opposite distribution.

When the global model aggregates updates from clients with skewed label distributions, it can become biased toward the over-represented classes. This phenomenon occurs because local models trained on data-rich in certain classes heavily influence global updates. Consequently, the federated model prioritizes learning these dominant classes and neglects the underrepresented ones, leading to decreased accuracy for minority classes and potentially even failing to recognize them altogether.

To address this challenge, exists a novel FL method called FedMGD [26]. FedMGD aims to mitigate the performance degradation caused by label distribution skew. The key innovation lies in introducing a global generative adversarial network (GAN). This GAN operates without access to the raw local datasets, preserving data privacy. However, it can still model the global data distribution by learning from the aggregated model updates received from participants. This allows the global model to be trained using information about the overall data distribution without compromising privacy.

3.5.2. Feature Distribution Skew

While label distribution skew focuses on class imbalance, this phenomenon arises when the distribution of feature values for the same class differs significantly across client datasets. Imagine client A possesses data primarily representing cats with long, white fur, while client B's cat data depicts mostly short-haired black cats. Even if the overall number of cat images (labels) is balanced, the underlying feature distributions (fur length, color) diverge. This disparity affects the model during the training phase.

The model struggles to learn a unified representation of the "cat" class due to the conflicting feature portrayals across clients. This can lead to increased training difficulty and ultimately result in a model with poorer generalization capabilities. The model might perform well on data that resembles the specific feature distributions it encountered during training, but it could struggle with unseen data that deviates from those distributions.

3.5.3. Quantity Skew

Quantity skew refers to the unequal distribution of data samples across participating clients. In this scenario, some clients possess significantly more data points compared to others.

Clients with abundant data exert a greater influence on the global model updates due to the sheer volume of local updates they contribute. This can lead to the model becoming biased toward the data distribution of clients holding more samples. Even if the label and feature distributions are balanced globally, the model might prioritize learning patterns specific to the dominant data source, potentially neglecting valuable information present in smaller datasets from other clients.

This results in a model that performs well on data resembling the dominant client's distribution but exhibits decreased performance on data from clients with less representation.

As presented in this section, a key obstacle in FL is training an effective model when devices possess heterogeneous data, which cannot be directly exchanged. This includes imbalances in label distribution (label skew), feature distribution (feature skew), and data quantity (quantity skew) across devices. To address this issue, a method with a hierarchical FL approach utilizing a hypernetwork (HN) aims to mitigate the negative influence of non-IID data [27]. This method is presented in a landscape of Digital Twin in Industrial IoT. The lower layer of this method leverages hypernetworks to generate local model parameters for each device. The upper layer then refines these hypernetworks by aggregating the model parameters from all devices. This approach decouples the number of parameters transmitted between the upper and lower layers, leading to improved communication efficiency, reduced computation costs, and ultimately, better model accuracy.

3.6. Model Aggregation

As highlighted, FL thrives in scenarios with heterogeneous data distributions across devices. While this protects data privacy, it also presents the challenge of effectively combining these diverse local models into a single, robust global model. This is where model aggregation techniques come into play. These techniques aim to intelligently combine the knowledge learned from individual devices, mitigating the negative effects of non-IID data and leading to a well-performing global model.

3.6.1. Synchronous Aggregation

Synchronous aggregation offers advantages in terms of convergence guarantees and ease of implementation. However, it can be susceptible to stragglers (devices that take significantly longer to train the model locally), delaying the entire update process and potentially hindering training efficiency. Additionally, communication overhead can be high due to the waiting periods before updates are uploaded.

3.6.2. Asynchronous Aggregation

Asynchronous aggregation techniques offer an alternative approach to synchronous aggregation, aiming to address limitations in scalability and efficiency. Unlike the coordinated update scheme of synchronous aggregation, asynchronous aggregation allows devices or institutions participating in FL training to upload their local model updates to the central server as soon as they become available, without waiting for others to finish. This eliminates delays caused by stragglers.

It avoids the communication bottlenecks associated with waiting periods in synchronous methods but introduces complexities in ensuring convergence of the global model, as participants contribute updates at varying times based on their local training speeds. FedTAR is an example of an FL model that uses asynchronous aggregation to minimize the sum energy consumption of all edge computing nodes of a wireless computing power network (WCPN) [28]. There is also the AMA-FES (adaptive-mixing aggregation, feature-extractor sharing) framework, which aims to mitigate the impact of the non-IID data and reduce computation load in a practical scenario where mobile UAVs act as FL training clients to conduct image classification tasks [29].

3.6.3. Hierarchical Aggregation

Hierarchical aggregation emerges as an optimization technique that addresses potential communication bottlenecks in scenarios with large numbers of participants or geographically distributed devices [30]. It also addresses privacy concerns by introducing a layered approach to update aggregation between user devices and the central server.

Hierarchical aggregation mitigates the privacy risk by having devices send their updates to intermediate servers first. These intermediate servers can then aggregate local updates before forwarding them to the central server, reducing the amount of individual data exposed. This approach is particularly valuable for the Industrial Internet of Things (IIoT) where sensitive data from various devices are involved [31].

Participants are organized into groups, forming a hierarchical structure. Local updates within a group are first aggregated, resulting in intermediate updates. These intermediate updates are then sent upwards in the hierarchy for further aggregation until they reach the central server.

Compared to directly sending individual updates to the central server, hierarchical aggregation significantly reduces communication costs. Only a condensed version of the updates travels through the network, alleviating bandwidth limitations and potentially accelerating the training process.

The specific structure of the hierarchy (number of layers, group sizes) can significantly impact efficiency. Additionally, techniques like selective aggregation, where only significant updates propagate through the hierarchy, can further optimize communication costs.

While hierarchical aggregation reduces communication overhead, it introduces an additional layer of information compression during the intermediate aggregation steps. This compression might lead to a certain loss of accuracy in the final global model.

A novel hierarchical FL framework is proposed for cloud–edge–robot collaborative training of deep learning models [32]. This framework allows robots to train the model for quality defect inspection of civil infrastructures without sharing sensitive data among themselves. The system is designed for resource-constrained robots, employing a lightweight model for efficient training and communication.

3.6.4. Robust Aggregation

As presented in Sections 3.2 and 3.5, FL models are susceptible to outliers within participant datasets and even malicious actors injecting poisoned data to manipulate the training process. Robust aggregation methods aim to detect and mitigate the influence of such anomalies on global model updates.

Various approaches can be employed for robust aggregation. These include clipping techniques that limit the magnitude of updates, outlier detection algorithms to identify and down-weight suspicious contributions, and median filtering to prioritize central tendencies within the updates [33,34].

A novel framework is secure and robust FL (SRFL), which is introduced to address security vulnerabilities in existing methods [35]. SRFL tackles the issue of model parameter leakage during aggregation using trusted execution environments (TEEs). This approach safeguards sensitive model components on resource-constrained IoT devices, even in situations with non-IID data. Evaluations demonstrate SRFL's effectiveness in improving accuracy and reducing backdoor attack success rates compared to traditional FL methods.

4. Main Practical Categories

Having explored the main theoretical trends across FL categories, we now delve into the applications driving this field forward. This section focuses on areas where FL is solving real-world problems. We will examine the distribution of research within these categories, including neural networks, classification, blockchain, Internet of Things, and edge computing. Through this analysis, we aim to identify the most promising and actively researched practical applications of FL technology.

4.1. Data Analysis

FL research shows a clear interest in leveraging powerful ML models for practical applications. The category of neural networks dominates the field, as Figures 5 and 6 depicted, with publications experiencing a staggering growth from 2019 to 2023. This highlights the focus on utilizing complex models to achieve superior performance in FL tasks. There is also a significant rise in classification, indicating a strong interest in using FL for tasks like image categorization. The emergence of blockchain and IoT (2019 onward) as prominent categories reflects the growing importance of integrating FL with secure and distributed data architectures. Similarly, edge computing has gained traction as researchers explore enabling FL on resource-constrained devices at the network edge.

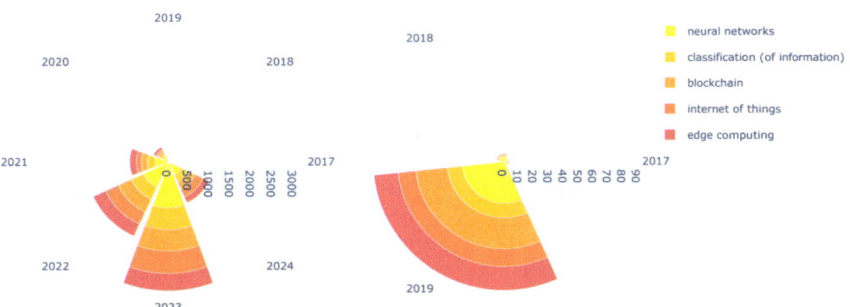

Figure 5. All practical keyword category groups.

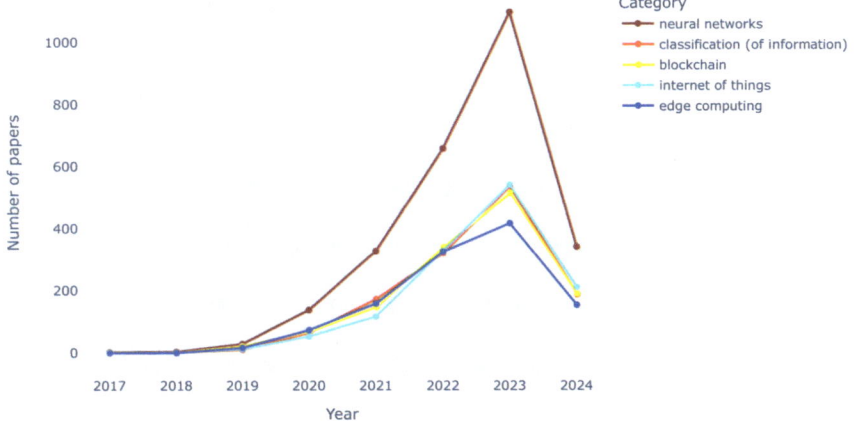

Figure 6. Tendencies of the practical keyword groups over the years.

4.2. Neural Networks

As FL continues its ascent as a privacy-preserving approach to training ML models, the role of neural networks (NNs) within this framework has become a focal point of research. With a growing number of papers dedicated to this topic, it is important to mention the relevant advancements in this field.

This section delineates NN architectures used under the FL framework, where data remain distributed across decentralized nodes while facilitating collaborative model training. Deep neural network (DNN) models tailored for FL encompass convolutional neural networks (CNNs), adept at feature extraction crucial for image processing tasks, and recurrent neural networks (RNNs), specialized in decoding sequential data and temporal dependencies. Furthermore, generative adversarial networks (GANs) demonstrate promise in generating realistic magnetic resonance imaging (MRI) images from undersampled data, while Transformers, initially developed for natural language processing (NLP) tasks, are repurposed to address image capture, information matching, and reconstruction challenges within the FL framework [36].

4.2.1. Traditional DNNs in FL

This section explores the application of CNNs and RNNs for collaborative training while preserving data privacy. We will explore specific use cases in domains like healthcare and cybersecurity, showcasing how FL empowers distributed learning on sensitive data.

CNN (Convolutional Neural Network)

CNNs excel in image processing tasks. Their core strength lies in capturing low-level to high-level features through convolutional operations. This makes them ideal for FL scenarios involving image data, such as medical imaging analysis [37] or object recognition in sensor networks [38]. CNNs can be trained on image data distributed across various devices without compromising privacy. For instance, FL with CNNs can be used to train models for disease detection in medical images without requiring hospitals to share the raw patient data [39].

Stacked CNNs (SCNNs) also excel in the cybersecurity field. A novel intrusion detection system (IDS) for wireless sensor networks (WSNs) based on the FL SCNN-Bi-LSTM model exists, which addresses limitations of traditional methods allowing sensor nodes to collaboratively train a central model without revealing their private data [40]. The SCNN-Bi-LSTM architecture analyzes both local and temporal network patterns to effectively identify even sophisticated and unknown cyber threats.

RNN (Recurrent Neural Network)

RNNs are adept at handling sequential data and capturing temporal dependencies. While not the primary choice for typical image processing tasks, RNNs can be valuable in FL settings where dynamic adjustments are needed. RNNs in healthcare are used for breast cancer detection, which allows hospitals to train an RNN on their mammogram data without sharing the raw images. This study proposes a hybrid approach combining FL with meta-heuristic optimization [41]. Another paper focuses on FL for pancreas segmentation, where data heterogeneity across institutions can hinder performance. To address this, their authors introduce FedRNN, a method that uses an RNN to adjust the aggregation weights based on the past performance of each participating site [42].

4.2.2. Emerging Applications of NN in FL

This section explores how emerging applications of NN in FL offer a revolutionary approach to training ML models while keeping data distributed across devices or servers. We will explore how GANs and Transformers are being leveraged to unlock new potential in FL applications.

GAN (Generative Adversarial Network)

A new approach called "federated synthesis" is emerging within FL. This technique aims to create synthetic data with the same properties as real data but without any privacy risks [43,44]. Researchers are exploring this method using GANs, to combine data from multiple sources while keeping it private. GANs consist of two competing NNs: a generator that creates new data, and a discriminator that tries to distinguish real data from generated data. This adversarial training allows GANs to generate highly realistic synthetic data.

Traditional GAN training requires sending large amounts of data to a central server. CAP-GAN is a novel framework that allows for collaborative training between cloud servers, edge servers, and even individual devices [44]. To address challenges caused by non-IID data, CAP-GAN incorporates a mix generator module that separates general and personalized features, improving performance on highly personalized datasets.

Transformers

Originally developed for NLP tasks, Transformers are powerful architectures based on the attention mechanism. This mechanism allows the model to focus on relevant parts of the input data, making it well-suited for tasks requiring long-range dependencies.

A recent research tackles challenges in medical image analysis with a Transformer-based FL framework. The method uses self-supervised pre-training with Transformers directly on individual institutions' data [45]. This approach overcomes limitations of data sharing and limited labeled data. The study shows significant improvements in accuracy on medical image classification tasks compared to traditional methods, even with variations in data across institutions.

4.3. Classification (of Information)

The field of FL is actively exploring its potential for various classification tasks, including image classification, object detection, and emotion recognition. This is particularly appealing due to the vast amount of labeled data often residing on private devices, which FL can leverage while preserving privacy.

A recent study [46] investigated a privacy-preserving approach to diagnosing skin lesions using FL. While the FL model achieved comparable performance to a traditional centralized model on data from a new hospital, it fell short when tested on data from a different source. Overall, the findings suggest that FL shows promise for melanoma classification while protecting patient privacy.

Another research proposes a new FL framework called FedCAE for fault diagnosis in industrial applications [47]. Traditional approaches require sharing large amounts of data, which can be impractical due to privacy concerns. FedCAE tackles this by using convolutional autoencoders (CAEs) on local devices to extract features from the data. These features are then uploaded to a central server for training a global fault diagnosis classifier, without revealing the raw data itself. The trained classifier is then downloaded to all devices for performing local diagnoses.

4.4. Blockchain

Blockchains enable secure, verifiable interactions between devices without a central authority [48]. The field of FL with blockchain integration, also known as blockchain-based FL (BCFL), is a rapidly evolving area [49]. Researchers are looking to leverage the strengths of both technologies to address limitations in traditional FL.

Recent research proposes a new FL method for blockchain named loosely coupled local differential privacy blockchain federated learning (LL-BCFL) that addresses data privacy and efficiency concerns on federated sharing methods for massive data in blockchain [50]. Traditional blockchain storage can be slow and unsuitable for private data. LL-BCFL tackles this by combining FL on user devices with blockchain storage. The system uses a client selection mechanism to ensure data integrity and participant honesty. Additionally, a local differential privacy mechanism protects against inference attacks during training.

To protect the FL process against poisoning attacks, two models have been developed under BCFL, namely, centralized aggregated BCFL (CA-BCFL) and fully decentralized BCFL (FD-BCFL) [24]. Both leverage secure off-chain computations to mitigate attacks without compromising performance. The study demonstrates that BCFL effectively defends against poisoning attacks while keeping operational costs low.

4.5. Internet of Things

FL has emerged as a powerful approach for the Internet of Things (IoT) domain. It tackles the challenge of training ML models on data generated by vast numbers of resource-constrained devices while preserving user privacy. The FL literature reflects this synergy, highlighting several key areas of advancement.

A major focus is on addressing the limitations of resource-constrained IoT devices. Traditional FL algorithms may not be suitable for devices with limited battery power, storage, and processing capabilities. Researchers are developing techniques like model compression (Section 5.3) and efficient communication (Section 3.3) mechanisms to reduce the computational burden on these devices. This ensures participation from a wider range of IoT devices in the FL algorithm process.

Another area of exploration is heterogeneity. IoT devices often generate data with varying formats and qualities [51]. This heterogeneity can negatively impact the performance of the model. Researchers are proposing data distribution (Section 3.5) techniques to improve the performance and model aggregation methods (Section 3.6) that can handle such inconsistencies. These techniques aim to improve the accuracy and robustness of the collaboratively learned model.

4.6. Edge Computing

While both IoT and edge computing are related to FL, they represent distinct concepts. IoT devices generate the data, while edge computing represents the layer of processing power located at the network's periphery, closer to the data source, and performs local computations [28].

One of the primary research areas is optimizing model performance and resource utilization in resource-constrained edge environments. In Section 5.3 techniques such as quantization and knowledge transfer are exposed, which are tailored to minimize the computational and memory requirements of FL models, making them suitable for deployment on low-power edge devices with limited processing capabilities. Furthermore, edge computing platforms with accelerators like GPUs and TPUs accelerate model inference and training, enhancing the efficiency and scalability of the systems.

FL is well-suited for edge devices, where data processing occurs locally [30]. It enables collaborative model training across devices at the network edge. CAP-GAN is a novel framework using GANs (presented in Section 4.2) in network edge [52]. This research tackles training GANs on devices at the network edge due to privacy and bandwidth limitations. However, traditional GAN training methods struggle with data that is not uniformly distributed across devices. To address this, CAP-GAN allows for parallel training of data and models across devices, cloud servers, and the network edge, overcoming isolated training issues. CAP-GAN introduced a mix generator module to handle highly personalized datasets that are common at the edge. Experiments show that this framework outperforms existing methods in handling non-uniformly distributed data.

5. Emerging Sub-Areas

Having explored the core theoretical categories and the practical application areas of FL research, we now turn our attention to emerging sub-areas. These sub-areas represent new lines of inquiry that have gained significant traction in recent years. Unlike the previously established categories, these sub-areas are distinguished by their later emergence and they are rapidly growing interest within the FL research community. This section delves into five such sub-areas: biological system modeling, model compression, speech recognition, real-time systems, and game theory.

5.1. Data Analysis

While all sub-areas show a clear rise in publications since 2019 and 2020, as Figure 7 depict, some demonstrate a more explosive growth trajectory. Figure 8 shows that biological system modeling exhibits the most dramatic increase, with publications nearly tripling from 2022 to 2023. This suggests a rapidly growing focus on applying FL to model complex biological systems like brain–computer interfaces (BCIs). Model compression also shows a steady and significant rise, highlighting the importance of reducing model size for deployment on resource-constrained devices in FL applications, like IoT or edge devices. Speech recognition and real-time systems show a more moderate but consistent growth, indicating a growing interest in integrating FL with these domains. Game theory, while experiencing a steady rise, has a slightly lower overall number of publications, suggesting it is a relatively new but promising sub-area exploring strategic interactions within FL systems.

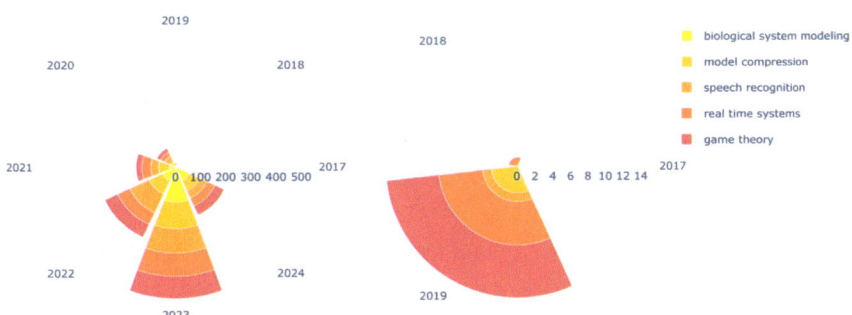

Figure 7. All emerging category groups over the years

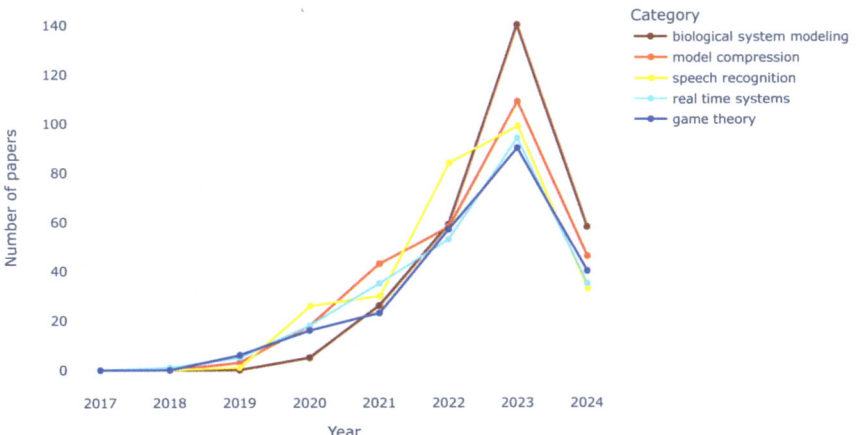

Figure 8. Tendencies of the emerging keyword groups over the years

5.2. Biological System Modeling

Brain–computer interfaces (BCIs) create a bridge between the brain and external devices by translating brain activity into commands [53]. These systems translate brain activity, captured through electroencephalogram (EEG) signals, into commands for external devices. However, a major hurdle in BCI development is the scarcity of data needed to train high-performance models. This is where FL steps in.

FL offers a privacy-preserving approach to training models on distributed datasets residing on individual devices. This eliminates the need for centralized data storage, addressing security and privacy concerns that plague biological datasets.

One recent paper proposes a novel framework called hierarchical personalized FL for EEG decoding (FLEEG) [54]. FLEEG tackles the challenge of device heterogeneity, where BCIs collect data from various sources with potentially different formats. This framework facilitates collaboration in model training across these diverse datasets, enabling knowledge sharing and boosting BCI performance. The studies presented by its researchers have shown that FLEEG can significantly improve classification accuracy, particularly for smaller datasets.

Another paper investigates the application of FL in classifying motor imagery (MI) EEG signals [37]. This approach utilizes a CNN on the PhysioNet dataset and compares two aggregation methods (FedAvg and FedProx) within the FL framework to traditional centralized ML approaches. The results demonstrate that FL can achieve classification accuracy comparable to centralized methods, while significantly reducing the risk of data leakage. This suggests that FL holds significant promise for MI-EEG signal classification in BCI systems.

5.3. Model Compression

FL allows collaborative machine learning without compromising data privacy. However, training these models across distributed devices or servers presents a challenge: model size. Large models can lead to slow communication and hinder the scalability of FL systems. To address this, researchers are actively exploring various model compression techniques.

5.3.1. Quantization

The distributed nature of FL can lead to communication bottlenecks due to the large size of model parameters. Here is where quantization emerges as a powerful technique to address this challenge. Quantization reduces the number of bits required to represent model parameters, significantly shrinking the model size. This translates to faster communication during the FL training process, making it more efficient and scalable. However, accuracy degradation can occur during the quantization process and researchers are actively developing methods to minimize this accuracy loss.

A recent study addresses communication efficiency in hierarchical FL, where model training is distributed across devices, edge servers, and a cloud server [55]. While existing approaches leverage hierarchical aggregation and model quantization to reduce communication costs, this study proposes an accurate convergence bound that considers model quantization. This bound informs practical strategies for client-edge and edge-cloud communication, such as dynamically adjusting aggregation intervals based on network delays. The effectiveness of these strategies is validated through simulations.

Another prominent and recent area of study is the 1-bit quantization. A study proposes a new scheme that uses 1-bit compressive sensing to significantly reduce the amount of data transmitted during model updates [56]. To optimize this method, they analyze the trade-off between communication efficiency and accuracy caused by data compression. The researchers then formulate a solution to minimize these errors through scheduling devices and adjusting transmission power. While an optimal solution exists, it is computationally expensive for large networks. To address this, they develop a more scalable method suitable for real-world applications with many devices. Simulations show this approach achieves comparable performance to traditional FL with significantly less communication, making it a promising technique for large-scale FL.

5.3.2. Knowledge Distillation

Another key approach is knowledge distillation. This technique involves training a smaller, student model to mimic the behavior of a larger, pre-trained teacher model. The student model learns from the teacher's predictions, resulting in a compressed model with comparable accuracy. Knowledge distillation is particularly useful in FL as it allows for transferring knowledge from a powerful trained model to smaller models deployed on user devices.

Recent research develops an intrusion detection method based on a semi-supervised FL scheme via knowledge distillation [57]. The study proposes an intrusion detection method in IoT devices. Existing FL methods for intrusion detection raise privacy concerns and struggle with non-private data distributions. To address this, the authors developed a method that leverages unlabeled data to improve detection accuracy while protecting privacy. Their approach uses a special NN model to both classify traffic data and assess the quality of labels generated by individual devices. This, combined with a hard-label strategy and voting mechanism, reduces communication overhead.

5.3.3. Pruning

Another promising direction is pruning. Pruning techniques identify and remove redundant or unimportant weights within a model. This process reduces the model's overall size without significantly impacting its performance. Advanced pruning algorithms can identify weights with minimal influence on the final output, allowing for compression while maintaining accuracy.

PruneFL is a new framework for FL that improves training efficiency on resource-constrained devices [58]. FL trains models on distributed data while protecting privacy, but edge devices often lack processing power and bandwidth. PruneFL tackles this by dynamically reducing model size during training through a distributed pruning approach. This reduces communication and computation requirements while maintaining accuracy. The method involves an initial pruning step and further pruning throughout the FL process, optimizing the model size for efficiency. Experiments on real-world datasets running on devices like the Raspberry Pi demonstrate that PruneFL significantly reduces training time compared to traditional FL and achieves comparable accuracy to the original model with a smaller size.

5.3.4. Sparsification

Researchers are also exploring sparsification techniques. Here, the focus is on converting model weights from dense matrices to sparse ones, containing mostly zeros. Sparse models require less memory and communication bandwidth, making them ideal for FL applications. Recent advancements involve combining sparsification with other compression methods like pruning to achieve even more compact models.

GossipFL is a novel framework that utilizes sparsification and gossiping to optimize bandwidth usage while ensuring training convergence. The authors designed a novel sparsification algorithm that enables each client to communicate with only one peer using a highly sparsified model [59]. Theoretical analysis and experiments using GossipFL demonstrate that this framework significantly reduces communication traffic and time compared to existing solutions while maintaining similar model accuracy.

5.4. Speech Recognition (SR)

SR is a technology that allows computers to translate spoken words into written text. This is achieved by analyzing speech's sound waves and identifying patterns corresponding to specific words or phonemes, which are the basic units of sound in a language.

Traditional SR models require vast datasets centralized in one location for training. This raises privacy concerns, especially for applications like forensic analysis, where data sensitivity is paramount.

The fight against online child exploitation is an example, where European law enforcement agencies (LEAs) require advanced tools to analyze the growing volume of audio data. Recent research explores FL as a solution for training SR models in this domain [60]. While the study compares the effectiveness of WAV2VEC2.0 and WHISPER models, the main focus lies in leveraging FL to overcome data privacy concerns.

The results show that FL models achieve word error rates (WERs) comparable to those trained in a traditional, centralized manner. This is particularly significant considering the challenges of non-IID data distribution, where the data used have unique characteristics due to languages, accents, or recording environments.

5.5. Real-Time Systems

The traditional approach of FL involves a central server aggregating updates from participants periodically. This raises limitations for applications demanding real-time performance. The research in FL delves into techniques for enabling real-time FL that ensure low latency.

Traditional periodic updates can introduce delays that hinder real-time responsiveness. Synchronous FL can lead to slow learning due to stragglers, which are devices that take longer to process information. A novel approach that breaks away from the limitations of synchronous FL uses scalable asynchronous FL for real-time surveillance systems [61]. Asynchronous FL allows devices to participate in the training process at their own pace, eliminating the bottleneck created by stragglers. This makes asynchronous FL a more suitable solution for large-scale, real-time applications where fast response is critical.

As it is presented in Section 3.2, security and privacy are paramount concerns in any FL system, and real-time settings pose additional challenges. Researchers are actively developing privacy-preserving communication protocols for real-time FL. Techniques like differential privacy [11] are being explored to achieve this balance between real-time performance and data security [62].

5.6. Game Theory

Game theory is a powerful mathematical field used to analyze situations where multiple parties (agents or players) interact and make decisions that can impact each other's outcomes [63]. Imagine a game of chess, where each agent considers not only their own possible moves but also how their opponent might respond. Game theory extends this concept to any situation where competing actors make strategic decisions in a setting with defined rules.

The core concept in game theory is the game itself, which acts as a model for the interactive situation. Each agent is a rational entity with well-defined preferences and a set of possible strategies they can employ. The key element is that an agent's success depends not only on their own choices but also on the strategies chosen by other players. A game will define the players, their available strategies, and how these strategies influence the final outcome for everyone involved.

Existing solutions based on game theory often assume perfect rationality in participants, so motivating participants to contribute to FL systems is an open field of research, which is used for collaborative training. A new model based on evolutionary game theory acknowledges participants' non-perfect decision-making in the long run [64]. By analyzing various scenarios, they identify strategies for parameter servers (coordinating the training) to maintain a sustainable FL system where participants are incentivized to contribute.

As commented, a limitation in existing game theory-related FL frameworks is the assumption of voluntary participation, but also it is the lack of defense against malicious actors. To address this, researchers propose a new scheme based on privacy-preserving techniques and game theory [62]. This scheme incentivizes participation through truthful mechanisms and limits the influence of malicious clients, all while achieving privacy guarantees.

6. Conclusions and Future Work

This study leverages advanced automated semantic keyword clustering techniques to analyze trends, tendencies, and emerging areas within the growing field of FL. By employing a transformer-based model, particularly the `all-mpnet-base-v2` model, the research identifies and groups 22,841 unique keywords of 7953 research articles based on their semantic meaning, providing a comprehensive view of the current state and future directions of the FL research landscape.

We present key research questions (RQs), revealing significant trends in security and communication as dominant areas of interest. The surge in publications related to these categories highlights the importance of addressing vulnerabilities and optimizing communication efficiency in FL systems. Furthermore, the analysis identifies the rising significance of coalitions, data distribution strategies, and model aggregation techniques, which are crucial for tackling challenges related to non-IID data and improving the performance of global models.

Emerging sub-areas such as biological system modeling, model compression, speech recognition, real-time systems, and the application of game theory present promising avenues for future research. These sub-areas show the field's dynamic nature and its potential for interdisciplinary applications.

To conclude, the answers to the RQs provide a structured understanding of the FL landscape: identifying the current trends (RQ1), examining their tendencies (RQ2), exploring practical application domains (RQ3), analyzing the tendencies within these domains (RQ4), uncovering emerging sub-areas (RQ5), investigating their tendencies (RQ6), and predicting potential future trends (RQ7).

6.1. RQ1: What Are the Current Trends in FL?

This question focuses on identifying the theoretical dominant areas of interest in FL research. FL is experiencing a period of significant research growth, as evidenced by the substantial increase in publications across all categories analyzed in Section 3.1. The data reveal several key trends that are shaping the current landscape of FL research: security, communication, coalitions, data distribution, and model aggregations.

The most prominent trend is the surge of interest in security, with 498 publications in 2023. This highlights a growing concern for addressing potential vulnerabilities in FL systems, as data privacy is paramount when training models collaboratively. Similarly, the rise in communication (382 publications in 2023) reflects the importance of optimizing communication efficiency, especially as the number of participating devices and the complexity of models increase.

6.2. RQ2: What Are the Tendencies of the Current Trends in FL?

This section delves deeper into RQ1 by analyzing the direction of the identified trends in FL. The analysis of publication trends across the theoretical FL categories reveals not only a surge in interest but also the trajectory of these trends.

The most notable observation is the explosive growth in both security and communication research since 2019. Coalitions show a consistent upward trend with a peak in 2023. This indicates a sustained interest in exploring how devices or institutions can group together to optimize FL speed convergence and accuracy of the trained models. For data distribution and model aggregation categories, the substantial rise suggests a growing interest in tackling challenges related to non-IID data and improving model aggregation techniques.

6.3. RQ3: What Are the Application Domains Where FL Techniques Are Applied?

This question explores the most relevant practical applications where FL techniques are being utilized. The data presented in Section 4.1 reveals a diverse range of application domains where FL techniques are finding utility. The most important key trends are the dominance of neural networks (NNs) and the emergence of secure and distributed architecture.

6.4. RQ4: What Are the Tendencies of the Application Domains?

As RQ2 delves deeper into RQ1, this RQ investigates the trends within the RQ3 identified domains.

Firstly, NNs stand out as the most prevalent category. This signifies a strong focus on leveraging powerful ML models to achieve superior performance in FL tasks. The significant and steady rise in publications suggests that researchers are actively exploring how to adapt and optimize complex NNs for collaborative learning in FL systems.

Beyond NNs, the data highlight a growing interest in integrating FL with secure and distributed data architectures. The rise of categories like blockchain and the Internet of Things (IoT) reflects this trend.

6.5. RQ5: What Are the Emerging Sub-Areas within FL?

Recognizing the potential for further exploration, we propose additional research questions that focus on under-researched areas of FL. This question aims to identify new or niche areas that have received less attention but hold promise for future development.

The analysis in Section 5.1 reveals several promising sub-areas that have garnered increasing attention in recent years. Among the most prominent emerging sub-areas are biological system modeling and model compression. Speech recognition and real-time systems are other emerging sub-areas with significant potential, with a close number of publications to model compression in 2023.

6.6. RQ6: What Are the Tendencies of the Emerging Sub-Areas?

We analyze the identified sub-areas in RQ5 to understand their growth trajectory and potential impact on the broader FL landscape.

Biological system modeling is the most rapidly growing sub-area with topics like bioinformatics and brain–computer interfaces. Game theory, while not exhibiting the most dramatic number of publications, also appears as an emerging sub-area with initial exploration beginning around 2019. This sub-area investigates strategic interactions within FL systems, which could be beneficial for areas like resource allocation or ensuring fairness among participants.

6.7. RQ7: What Are the Potential Future Trends of FL?

Finally, to provide a more comprehensive picture, we introduce this additional question, which looks ahead to predict potential future directions and areas of growth in FL research.

The consistently increasing number of publications in NNs suggests a continued focus on leveraging powerful models for FL tasks. Also, we can expect sustained research efforts in core areas like security and communication efficiency, as the significant rise in publications until 2023 highlights their importance. Researchers might focus on developing more robust security mechanisms to address evolving threats and optimizing communication protocols for specific federated learning applications.

Another core area is data distribution, which is likely to see continued growth. With the increasing interest in applying FL to real-world scenarios involving non-IID data, researchers will likely explore more sophisticated techniques to handle data heterogeneity and improve model performance.

6.8. Future Work

Future work will focus on expanding the software developed to include other database sources and utilizing the software to experiment with the linkage method and the distance metric of the agglomerative clustering algorithm and explore different clustering algorithms.

The results using Euclidean distance and Ward's linkage are used in this research article to group the keywords by their semantic meaning, offering significant insights into FL research trends. In future work, experimenting with different parameter values will enable us to assess the impact of different distance metrics, such as cosine similarity and Manhattan distance, on the clustering results. Additionally, experimenting with various linkage methods, including single linkage, complete linkage, and average linkage, will allow us to compare strategies for forming thematic clusters.

Funding: This research was funded by MCIN/AEI/10.13039/501100011033 and "ERDF A way of making Europe" grant number PID2021-123673OB-C31 and funded by VAE-VADEN UPV grant number TED2021-131295B-C32 and funded by GUARDIA grant number PROMETEO CIPROM/2021/077 and funded by Ayudas del Vicerrectorado de Investigacion de la UPV grant number PAID-PD-22.

Conflicts of Interest: The authors declare no conflicts of interest. The funders had no role in the design of the study; in the collection, analyses, or interpretation of data; in the writing of the manuscript; or in the decision to publish the results.

Appendix A

Table A1. The first 50 keyword groups ordered by the overall number of papers.

Rank	Category	Total	2017	2018	2019	2020	2021	2022	2023	2024
0	federated learning	7953	2	6	85	393	964	2027	3394	1082
1	learning systems	5028	2	3	42	238	571	1266	2184	722
2	privacy	4175	2	3	47	245	521	1035	1754	568
3	machine learning	3458	1	2	46	183	425	902	1446	453
4	neural networks	2592	2	3	28	137	327	657	1097	341
5	global models	1568	0	2	16	63	211	402	679	195
6	data models	1551	0	3	29	101	202	382	581	253
7	computational modeling	1460	0	1	9	74	145	336	645	250
8	classification (of information)	1292	0	1	10	65	172	321	536	187
9	blockchain	1281	1	0	21	68	147	340	515	189
10	modeling accuracy	1269	0	1	16	63	154	294	539	202
11	Internet of Things	1262	0	1	12	53	116	328	541	211
12	artificial intelligence	1262	1	2	18	80	154	318	509	180
13	decentralized	1233	1	2	19	78	145	308	523	157
14	performance	1183	0	0	8	57	155	302	495	166
15	state of the art	1177	0	1	19	80	183	296	473	125
16	learning frameworks	1163	0	2	24	73	188	302	442	132
17	edge computing	1142	0	0	16	73	158	325	417	153
18	personalizations	1115	0	2	18	59	148	307	461	120
19	communication	1110	1	1	20	84	182	300	382	140
20	poisoning attacks	1095	0	1	6	41	115	270	485	177
21	security	1076	1	1	7	43	110	255	498	161
22	job analysis	1065	0	2	18	46	126	271	416	186
23	large amounts	1055	0	1	15	55	139	276	417	152
24	computational efficiency	1014	1	1	18	68	145	257	395	129
25	distributed machine learning	1013	0	3	13	91	171	264	364	107
26	over the airs	967	0	1	7	35	81	255	448	140
27	coalition	942	1	1	8	77	104	297	355	99
28	centralized	934	0	1	7	27	108	263	393	135
29	servers	929	0	0	2	39	95	227	390	176
30	commerce	908	0	2	19	53	133	243	359	99
31	wireless networks	899	0	2	11	48	106	234	378	120
32	information management	894	0	1	12	46	99	217	397	122
33	optimizations	735	0	1	7	48	85	190	293	111
34	network architecture	704	0	1	10	51	93	182	299	68
35	numerical methods	687	0	1	5	47	88	177	267	102
36	budget control	673	0	0	10	31	80	185	267	100
37	data distribution	671	0	0	6	27	72	170	297	99
38	iterative methods	664	0	1	7	38	67	149	296	106
39	smart city	656	0	1	11	42	100	168	257	77
40	benchmarking	637	1	1	9	53	99	142	255	77
41	energy utilization	627	0	1	6	26	86	173	252	83
42	human	613	0	0	4	20	48	156	293	92
43	forecasting	611	0	0	7	26	82	159	239	98
44	cloud computing	605	0	1	10	35	68	165	244	82
45	transfer learning	596	0	1	6	21	56	153	253	106
46	distillation	596	0	1	5	25	69	135	263	98
47	health care	594	0	0	4	21	77	139	280	73
48	diseases	588	0	0	0	14	60	123	296	95
49	model aggregations	574	0	0	5	34	92	139	232	72

Table A2. The last 50 keyword groups ordered by the overall number of papers.

Rank	Category	Total	2017	2018	2019	2020	2021	2022	2023	2024
50	computer vision	572	0	1	6	32	64	140	251	78
51	task analysis	567	0	0	3	20	49	143	245	107
52	5g mobile communication systems	545	0	1	7	36	75	141	216	69
53	bandwidth	529	0	1	7	50	71	123	207	70
54	current	523	0	0	6	15	62	142	229	69
55	signal processing	510	0	0	5	36	78	129	190	72
56	antennas	499	0	0	3	29	71	130	197	69
57	quality of service	495	0	0	13	41	57	134	184	66
58	diagnosis	491	0	0	1	8	59	83	265	75
59	stochastic systems	487	0	1	5	28	67	118	201	67
60	image enhancement	487	0	0	3	20	51	132	208	73
61	decision making	486	0	0	7	24	59	134	199	63
62	resource allocation	480	0	2	6	22	63	128	190	69
63	inference attacks	476	0	0	1	28	60	101	209	77
64	convergence	468	0	0	3	20	54	101	210	80
65	vehicles	466	0	1	4	23	55	121	181	81
66	intelligent vehicle highway systems	458	0	0	2	18	54	110	203	71
67	digital storage	446	0	0	6	29	47	127	185	52
68	cryptography	438	1	0	9	31	48	92	181	76
69	matrix algebra	433	1	0	7	24	51	116	166	68
70	reinforcement learning	427	0	0	5	20	49	111	178	64
71	risk assessment	424	0	0	6	23	60	94	199	42
72	intrusion detection	416	0	0	1	13	46	108	189	59
73	iid data	403	0	0	2	12	44	114	180	51
74	large scales	397	0	0	3	18	52	112	160	52
75	medical imaging	369	0	0	1	12	34	85	172	65
76	incentive mechanism	352	0	0	7	32	47	92	118	56
77	clustering	350	1	0	2	11	44	92	153	47
78	channel state information	342	0	0	4	21	55	104	118	40
79	gradient methods	334	0	0	6	24	49	87	129	39
80	Industrial Internet of Things	301	0	0	2	19	51	71	121	37
81	biological system modeling	288	0	0	0	5	26	59	140	58
82	speech recognition	273	0	0	1	26	30	84	99	33
83	real-time systems	241	0	1	5	18	35	53	94	35
84	game theory	232	0	0	6	16	23	57	90	40
85	graph neural networks	195	0	0	1	1	18	41	97	37
86	machine design	180	0	1	1	27	34	37	55	25
87	unmanned aerial vehicles (UAV)	176	0	0	0	7	30	43	70	26
88	spatial-temporal	174	0	0	1	10	15	49	73	26
89	labeled data	174	0	0	2	9	25	41	75	22
90	traffic congestion	166	0	0	1	9	24	40	65	27
91	quantization	164	0	0	0	6	24	40	61	33
92	sensor nodes	156	0	0	3	10	21	27	70	25
93	model compression	143	0	0	3	15	23	24	60	18
94	data sample	138	0	0	2	13	20	28	59	16
95	tumors	132	0	0	0	5	11	32	67	17
96	hyperparameter	128	0	0	4	12	20	30	50	12
97	synchronization	121	0	0	1	4	20	28	55	13
98	leaf disease	118	0	0	1	0	3	16	85	13
99	web services	90	0	0	3	11	9	36	25	6

References

1. Konečný, J.; McMahan, H.B.; Yu, F.X.; Richtárik, P.; Suresh, A.T.; Bacon, D. Federated Learning: Strategies for Improving Communication Efficiency. *arXiv* **2016**, arXiv:1610.05492.
2. Vaswani, A.; Shazeer, N.; Parmar, N.; Uszkoreit, J.; Jones, L.; Gomez, A.N.; Kaiser, L.; Polosukhin, I. Attention is all you need. In Proceedings of the 31st International Conference on Neural Information Processing Systems, NIPS'17, Red Hook, NY, USA, 4–9 December 2017; pp. 6000–6010.

3. Lo, K.; Wang, L.L.; Neumann, M.; Kinney, R.; Weld, D. S2ORC: The Semantic Scholar Open Research Corpus. In Proceedings of the 58th Annual Meeting of the Association for Computational Linguistics, Online, 5–10 July 2020; pp. 4969–4983. [CrossRef]
4. Hashimoto, T.B.; Alvarez-Melis, D.; Jaakkola, T.S. Word embeddings as metric recovery in semantic spaces. *Trans. Assoc. Comput. Linguist.* **2016**, *4*, 273–286. [CrossRef]
5. Fredrikson, M.; Jha, S.; Ristenpart, T. Model Inversion Attacks that Exploit Confidence Information and Basic Countermeasures. In Proceedings of the 22nd ACM SIGSAC Conference on Computer and Communications Security, CCS '15, New York, NY, USA, 12–16 October 2015; pp. 1322–1333. [CrossRef]
6. Zhang, L.; Xu, J.; Vijayakumar, P.; Sharma, P.K.; Ghosh, U. Homomorphic Encryption-Based Privacy-Preserving Federated Learning in IoT-Enabled Healthcare System. *IEEE Trans. Netw. Sci. Eng.* **2023**, *10*, 2864–2880. [CrossRef]
7. Schlegel, R.; Kumar, S.; Rosnes, E.; Amat, A.G.i. CodedPaddedFL and CodedSecAgg: Straggler Mitigation and Secure Aggregation in Federated Learning. *IEEE Trans. Commun.* **2023**, *71*, 2013–2027. [CrossRef]
8. Asad, M.; Shaukat, S.; Javanmardi, E.; Nakazato, J.; Bao, N.; Tsukada, M. Secure and Efficient Blockchain-Based Federated Learning Approach for VANETs. *IEEE Internet Things J.* **2024**, *11*, 9047–9055. [CrossRef]
9. Qiao, F.; Li, Z.; Kong, Y. A Privacy-Aware and Incremental Defense Method Against GAN-Based Poisoning Attack. *IEEE Trans. Comput. Soc. Syst.* **2024**, *11*, 1708–1721. [CrossRef]
10. Zhou, J.; Wu, N.; Wang, Y.; Gu, S.; Cao, Z.; Dong, X.; Choo, K.K.R. A Differentially Private Federated Learning Model Against Poisoning Attacks in Edge Computing. *IEEE Trans. Dependable Secur. Comput.* **2023**, *20*, 1941–1958. [CrossRef]
11. Dwork, C. Differential privacy. In Proceedings of the International Colloquium on Automata, Languages, and Programming, Venice, Italy, 10–14 July 2006; Springer: Berlin/Heidelberg, Germany, 2006; pp. 1–12.
12. Jiang, W.; Li, H.; Liu, S.; Ren, Y.; He, M. A flexible poisoning attack against machine learning. In Proceedings of the ICC 2019—2019 IEEE International Conference on Communications (ICC), Shanghai, China, 20–24 May 2019; IEEE: Piscataway, NJ, USA, 2019; pp. 1–6.
13. Gupta, P.; Yadav, K.; Gupta, B.B.; Alazab, M.; Gadekallu, T.R. A Novel Data Poisoning Attack in Federated Learning based on Inverted Loss Function. *Comput. Secur.* **2023**, *130*, 103270. [CrossRef]
14. Omran, A.H.; Mohammed, S.Y.; Aljanabi, M. Detecting Data Poisoning Attacks in Federated Learning for Healthcare Applications Using Deep Learning. *Iraqi J. Comput. Sci. Math.* **2023**, *4*, 225–237. [CrossRef]
15. Li, S.; Ngai, E.; Voigt, T. Byzantine-Robust Aggregation in Federated Learning Empowered Industrial IoT. *IEEE Trans. Ind. Inform.* **2023**, *19*, 1165–1175. [CrossRef]
16. Yang, M.; Cheng, H.; Chen, F.; Liu, X.; Wang, M.; Li, X. Model poisoning attack in differential privacy-based federated learning. *Inf. Sci.* **2023**, *630*, 158–172. [CrossRef]
17. Kalapaaking, A.P.; Khalil, I.; Yi, X. Blockchain-Based Federated Learning with SMPC Model Verification against Poisoning Attack for Healthcare Systems. *IEEE Trans. Emerg. Top. Comput.* **2024**, *12*, 269–280. [CrossRef]
18. Wang, Z.; Huang, Y.; Song, M.; Wu, L.; Xue, F.; Ren, K. Poisoning-Assisted Property Inference Attack against Federated Learning. *IEEE Trans. Dependable Secur. Comput.* **2023**, *20*, 3328–3340. [CrossRef]
19. Zhao, P.; Cao, Z.; Jiang, J.; Gao, F. Practical Private Aggregation in Federated Learning against Inference Attack. *IEEE Internet Things J.* **2023**, *10*, 318–329. [CrossRef]
20. Gong, X.; Chen, Y.; Wang, Q.; Kong, W. Backdoor Attacks and Defenses in Federated Learning: State-of-the-Art, Taxonomy, and Future Directions. *IEEE Wirel. Commun.* **2023**, *30*, 114–121. [CrossRef]
21. Lyu, X.; Han, Y.; Wang, W.; Liu, J.; Wang, B.; Liu, J.; Zhang, X. Poisoning with Cerberus: Stealthy and Colluded Backdoor Attack against Federated Learning. *Proc. AAAI Conf. Artif. Intell.* **2023**, *37*, 9020–9028. [CrossRef]
22. Lai, Y.C.; Lin, J.Y.; Lin, Y.D.; Hwang, R.H.; Lin, P.C.; Wu, H.K.; Chen, C.K. Two-phase Defense against Poisoning Attacks on Federated Learning-based Intrusion Detection. *Comput. Secur.* **2023**, *129*, 103205. [CrossRef]
23. Carrascosa, C.; Rincón, J.; Rebollo, M. Co-Learning: Consensus-based Learning for Multi-Agent Systems. In Proceedings of the Advances in Practical Applications of Agents, Multi-Agent Systems, and Complex Systems Simulation. The PAAMS Collection, L'Aquila, Italy, 13–15 July 2022; Dignum, F., Mathieu, P., Corchado, J.M., De La Prieta, F., Eds.; Springer: Cham, Switzerland, 2022; pp. 63–75.
24. Thennakoon, R.; Wanigasundara, A.; Weerasinghe, S.; Seneviratne, C.; Siriwardhana, Y.; Liyanage, M. Decentralized Defense: Leveraging Blockchain against Poisoning Attacks in Federated Learning Systems. In Proceedings of the 2024 IEEE 21st Consumer Communications & Networking Conference (CCNC), Las Vegas, NV, USA, 6–9 January 2024; pp. 950–955. [CrossRef]
25. Rebollo, M.; Rincon, J.A.; Hernández, L.; Enguix, F.; Carrascosa, C. Extending the Framework for Developing Intelligent Virtual Environments (FIVE) with Artifacts for Modeling Internet of Things Devices and a New Decentralized Federated Learning Based on Consensus for Dynamic Networks. *Sensors* **2024**, *24*, 1342. [CrossRef]
26. Sheng, T.; Shen, C.; Liu, Y.; Ou, Y.; Qu, Z.; Liang, Y.; Wang, J. Modeling global distribution for federated learning with label distribution skew. *Pattern Recognit.* **2023**, *143*, 109724. [CrossRef]
27. Yang, J.; Jiang, W.; Nie, L. Hypernetworks-Based Hierarchical Federated Learning on Hybrid Non-IID Datasets for Digital Twin in Industrial IoT. *IEEE Trans. Netw. Sci. Eng.* **2024**, *11*, 1413–1423. [CrossRef]
28. Sun, W.; Li, Z.; Wang, Q.; Zhang, Y. FedTAR: Task and Resource-Aware Federated Learning for Wireless Computing Power Networks. *IEEE Internet Things J.* **2023**, *10*, 4257–4270. [CrossRef]

29. Li, J.; Liu, X.; Mahmoodi, T. Federated Learning in Heterogeneous Wireless Networks with Adaptive Mixing Aggregation and Computation Reduction. *IEEE Open J. Commun. Soc.* **2024**, *5*, 2164–2182. [CrossRef]
30. Wu, Q.; Chen, X.; Ouyang, T.; Zhou, Z.; Zhang, X.; Yang, S.; Zhang, J. HiFlash: Communication-Efficient Hierarchical Federated Learning with Adaptive Staleness Control and Heterogeneity-Aware Client-Edge Association. *IEEE Trans. Parallel Distrib. Syst.* **2023**, *34*, 1560–1579. [CrossRef]
31. Chen, J.; Xue, J.; Wang, Y.; Huang, L.; Baker, T.; Zhou, Z. Privacy-Preserving and Traceable Federated Learning for data sharing in industrial IoT applications. *Expert Syst. Appl.* **2023**, *213*, 119036. [CrossRef]
32. Wu, H.T.; Li, H.; Chi, H.L.; Kou, W.B.; Wu, Y.C.; Wang, S. A hierarchical federated learning framework for collaborative quality defect inspection in construction. *Eng. Appl. Artif. Intell.* **2024**, *133*, 108218. [CrossRef]
33. Uddin, M.P.; Xiang, Y.; Cai, B.; Lu, X.; Yearwood, J.; Gao, L. ARFL: Adaptive and Robust Federated Learning. *IEEE Trans. Mob. Comput.* **2024**, *23*, 5401–5417. [CrossRef]
34. Yang, H.; Gu, D.; He, J. A Robust and Efficient Federated Learning Algorithm against Adaptive Model Poisoning Attacks. *IEEE Internet Things J.* **2024**, *11*, 16289–16302. [CrossRef]
35. Cao, Y.; Zhang, J.; Zhao, Y.; Su, P.; Huang, H. SRFL: A Secure & Robust Federated Learning framework for IoT with trusted execution environments. *Expert Syst. Appl.* **2024**, *239*, 122410. [CrossRef]
36. Hossain, M.B.; Shinde, R.K.; Oh, S.; Kwon, K.C.; Kim, N. A Systematic Review and Identification of the Challenges of Deep Learning Techniques for Undersampled Magnetic Resonance Image Reconstruction. *Sensors* **2024**, *24*, 753. [CrossRef]
37. Ghader, M.; Farahani, B.; Rezvani, Z.; Shahsavari, M.; Fazlali, M. Exploiting Federated Learning for EEG-based Brain-Computer Interface System. In Proceedings of the 2023 IEEE International Conference on Omni-Layer Intelligent Systems (COINS), Berlin, Germany, 23–25 July 2023; pp. 1–6. [CrossRef]
38. Mehta, S.; Kukreja, V.; Gupta, A. Next-Generation Wheat Disease Monitoring: Leveraging Federated Convolutional Neural Networks for Severity Estimation. In Proceedings of the 2023 4th International Conference for Emerging Technology (INCET), Belgaum, India, 26–28 May 2023; pp. 1–6. [CrossRef]
39. Pandianchery, M.S.; Sowmya, V.; Gopalakrishnan, E.A.; Ravi, V.; Soman, K.P. Centralized CNN–GRU Model by Federated Learning for COVID-19 Prediction in India. *IEEE Trans. Comput. Soc. Syst.* **2024**, *11*, 1362–1371. [CrossRef]
40. Bukhari, S.M.S.; Zafar, M.H.; Houran, M.A.; Moosavi, S.K.R.; Mansoor, M.; Muaaz, M.; Sanfilippo, F. Secure and privacy-preserving intrusion detection in wireless sensor networks: Federated learning with SCNN-Bi-LSTM for enhanced reliability. *Ad Hoc Netw.* **2024**, *155*, 103407. [CrossRef]
41. Kumbhare, S.; Kathole, A.B.; Shinde, S. Federated learning aided breast cancer detection with intelligent Heuristic-based deep learning framework. *Biomed. Signal Process. Control* **2023**, *86*, 105080. [CrossRef]
42. Deng, Z.; Qureshi, T.A.; Javed, S.; Wang, L.; Christodoulou, A.G.; Xie, Y.; Gaddam, S.; Pandol, S.J.; Li, D. FedRNN: Federated Learning with RNN-Based Aggregation on Pancreas Segmentation. In Proceedings of the Medical Imaging and Computer-Aided Diagnosis, San Diego, CA, USA, 19–23 February 2023; Su, R., Zhang, Y., Liu, H., Frangi, A.F., Eds.; Springer: Singapore, 2023; pp. 453–464.
43. Little, C.; Elliot, M.; Allmendinger, R. Federated learning for generating synthetic data: A scoping review. *Int. J. Popul. Data Sci.* **2023**, *8*. [CrossRef] [PubMed]
44. Cai, X.; Lan, Y.; Zhang, Z.; Wen, J.; Cui, Z.; Zhang, W. A Many-Objective Optimization Based Federal Deep Generation Model for Enhancing Data Processing Capability in IoT. *IEEE Trans. Ind. Inform.* **2023**, *19*, 561–569. [CrossRef]
45. Yan, R.; Qu, L.; Wei, Q.; Huang, S.C.; Shen, L.; Rubin, D.L.; Xing, L.; Zhou, Y. Label-Efficient Self-Supervised Federated Learning for Tackling Data Heterogeneity in Medical Imaging. *IEEE Trans. Med Imaging* **2023**, *42*, 1932–1943. [CrossRef] [PubMed]
46. Haggenmüller, S.; Schmitt, M.; Krieghoff-Henning, E.; Hekler, A.; Maron, R.C.; Wies, C.; Utikal, J.S.; Meier, F.; Hobelsberger, S.; Gellrich, F.F.; et al. Federated Learning for Decentralized Artificial Intelligence in Melanoma Diagnostics. *JAMA Dermatol.* **2024**, *160*, 303–311. [CrossRef] [PubMed]
47. Yu, Y.; Guo, L.; Gao, H.; He, Y.; You, Z.; Duan, A. FedCAE: A New Federated Learning Framework for Edge-Cloud Collaboration Based Machine Fault Diagnosis. *IEEE Trans. Ind. Electron.* **2024**, *71*, 4108–4119. [CrossRef]
48. Christidis, K.; Devetsikiotis, M. Blockchains and Smart Contracts for the Internet of Things. *IEEE Access* **2016**, *4*, 2292–2303. [CrossRef]
49. Tang, Y.; Zhang, Y.; Niu, T.; Li, Z.; Zhang, Z.; Chen, H.; Zhang, L. A Survey on Blockchain-Based Federated Learning: Categorization, Application and Analysis. *Comput. Model. Eng. Sci.* **2024**, *139*, 2451–2477. [CrossRef]
50. Wu, B.; Kang, H. Research on Federated Sharing Methods for Massive Data in Blockchain. In *Proceedings of the Smart Grid and Internet of Things*; Deng, D.J., Chen, J.C., Eds.; Springer: Cham, Switzerland, 2024; pp. 12–27.
51. Sumitra; Shenoy, M.V. HFedDI: A novel privacy preserving horizontal federated learning based scheme for IoT device identification. *J. Netw. Comput. Appl.* **2023**, *214*, 103616. [CrossRef]
52. Zhang, J.; Zhao, L.; Yu, K.; Min, G.; Al-Dubai, A.Y.; Zomaya, A.Y. A Novel Federated Learning Scheme for Generative Adversarial Networks. *IEEE Trans. Mob. Comput.* **2024**, *23*, 3633–3649. [CrossRef]
53. Nicolas-Alonso, L.F.; Gomez-Gil, J. Brain computer interfaces, a review. *Sensors* **2012**, *12*, 1211–1279. [CrossRef]
54. Liu, R.; Chen, Y.; Li, A.; Ding, Y.; Yu, H.; Guan, C. Aggregating intrinsic information to enhance BCI performance through federated learning. *Neural Netw.* **2024**, *172*, 106100. [CrossRef]

55. Liu, L.; Zhang, J.; Song, S.; Letaief, K.B. Hierarchical Federated Learning with Quantization: Convergence Analysis and System Design. *IEEE Trans. Wirel. Commun.* **2023**, *22*, 2–18. [CrossRef]
56. Fan, X.; Wang, Y.; Huo, Y.; Tian, Z. 1-Bit Compressive Sensing for Efficient Federated Learning over the Air. *IEEE Trans. Wirel. Commun.* **2023**, *22*, 2139–2155. [CrossRef]
57. Zhao, R.; Wang, Y.; Xue, Z.; Ohtsuki, T.; Adebisi, B.; Gui, G. Semisupervised Federated-Learning-Based Intrusion Detection Method for Internet of Things. *IEEE Internet Things J.* **2023**, *10*, 8645–8657. [CrossRef]
58. Jiang, Y.; Wang, S.; Valls, V.; Ko, B.J.; Lee, W.H.; Leung, K.K.; Tassiulas, L. Model Pruning Enables Efficient Federated Learning on Edge Devices. *IEEE Trans. Neural Netw. Learn. Syst.* **2023**, *34*, 10374–10386. [CrossRef]
59. Tang, Z.; Shi, S.; Li, B.; Chu, X. GossipFL: A Decentralized Federated Learning Framework with Sparsified and Adaptive Communication. *IEEE Trans. Parallel Distrib. Syst.* **2023**, *34*, 909–922. [CrossRef]
60. Vásquez-Correa, J.C.; Álvarez Muniain, A. Novel Speech Recognition Systems Applied to Forensics within Child Exploitation: Wav2vec2.0 vs. Whisper. *Sensors* **2023**, *23*, 1843. [CrossRef]
61. Hagos, D.H.; Tankard, E.; Rawat, D.B. A Scalable Asynchronous Federated Learning for Privacy-Preserving Real-Time Surveillance Systems. In Proceedings of the IEEE INFOCOM 2023—IEEE Conference on Computer Communications Workshops (INFOCOM WKSHPS), New York, NY, USA, 17–20 May 2023; pp. 1–6. [CrossRef]
62. Zhang, L.; Zhu, T.; Xiong, P.; Zhou, W.; Yu, P.S. A Robust Game-Theoretical Federated Learning Framework with Joint Differential Privacy. *IEEE Trans. Knowl. Data Eng.* **2023**, *35*, 3333–3346. [CrossRef]
63. Shoham, Y.; Leyton-Brown, K. *Multiagent Systems: Algorithmic, Game-Theoretic, and Logical Foundations*; Cambridge University Press: Cambridge, UK, 2008.
64. Luo, X.; Zhang, Z.; He, J.; Hu, S. Strategic Analysis of the Parameter Servers and Participants in Federated Learning: An Evolutionary Game Perspective. *IEEE Trans. Comput. Soc. Syst.* **2024**, *11*, 132–143. [CrossRef]

Disclaimer/Publisher's Note: The statements, opinions and data contained in all publications are solely those of the individual author(s) and contributor(s) and not of MDPI and/or the editor(s). MDPI and/or the editor(s) disclaim responsibility for any injury to people or property resulting from any ideas, methods, instructions or products referred to in the content.

Article

Prototype Selection for Multilabel Instance-Based Learning [†]

Panagiotis Filippakis [1],*, Stefanos Ougiaroglou [1] and Georgios Evangelidis [2]

[1] Department of Information and Electronic Engineering, School of Engineering, International Hellenic University, 57400 Thessaloniki, Greece; stoug@ihu.gr
[2] Department of Applied Informatics, School of Information Sciences, University of Macedonia, 156 Egnatia Street, 54636 Thessaloniki, Greece; gevan@uom.gr
* Correspondence: filipana1@iee.ihu.gr
[†] This paper is an extended version of our paper published in 27th International Database Engineered Application Symposium, IDEAS 2023, Heraklion, Greece, 5–7 May 2023.

Abstract: Reducing the size of the training set, which involves replacing it with a condensed set, is a widely adopted practice to enhance the efficiency of instance-based classifiers while trying to maintain high classification accuracy. This objective can be achieved through the use of data reduction techniques, also known as prototype selection or generation algorithms. Although there are numerous algorithms available in the literature that effectively address single-label classification problems, most of them are not applicable to multilabel data, where an instance can belong to multiple classes. Well-known transformation methods cannot be combined with a data reduction technique due to different reasons. The Condensed Nearest Neighbor rule is a popular parameter-free single-label prototype selection algorithm. The IB2 algorithm is the one-pass variation of the Condensed Nearest Neighbor rule. This paper proposes variations of these algorithms for multilabel data. Through an experimental study conducted on nine distinct datasets as well as statistical tests, we demonstrate that the eight proposed approaches (four for each algorithm) offer significant reduction rates without compromising the classification accuracy.

Keywords: data reduction techniques; instance reduction; multilabel classification; prototype selection; instance-based classification; binary relevance; CNN; IB2; BRkNN

Citation: Filippakis, P.; Ougiaroglou, S.; Evangelidis, G. Prototype Selection for Multilabel Instance-Based Learning. *Information* **2023**, *14*, 572. https://doi.org/10.3390/info14100572

Academic Editor: Peter Revesz

Received: 31 July 2023
Revised: 13 October 2023
Accepted: 17 October 2023
Published: 19 October 2023

Copyright: © 2023 by the authors. Licensee MDPI, Basel, Switzerland. This article is an open access article distributed under the terms and conditions of the Creative Commons Attribution (CC BY) license (https:// creativecommons.org/licenses/by/ 4.0/).

1. Introduction

Multilabel classification [1] involves predicting multiple potential classes or labels for a single instance, while single-label classification focuses on assigning only one class to each instance. Multilabel classification is commonly employed to classify diverse forms of data, such as images, books, artists, music, videos and movies. For instance, a movie can be classified as both "crime" and "adventure", a text can cover many different topics and a music track may encompass multiple genres or moods. Multilabel classification can be characterized as a generalization of single-label classification, where a classifier is capable of handling scenarios where multiple labels may be applicable to a single instance. This extension allows for more versatile predictions as it accommodates cases where instances can belong to multiple classes at the same time.

The k-Nearest Neighbors (k-NN) [2] classification algorithm serves as a typical illustration of a lazy or instance-based classifier. It operates by retrieving the k-nearest neighbors of an unclassified instance and employing a majority voting approach to assign a classification. In simpler terms, the unclassified instance is assigned to the most prevalent class among the classes of the retrieved k-nearest neighbors. This classifier is renowned for its simplicity, ease of implementation and robust classification performance, making it valuable for both single-label and multilabel classification tasks. Nonetheless, it comes with a drawback of high computational cost due to the need to calculate the distances between each unclassified instance and all instances in the training set.

Hence, the size of the training set plays a vital role in instance-based classification. While a large training set yields higher classification accuracy, it also entails increased computational costs. To expedite the k-NN classifier, it becomes necessary to mitigate its memory and CPU requirements by reducing the training set's size. In single-label classification tasks, one approach is to employ a data reduction technique (DRT) capable of reducing either the number of training instances or attributes [3]. This paper specifically focuses on DRTs from the perspective of instance reduction. The objective of this paper is to achieve efficient k-NN classification on multilabel data by decreasing the training set's size without compromising accuracy.

DRTs encompass two categories, namely prototype selection (PS) [4] and prototype generation (PG) [5]. In practical terms, these techniques serve as data pre-processing tasks aimed at replacing the original training dataset with a smaller subset known as the "condensing set". Utilizing the condensing set enables the k-NN classifier to achieve comparable accuracy to using the full training dataset but with significantly reduced computational costs. PS algorithms choose specific instances, or prototypes, from the original training set, whereas PG algorithms generate prototypes by summarizing similar training instances belonging to the same class. The fundamental concept underlying many DRTs is that only training instances in close proximity to class decision boundaries, in terms of a Euclidean metric space, are crucial for classification tasks. Those training instances situated within the "internal" area of a class, far away from decision boundaries, can be safely removed without compromising classification accuracy. Consequently, DRTs aim to select or generate an adequate number of prototypes that reside near the decision boundary areas for each class. The majority of DRTs primarily focus on single-label classification problems.

It is worth mentioning that a subcategory of PS algorithms focuses on noise removal and operates differently from other PS and PG algorithms. These algorithms are designed to eliminate noise and smooth the decision boundaries between discrete classes. As a result, they create an edited training set that leads to accuracy gains in the classification process.

The label powerset (LP) transformation technique [1] offers a straightforward solution for employing a DRT in multilabel classification tasks. LP transforms a multilabel dataset into a single-label dataset by considering each label combination, or labelset, as a separate class. However, it is important to acknowledge that LP is suitable only when the number of labels and potential labelsets is limited and there are ample instances available for each labelset. In situations where the number of label combinations becomes excessively large, the reduction rate may not be sufficient, resulting in inadequate representation of certain combinations. Moreover, the total count of distinct label combinations can grow exponentially, giving rise to scalability issues.

Binary relevance (BR) is a popular transformation technique that addresses multilabel classification problems by transforming them into single-label classification problems. Essentially, BR involves converting the original multilabel problem into multiple independent binary classification problems. In order to predict the labels associated with an unclassified instance, a separate classifier is required for each available label. When BR is combined with the k-NN classifier, it is referred to as BRkNN [6]. This combination proves effective because the k-NN classifier is a lazy classifier that does not construct a classification model. During the classification of an instance x using BRkNN, the algorithm searches for the k-nearest neighbors to x, just like in the case of single-label k-NN. Subsequently, the voting procedure of the nearest neighbors is repeated individually for each label.

When a data reduction technique (DRT) is applied before utilizing the k-NN classifier, the classifier transitions from being lazy to eager. In such scenarios, the condensing set becomes the classification model. However, in the context of multilabel classification, the application of BR (binary relevance) to construct a condensing set for each label undermines the objective of data reduction. On one hand, the goal of data reduction is not achieved. On the other hand, since the k-NN classifier must search for nearest neighbors within each specific condensing set to make predictions for individual labels, the computational cost remains high. Hence, combining BR with a DRT becomes infeasible due to the presence of multiple binary condensing sets. This highlights the necessity of modifying DRTs to effectively handle multilabel datasets, which serves as the motivation for the current research work.

The Condensed Nearest Neighbor (CNN) rule [7] stands as the oldest prototype selection (PS) algorithm for single-label classification tasks. It operates by eliminating training instances that reside far from the decision boundaries, and this is achieved by running over the training data multiple times. Instance-Based Learning 2 (IB2) [8] is the one-pass variation of CNN and has the same motivation with that of CNN. IB2 involves an extremely low pre-processing computational cost to build the condensing set. Both CNN and IB2 are popular parameter-free PS algorithms. However, they are inappropriate for multilabel data.

The objective of the present paper is to extend the applicability of CNN and IB2 by introducing variations suitable for multilabel classification problems. In [9], we proposed three multilabel variations of CNN. Here, we extend the previous work by adding one more variation of CNN, which is based on Levenshtein distance [10], and by introducing the multilabel version of IB2 with the four variations. Thus, the contributions of the paper are summarized as follows:

- We propose four variations of CNN and IB2 that are suitable for multilabel classification problems.
- One of the variations uses a novel adaptation of Levenshtein distance for multilabel instances.
- We conduct an experimental study using nine multilabel datasets and complement it with corresponding statistical tests of significance. The study reveals that the proposed variations offer significant reduction rates without compromising the classification accuracy.

The rest of this paper is organized as follows. Section 2 presents the related work in data reduction on multilabel datasets. Section 3 presents the CNN algorithm for single-label classification problems. Section 4 presents the Instance-Based Two-Step (IB2) algorithm for single-label classification problems. Section 5 describes the proposed variations of CNN and IB2 for multilabel classification problems that use the BRkNN multilabel classifier. Section 6 presents the experimental study that compares the proposed algorithms. Finally, Section 7 concludes the paper.

2. Related Work

The majority of publications focused on multilabel problems tend to discuss classification algorithms rather than approaches aimed at reducing computational costs associated with large multilabel training sets. There are also many attempts to offer programming APIs [11] and environments [12] specific for multilabel classification. Limited research is available on data reduction techniques specifically tailored for such datasets. In this section, we examine the scarce relevant literature.

In [13], the authors propose a PS algorithm designed for multilabel datasets. This algorithm aims to eliminate noise during the editing process and achieve balanced class decision boundaries. Inspired by the Edited Nearest Neighbor rule (ENN-rule) [14], the authors suggest an under-sampling method for addressing imbalanced training sets.

Another article, [15], introduces a prototype selection editing algorithm based on ENN-rule. The algorithm utilizes the Hamming loss metric to identify noisy training instances. The concept is straightforward: instances with high Hamming loss are considered for elimination due to their proximity to decision boundaries, similar to ENN-rule.

In [16], the authors make the first attempt to adapt PS algorithms to multilabel problems. Their proposed algorithms are based on local sets [17] and the LP transformation technique [1]. In single-label problems, the local set of an instance refers to the largest set of instances centered around it, all belonging to the same class. The authors argue that, in multilabel datasets, a local set does not necessarily need to contain instances with the exact same labelsets; they may have slightly different labelsets. The authors calculate the Hamming loss over the labelsets to measure the differences, and the distance between two instances is determined using the Hamming loss of their labelsets. If the distance exceeds a specified threshold, the instances are classified as belonging to different "classes".

The multilabel prototype selection with Co-Occurrence and Generalized Condensed Nearest Neighbor (CO-GCNN) was proposed in [18]. CO-GCNN captures label correlation by computing the co-occurrence frequency of label pairs and subsequently segregating the initial data into positive and negative categories. The single-label generalized CNN [19] is performed in order to produce the condensing set. In effect, CO-GCNN transforms the multilabel classification problem into a single-label classification problem. It leverages the benefit of incorporating pairwise label correlations as a constraint during the data transformation step. The authors state that the incorporation of pairwise label correlations enables the chosen prototypes to more accurately represent the original dataset.

In Ref. [20], a simple multilabel prototype selection algorithm based on clustering is proposed. The proposed methodology uses a clustering algorithm as a PS algorithm and then the well-known Multilabel k-Nearest Neighbor algorithm (ML-KNN) [21] performs the labels prediction.

The article [22] explores the use of single-label prototype selection algorithms along with binary relevance (BR), label powerset (LP) and other transformation techniques. For BR and its variants, the proposed strategy generates individual single-label training sets for different label types. Each training set undergoes a PS algorithm to create a condensing set specific to each label. Instances receive votes each time they are selected, and the accumulated votes form a single vector. Instances with votes surpassing a predefined threshold are selected, resulting in a complete condensing set.

The work presented in [23] does not introduce a DRT. However, the proposed method uses a PS algorithm as an intermediate step. The authors argue that PS leads to accuracy loss. Their method aims to combine the classification accuracy of retaining the original training set with the time efficiency of a PS method. More specifically, the authors propose a three-phase strategy for multilabel classification: initially, a PS algorithm is employed on the complete training set, generating a single-label condensed set. This operation is performed once as a pre-processing step. When an instance x is presented and must be classified, the proposed method selects a reduced set of labels as potential hypotheses, considering only the condensed set. In effect, the condensing set works as a recommender system that recommends the labels where x belongs to. The authors suggest picking the c nearest classes to input x, where c is a user-specified parameter. The final classification is performed by k-NN, employing a dynamically formed subset of the original training set limited to the c labels identified in the previous step.

In [24], a data pre-processing technique to improve label distribution learning (LDL) [25] algorithms is proposed. LDL is a general learning framework that assigns an instance to a distribution over a set of labels. Specifically, the proposed method is called ProLSFEO-LDL and combines prototype selection and label-specific feature learning. The paper proposes an evolutionary algorithm adapted to the specificities of LDL, aiming to optimize the initial solution to meet desired expectations.

In [26], the authors introduce an attempt to reduce multilabel datasets using homogeneous clustering. The algorithm, known as Multilabel Reduction through Homogeneous Clusters (MLRHC), is an adaptation of the single-label prototype generation algorithm RHC. MLRHC applies K-means clustering iteratively to produce homogeneous clusters, which are then replaced by their center. In MLRHC, a cluster is considered homogeneous if all instances within the cluster share at least one common label. The initial dataset is clustered using the existing labels as initial means, with the number of clusters matching the number of existing labels. Homogeneous clusters are replaced by their center, which is assigned the common label along with any label appearing in at least half of the cluster's instances. Similarly, in [27], the authors extend their previous work by proposing a variant called Multilabel Reduction by Space Partitioning (MLRSP3), which also relies on the concept of homogeneity based on instances sharing at least one common label. MLRSP3 is based on the RSP3 PG algorithm [28] and starts with a non-homogeneous cluster, selects the two farthest instances and divides the training set into two clusters by assigning instances to the closest farthest instance. This process continues until all clusters become homogeneous. Like MLRHC, the center of each homogeneous cluster in MLRSP3 becomes a multilabel prototype labeled with the common label and any label appearing in at least half of the instances within the cluster.

Quite similar work is presented in [29]. More specifically, the authors propose adaptations of the Chen and Jozwik multiclass PG algorithm [30] and of its descendants, RSP1, RSP2 and RSP3 [28], to the multilabel case. The proposed adaptations are evaluated on three multilabel NN-based classifiers using 12 datasets of varying domains and sizes, along with artificially induced noise scenarios. The results show high efficiency and classification performance, especially in noisy scenarios.

The paper presented in [31] does not introduce a PS or a PG algorithm. However, it deals with the high computational cost in multilabel classification. Specifically, the paper introduces a novel approach for multilabel classification by leveraging hypergraph spectral learning. Hypergraphs, which generalize traditional graphs by allowing edges to be arbitrary non-empty subsets of vertices, are employed to capture high-order relations among labels. The proposed formulation leads to an eigenvalue problem, which may be computationally expensive for large-scale datasets. To address this, the paper presents an approximate formulation that reduces computational complexity while maintaining competitive classification performance.

Figure 1 summarizes the presented works in a form of a hierarchy. More specifically, the presented works can be categorized into three main categories. The first category includes the paper related with the PS algorithms and the adaption of them. The works related to adaptation of PG algorithms belong to the second category. The methods that belong to the third category cannot be characterized as either PS or PG. However, they are able to speed up the multilabel classification tasks, which is the goal of the PS and PG algorithms.

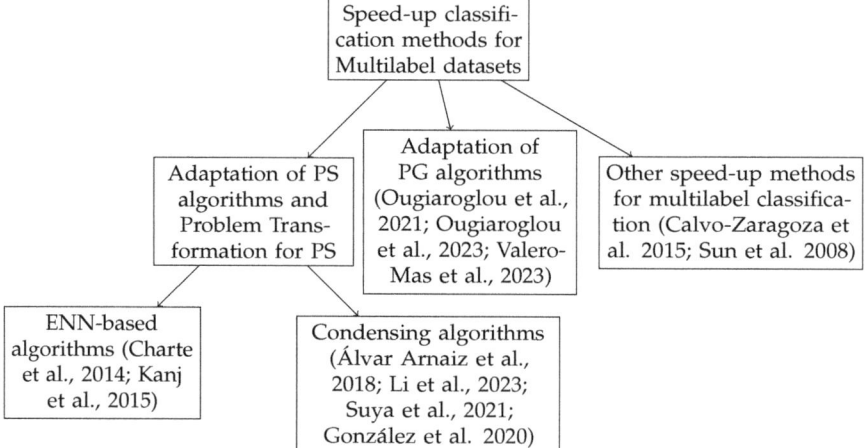

Figure 1. Hierarchy of the presented algorithms and methods: Charte et al., 2014 [13]; Kanj et al., 2015 [15]; Álvar Arnaiz et al., 2018 [16,22]; Li et al., 2023 [18]; Suya et al., 2021 [20]; González et al., 2020 [24]; Ougiaroglou et al., 2021 [26]; Ougiaroglou et al., 2023 [27]; Valero-Mas et al., 2023 [29]; Calvo-Zaragoza et al., 2015 [23]; Sun et al., 2008 [31].

3. The Single-Label Condensed Nearest Neighbor Rule

In Section 1, the Condensed Nearest Neighbor (CNN) rule [7] is discussed as the first and most widely utilized PS algorithm. CNN is a parameter-free single-label PS algorithm that constructs its condensing set by iteratively examining the training data.

CNN involves two storage areas, namely the Condensing Set (CS) and the Training Set (TS). Initially, the TS contains the entire training set while the CS is empty. The process begins by randomly selecting an instance from the TS and transferring it to the CS. Each instance $x \in TS$ is then compared to the instances currently stored in the CS.

Specifically, for each instance $x \in TS$, CNN identifies its nearest neighbor (1-NN) within the current CS using the Euclidean distance. If x is correctly classified by its nearest neighbor in the CS, it remains in the TS. However, if x is misclassified, it is removed from the TS and added to the CS. This process continues until all instances $x \in TS$ have been considered. Subsequently, the algorithm proceeds to the next scan of the TS.

The algorithm terminates when, during a complete scan of the TS, no instances are transferred from the TS to the CS, indicating that all instances in the TS are correctly classified based on the content of the CS. Algorithm 1 provides the pseudo-code representation of the CNN algorithm.

It is worth mentioning that, during the first algorithm iteration, CS is not empty. It contains a randomly selected instance (see line 2 in Algorithm 1). Therefore, for all examined instances of TS, there is always a nearest neighbor in CS.

Algorithm 1 CNN

Input: TS
Output: CS
1: $CS \leftarrow \emptyset$
2: randomly pick an instance of TS and move it to CS
3: **repeat**
4:    $stop \leftarrow TRUE$
5:    **for** each $x \in TS$ **do**
6:      $NN \leftarrow$ nearest neighbor of x in CS using Euclidean distance
7:      **if** $NN_{class} \neq x_{class}$ **then**
8:        $CS \leftarrow CS \cup \{x\}$
9:        $TS \leftarrow TS - \{x\}$
10:       $stop \leftarrow FALSE$
11:     **end if**
12:    **end for**
13: **until** $stop == TRUE$ {no move during a pass of TS}
14: discard TS
15: **return** CS

The fundamental principle of the CNN algorithm is to include incorrectly classified instances in the Condensing Set (CS) since they are considered border instances located near decision boundaries. CNN ensures that each removed instance from the Training Set (TS) can be correctly classified using the information contained in the CS set. CNN's lack of parameters is a significant advantage. The condensing set is built without the need of any user-specified input parameter. Thus, costly computational parameter tuning procedures are avoided. On the other hand, there are some disadvantages to consider:

- Variability in Results: Running the CNN algorithm multiple times on the same TS may produce different condensing sets due to variations in the randomly selected initial instance (line 2 of Algorithm 1) or differences in the order in which TS instances are examined (line 5 of Algorithm 1).
- Memory Requirements: CNN is a memory-based algorithm, meaning that all instances need to reside in main memory during its execution.

- Computational Cost: The CNN algorithm requires multiple passes over the training set.

In terms of quality, the CNN algorithm operates as follows: if the underlying densities of different classes have minimal overlap, indicating a low Bayes risk, CNN tends to select instances located close to the possibly fuzzy boundary between classes. Instances deeply embedded within a class are unlikely to be transferred to the CS since they are correctly classified. However, if the Bayes risk is high, the CS will essentially contain almost every instance from the initial TS set, resulting in negligible sample size reduction.

4. The IB2 Algorithm

Aha et al. [8] introduced a set of instance-based learning algorithms. Among these algorithms, IB1 (Instance-Based Learning) served as a baseline and was essentially equivalent to the 1-NN algorithm.

The IB2 algorithm operates incrementally by initially having an empty set, CS, and adding each instance from TS to CS if it is misclassified by the instances already present in CS. Algorithm 2 provides the pseudo-code representation of the IB2 algorithm.

Algorithm 2 IB2
Input: TS
Output: CS
1: $CS \leftarrow \emptyset$
2: randomly pick an instance of TS and move it to CS
3: **for** each $x \in TS$ **do**
4: $\quad NN \leftarrow$ nearest neighbor of x in CS using Euclidean distance
5: $\quad$ **if** $NN_{class} \neq x_{class}$ **then**
6: $\quad\quad CS \leftarrow CS \cup \{x\}$
7: $\quad\quad TS \leftarrow TS - \{x\}$
8: $\quad$ **end if**
9: **end for**
10: discard TS
11: **return** CS

IB2 bears similarities to CNN-rule but it does not repeat the process after the first pass through the training set. As a result, IB2 does not guarantee the correct classification of all remaining instances in the TS. In effect, IB2 is a one-pass version of CNN.

Like CNN, IB2 aims to retain border instances in CS while eliminating internal instances that are surrounded by instances belonging to the same class. Similar to the CNN algorithm, IB2 is highly sensitive to noise because erroneous instances are often misclassified, resulting in the preservation of noisy instances, while more reliable instances are removed. The benefits and characteristics of IB2 algorithm are

- Since IB2 avoids multiple passes over training data, is quite faster than CNN.
- The condensing set obtained from IB2 is generally smaller than that of CNN, leading to faster classification and reduced storage requirements.
- IB2 supports incremental learning, where new instances can be added to the condensing set without requiring complete retraining of the algorithm.
- The decision boundary revision step allows IB2 to adapt to new instances and adjust the reduced training set accordingly.

5. The Proposed Algorithms

As discussed in Section 1, traditional data reduction algorithms are not suitable for use with the binary relevance transformation method in multilabel data. Applying data reduction in conjunction with the binary relevance transformation would lead to the creation of numerous condensing sets, one for each label.

In this section, we introduce variations of the CNN and IB2 algorithms that are designed for multilabel datasets. These proposed algorithms are named

- Multilabel CNN Hamming Distance and Multilabel IB2 Hamming Distance (MLCNN-H and MLIB2-H);
- Multilabel CNN Jaccard Distance and Multilabel IB2 Jaccard Distance (MLCNN-J and MLIB2-J);
- Multilabel CNN Levenshtein Distance and Multilabel IB2 Levenshtein Distance (MLCNN-L and MLIB2-L) and
- Multilabel CNN Binary Relevance and Multilabel IB2 Binary Relevance (MLCNN-BR and MLIB2-BR).

5.1. MLCNN-H and MLIB2-H

The MLCNN-H algorithm is based on a similar principle to CNN, with the idea that an instance with a significantly different labelset compared to its nearest neighbor should be included in the multilabel condensing set. To achieve this, MLCNN-H requires a method to measure the distance or difference between multilabel instances and a mechanism to determine when two instances are considered different or similar.

For that purpose, MLCNN-H uses Hamming distance. The Hamming distance between two labelsets is the number of positions at which the corresponding labels do not match.

Definition 1 (Hamming Distance). *Given two labelsets u and v, each of length n, their Hamming distance is the total number of positions where their labels do not match:*

$$HD(u,v) = Cardinality(\{i : u_i \neq v_i, i = 1, \ldots, n\})$$

In MLCNN-H, the labelset of an instance is represented as a sequence of binary values (0 or 1), where 0 indicates that the instance does not belong to the corresponding label and 1 indicates that it does. To compute the Hamming distance (HD) between two instances, the labelsets are compared using an XOR operation to count the number of differing labels. To express HD in the [0,1] interval, this count is then divided by the length of a labelset, which represents the total number of labels in the dataset. Consequently, when HD is zero, the two instances have identical labelsets, while an HD of one indicates completely different labelsets. For example:

- $HD(110001, 110001) = 0/6 = 0$
- $HD(110001, 001110) = 6/6 = 1$
- $HD(110001, 000010) = 4/6 = 0.33$
- $HD(110001, 100010) = 3/6 = 0.5$

MLCNN-H incorporates the concept of label density in the dataset. The label density is calculated as the average number of labels per instance divided by the number of distinct labels in the dataset [32]. MLCNN-H considers the labelsets of examined instances to be significantly different if their HD is greater than the dataset density.

Similar to the single-label CNN algorithm, MLCNN-H utilizes two sets: the Condensing Set (CS) and the Training Set (TS). Initially, TS contains the complete training set, while CS is empty. MLCNN-H randomly selects an instance from TS and transfers it to CS. For each instance $x \in TS$, the algorithm finds the nearest neighbor (e.g., y) within the CS. Then, MLCNN-H calculates the Hamming distance (HD) metric between x and y, quantifying the difference in their labelsets. If HD is greater than the dataset density, x is removed from TS and added to CS; otherwise, it remains in TS. Once all instances $x \in TS$ have been examined, the process continues with subsequent scans of the remaining instances in TS. MLCNN-H terminates when no transfers from TS to CS are made during a complete pass over TS. In each next scan over the remaining instances in TS, more instances move from TS to CS.

Similar to MLCNN-H, the MLIB2-H algorithm follows exactly the same process as the previous one, with the only difference being that the process stops after examining all instances in TS in one pass. In effect, MLIB2-H is the one-pass version of MLCNN-H. Therefore, MLIB2-H is quite faster and achieves higher reduction rates than MLCNN-H.

5.2. MLCNN-J and MLIB2-J

MLCNN-J and MLIB2-J employ the Jaccard distance for asymmetric binary attributes, where the presence of a label is considered more important than its absence.

Definition 2 (Jaccard Distance for asymmetric binary attributes). *Given two labelsets u and v, each of length n, we define:*

$$A = Cardinality(\{i : u_i = v_i = 1, i = 1, \ldots, n\})$$

$$B = Cardinality(\{i : u_i = 1 \wedge v_i = 0, i = 1, \ldots, n\})$$

$$C = Cardinality(\{i : u_i = 0 \wedge v_i = 1, i = 1, \ldots, n\})$$

$$D = Cardinality(\{i : u_i = v_i = 0, i = 1, \ldots, n\})$$

Then, their Jaccard distance for asymmetric binary attributes is the percentage of the number of positions where their labels do not match over the total number of positions where at least one of the labelsets has an appearing label. In other words, D is not taken into consideration:

$$JD(u,v) = \frac{B+C}{A+B+C}$$

Thus, matching zeros (absent labels) are disregarded when calculating the distance between two labelsets. For example:

- $JD(110001, 110001) = 0/3 = 0$
- $JD(110001, 001110) = 6/6 = 1$
- $JD(110001, 000010) = 4/4 = 1$
- $JD(110001, 100010) = 3/4 = 0.75$

MLCNN-J and MLIB2-J are variations of MLCNN-H and MLIB2-H, respectively. Both are based on the idea of selecting instances with labelsets that significantly differ from their nearest neighbors in the CS as prototypes. However, MLCNN-J and MLIB2-J differ from MLCNN-H and MLIB2-H in two key aspects. Firstly, instead of using Hamming distance, MLCNN-J and MLIB2-J employ Jaccard distance (JD) as the dissimilarity metric. Secondly, MLCNN-J and MLIB2-J introduce a different JD threshold to determine the extent to which two instances differ.

The JD threshold plays a crucial role in determining which instances are considered different enough to be included in the CS. Initially, MLCNN-J and MLIB2-J consider instances with fewer than half the labels in common (JD threshold of 0.5) as different. This means that instances sharing at least half the labels are deemed similar and not added to the CS. However, in order to increase the reduction rates, higher JD threshold values were explored during experimentation. Ultimately, two JD threshold values were chosen for testing: 0.5 and 0.75.

By employing Jaccard distance and adjusting the JD threshold, MLCNN-J and MLIB2-J aim to identify instances that significantly differ from their nearest neighbors and include them as prototypes in the CS. The choice of the JD threshold affects the reduction rates, with higher thresholds potentially leading to greater reductions by considering more instances as similar to their nearest neighbors and excluding them from the CS.

The MLIB2-J algorithm follows the exact same procedure as the MLCNN-J but with a distinct difference: it conducts a single complete iteration on the training set and subsequently terminates. Thus, it achieves higher reduction rates than MLCNN-J.

5.3. MLCNN-L and MLIB2-L

MLCNN-L and MLIB2-L are the third pair of multilabel variations of CNN. MLCNN-L and MLIB2-L utilize the Levenshtein distance metric. The Levenshtein distance serves as a quantification of dissimilarity between two sets.

The Levenshtein distance represents the minimum number of edit operations needed to convert one string into another. These edit operations encompass insertions, deletions and substitutions. Among the family of distance metrics known as edit distance, the Levenshtein distance stands out as one of the most widely used and popular metrics.

For instance, for the transformation of the string "COVID" to the string "MOVING", three operations are required. Hence, the Levenshtein distance between these two strings is three. More specifically, C is substituted by M, D is substituted by N and, finally, G is inserted.

Properties of the Levenshtein distance include:

1. Non-Negativity: The Levenshtein distance is always non-negative.
2. Symmetry: The distance between "S" and "T" is the same as the distance between "T" and "S".
3. Identity: The distance between a string and itself is always zero.
4. Triangle Inequality: For any three strings "S", "T" and "W", the distance from "S" to "W" is no greater than the sum of the distances from "S" to "T" and from "T" to "W".
5. Substructure Optimality: The optimal solution for the overall Levenshtein distance can be obtained by combining optimal solutions to the subproblems (i.e., the prefix substrings) of "S" and "T" [33].

MLCNN-L and MLIB2-L are modified versions of MLCNN-J and MLIB2-J that focus on selecting instances with labelsets that exhibit substantial differences from their nearest neighbors in the CS as prototype examples. However, MLCNN-L and MLIB2-L distinguish themselves from MLCNN-J and MLIB2-J in two ways. Firstly, they replace the use of Jaccard distance with Levenshtein distance as the dissimilarity metric. Secondly, MLCNN-L and MLIB2-L introduce a distinct Levenshtein distance threshold to determine the degree of dissimilarity between two instances. In MLCNN-L and MLIB2-L, the concept of label cardinality is incorporated. Label cardinality (LC) of a dataset refers to the average number of labels per instance in the dataset.

In MLCNN-L and MLIB2-L, the labelsets of examined instances are considered significantly different if the Levenshtein distance between them exceeds half of the label cardinality (LC). Therefore, MLCNN-L and MLIB2-L utilize the Levenshtein distance (LV) and consider instances for which $LV > \frac{LC}{2}$ to differ significantly. The goal is to include these instances as prototypes in the CS.

Let us now illustrate how the Levenshtein distance is computed for two instances. Using the binary representation of the labelsets of the instances, we perform an on-the-fly mapping to an ASCII string. This is accomplished by mapping label positions to a fixed sorted sequence of ASCII characters. For example, assuming that there are six labels in total, Positions 1–6 are mapped to characters A through F. Hence, 011010 is mapped to BCE, whereas 111001 is mapped to ABCF. The examples below demonstrate the use of Levenshtein distance on the mapped labelsets:

- LEV(110001, 110001) = LEV(ABF, ABF) = 0
- LEV(110001, 001110) = LEV(ABF, CDE) = 3
- LEV(110001, 000010) = LEV(ABF, E) = 3
- LEV(110001, 100010) = LEV(ABF, AE) = 2

We can take advantage of the fact that, by design, the resulting strings are sorted sequences of characters and compute Levenshtein distances directly on the binary labelsets, i.e., without mapping them to strings.

Like in the case of Jaccard distance for asymmetric binary attributes, our method for mapping binary labelsets to strings disregards non-appearing labels, i.e., 0 to 0 matches.

We define a matching substring pair of two labelsets to consist of 1s only, whereas non-matching substring pairs of two labelsets are all the remaining cases. For example, let us take labelsets x = 11001001 and y = 11110001. We can express these two labelsets as a sequence of matching and non-matching substring pairs as follows: (11, 11), (00100, 11000) and (1, 1). The first and third substring pairs are matching, whereas the second substring pair is non-matching. It is obvious that matching substring pairs correspond to identical strings, while non-matching substring pairs correspond to strings without a single common character.

In our example, since mapped(x) = ABEH and mapped(y) = ABCDH, the corresponding mapped substring pairs are (AB, AB), (E, CD) and (H, H). Observe that F and G are missing from both mapped labelsets. By definition, the Levenshtein distance of identical strings is zero and of strings without any common characters is the length of the longest string. Thus, to calculate the Levenshtein distance of the mapped labelsets of our example, we sum the Levenshtein distances of their non-matching substring pairs. In our example, this is pair (E, CD) and the distance is 2. Using the original binary labelsets, the Levenshtein distance of a non-matching substring pair is the maximum number of 1s among the two substrings. In our example, the Levenshtein distance of (00100, 11000) is 2.

Like the previous presented variations, the MLIB2-L algorithm adheres to the same procedure as MLCNN-L but with a distinction: it carries out a single full iteration on the training set and then terminates.

5.4. MLCNN-BR and MLIB2-BR

MLCNN-BR and MLIB2-BR take a different approach compared to MLCNN-H and MLIB2-H, MLCNN-J and MLIB2-J, MLCNN-L and MLIB2-L. Both MLCNN-BR and MLIB2-BR begin by using the binary relevance transformation method to transform the multilabel problem with |L| labels into |L| single-label problems. Each label of the training set is processed separately using the CNN or IB2 algorithm. This results in the creation of multiple CSs, with each set corresponding to a specific label. For example, if the initial training set has ten labels, MLCNN-BR and MLIB2-BR generate ten CSs.

Each CS contains prototypes labeled as 1 (indicating that the instance belongs to the corresponding label) or 0 (indicating that the instance does not belong to the corresponding label). From each condensing set, MLCNN-BR and MLIB2-BR select only the prototypes with a label value of 1, discarding the prototypes with a label value of 0. Finally, MLCNN-BR and MLIB2-BR merge all the CS to create the final multilabel CS.

To illustrate how MLCNN-BR works, consider the example of running CNN or IB2 for each label on a two-dimensional training dataset with two labels. Suppose two CSs, as shown in Tables 1 and 2, are derived. The first CS contains six prototypes labeled as "1", while the second CS contains four prototypes labeled as "1". The final multilabel CS constructed by MLCNN-BR or MLIB2-BR, as shown in Table 3, contains eight prototypes.

In the final multilabel CS, prototypes (1, 1), (1, 8), (4, 5) and (9, 1) originate exclusively from the CS of the first label. Prototypes (5, 6) and (9, 9) originate exclusively from the CS of the second label. Prototypes (3, 8) and (8, 4) are common in both CSs.

By applying the binary relevance transformation and utilizing CNN on each label separately, MLCNN-BR constructs a final multilabel CS that includes prototypes with the corresponding labels. The merging process ensures that the final CS captures relevant prototypes from each label's CS.

Table 1. Condensing set for the first label.

Instances	First Label
(1, 1)	1
(1, 8)	1
(2, 7)	0
(3, 8)	1
(4, 5)	1
(7, 1)	0
(8, 4)	1
(9, 1)	1

Table 2. Condensing set for the second label.

Instances	Second Label
(1, 1)	0
(2, 7)	0
(3, 8)	1
(5, 6)	1
(7, 5)	0
(8, 4)	1
(9, 9)	1

Table 3. Final merged condensing set.

Instances	First Label	Second Label
(1, 1)	1	0
(1, 8)	1	0
(3, 8)	1	1
(4, 5)	1	0
(5, 6)	0	1
(8, 4)	1	1
(9, 1)	1	0
(9, 9)	0	1

The MLIB2-BR algorithm shares the same procedure as the MLCNN-BR algorithm, with the only difference being that it performs a single complete iteration on the training set before terminating.

6. Experimental Study

6.1. Experimental Setup

In our experimentation, we utilized nine multilabel datasets provided by Mulan dataset repository [11]. These datasets consisted of numeric features and contained a minimum of five hundred (500) instances. The key characteristics of the used datasets are summarized in Table 4. The last two columns of the table present the dataset cardinality and density. Cardinality represents the average number of labels per instance, while density is calculated by dividing the cardinality by the total number of labels. The domain of each dataset is listed in the second column of Table 4.

Since the datasets encompass features with different value ranges, this can impact the classification process as higher-valued features may dominate the distance calculation between instances. To address this, we normalized the values of all features to the [0,1] range. The normalization was performed using the MinMaxScaler from the scikit-learn Python library [34].

Subsequently, the datasets were split into training and test sets using the stratified 5-fold cross-validation method [35]. This approach ensures that the estimates have reduced variance and improves the generalization performance estimation of classification algorithms. Stratified cross-validation guarantees that the proportion of the feature of interest remains the same in the original data, training set and test set. This ensures that no values are over- or under-represented in the training and test sets, providing a more accurate evaluation of performance and error.

For the implementation of CNN and IB2 and the proposed variations of them, we used Python 3.12.0 and employed Multiprocessing. The utilization of Multiprocessing allows for the concurrent processing of multiple distinct parts of the same Python script by two or more CPU threads. This not only improves processing speed but also enables handling larger volumes of data.

Table 4. Dataset characteristics.

Datasets	Domain	Size	Attributes	Labels	Cardinality	Density
CAL500 (CAL)	Music	502	68	174	26.044	0.150
Emotions (EMT)	Music	593	72	6	1.869	0.311
Water quality (WQ)	Chemistry	1060	16	14	5.073	0.362
Scene (SC)	Image	2407	294	6	1.074	0.179
Yeast (YS)	Biology	2417	103	14	4.237	0.303
Birds (BRD)	Sounds	645	260	19	1.014	0.053
CHD49 (CHD)	Medicine	555	49	6	2.580	0.430
Image (IMG)	Image	2000	294	5	1.236	0.247
Mediamill (MDM)	Video	43,907	120	101	4.376	0.043

The objective of data reduction is to selectively choose training instances to be used as input for instance-based classifiers. This process involves identifying and potentially eliminating redundant instances. The ultimate goal is to obtain smaller datasets that effectively represent the original dataset. The aim is to simplify the dataset, improve computational efficiency and potentially enhance classification performance [36]. It is essential to ensure that the condensing set retains an acceptable amount of information compared to the original dataset.

Therefore, we evaluate the performance of BRkNN when it runs over the initial training set and over the condensing sets generated by the proposed algorithms. To measure the effectiveness, we obtained the reduction rate and Hamming loss through a five-fold stratified cross-validation framework. The reduction rate is calculated by comparing the number of instances before and after the reduction process. Thus, a reduction rate of 90% means that 90% of the original training set instances are discarded and the final condensing set consists of 10% of the original instances.

The utilization of eager multilabel classifiers in the experimental study does not align with the objective of the proposed variations, which aim to enhance the speed of instance-based classifiers in multilabel domains. Thus, we do not include eager multilabel classifiers in the experimental study.

As the computational cost of the BR*k*NN classifier is dependent on the size of the training set utilized, our study does not report the CPU time required for classification. Essentially, a higher reduction rate corresponds to a lower computational cost for the BR*k*NN classifier during the classification process. The effectiveness of the predictions is evaluated by computing the Hamming loss, which measures the ratio of incorrectly predicted labels to the total number of labels. The Hamming loss is computed as follows:

$$HL = \frac{1}{m} \sum_{i=1}^{m} \frac{|Y_i \Delta Z_i|}{|L|}$$

Y_i represents the set of actual labels for each instance, while Z_i represents the set of predicted labels for each instance. The total number of instances in the dataset is denoted as m, and $|L|$ refers to the total number of labels. The symmetric difference between two sets, denoted as Δ, can be visualized as the XOR operation. In other words, $|Y_i \Delta Z_i|$ is the number of non-matching labels between the labelsets of the two instances.

For instance, let us consider a multilabel dataset with five labels and a testing instance $x1$ with an actual label set of 11001 and a predicted label set of 11010. In this case, the Hamming loss is calculated as $\frac{2}{5} = 0.4$ because two of the labels do not match (these are fourth and fifth labels). Similarly, if another testing instance $x2$ has an actual label set of 00001 and a predicted label set of 11010, the Hamming loss is $\frac{4}{5} = 0.8$ because all but the third label do not match. To calculate the Hamming loss for a testing set comprising these two instances, we compute the average as $\frac{1}{2} \times (0.4 + 0.8) = 0.6$.

Lastly, it is worth mentioning that, in accordance with established conventions in the relevant literature (e.g., [4,5]), all experiments were conducted using $k = 1$.

6.2. Experimental Results

Table 5 presents the measurements of the experimental study, reporting the Hamming loss (HL) and reduction rates (RR) achieved by MLCNN-H, MLCNN-J, MLCNN-BR, MLCNN-L, MLIB2-H, MLIB2-J, MLIB2-BR and MLIB2-L for each dataset.

MLCNN-H achieved reduction rates ranging from 8.28% to 55.70%, MLCNN-J ($JD > 0.5$) from 0.75% to 51.93%, MLCNN-J ($JD > 0.75$) from 19.45% to 60.62%, MLCNN-BR from 0% to 58.18% and MLCNN-L from 0.10% to 43.05%. MLIB2-H achieved reduction rates ranging from 14.26% to 73.23%, MLIB2-J ($JD > 0.5$) from 1.20% to 73.23%, MLIB2-J ($JD > 0.75$) from 26.28% to 84.67%, MLIB2-BR from 0% to 75.69% and MLIB2-L from 0.35% to 58.85%.

Figure 2 illustrates the reduction rates achieved on each dataset by each algorithm. Instead of the reduction rate, we report the percentage of retained instances for each dataset (in other words, we report 1—*reduction_rate*.

On average, we observe that MLIB2-J ($JD > 0.75$) achieves the highest reduction rate, followed by MLIB2-H, MLCNN-J ($JD > 0.75$) and MLIB2-J ($JD > 0.75$). It is important to note that the distribution of instances within the dataset greatly influence the reduction rate. As the data are complex and not uniformly distributed in space, we observe significant fluctuations in the reduction rate across different datasets. However, in general, we notice that MLCNN-BR and MLIB2-BR exhibit less stable behavior in terms of the reduction rate achieved for each dataset compared to other algorithms.

Additionally, in Table 5, we observe only small differences in Hamming loss between the BR*k*NN classifier that utilizes the initial training set and the BR*k*NN classifier that employs the multilabel condensing sets created by MLCNN-H, MLCNN-J ($JD > 0.5$), MLCNN-J ($JD > 0.75$), MLCNN-BR, MLCNN-L, MLIB2-H, MLIB2-J (0.5), MLIB2-J ($JD > 0.75$), MLIB2-BR and MLIB2-L. In the following section, we provide a statistical analysis to further explore the performance of the algorithms.

Furthermore, as expected, the MLIB2 algorithm achieves higher reduction rates in all its versions than the corresponding versions of MLCNN. This is because IB2 performs one pass on the data compared to the multiple passes performed by CNN, which move more instances in the CS.

The proposed algorithms achieve instance reduction while maintaining high levels of accuracy.

6.3. Statistical Comparisons

In line with the commonly employed approach in the domain of PS and PG algorithms [4,5,37–41], we have supplemented the experimental study with a Wilcoxon signed rank test [42]. This test serves to statistically validate the accuracy of the measurements presented in Table 5. By comparing all the algorithms in pairs based on their performance on each dataset, the Wilcoxon signed rank test confirms their relative rankings. We performed the Wilcoxon signed rank test using the PSPP 2.0.0 statistical software.

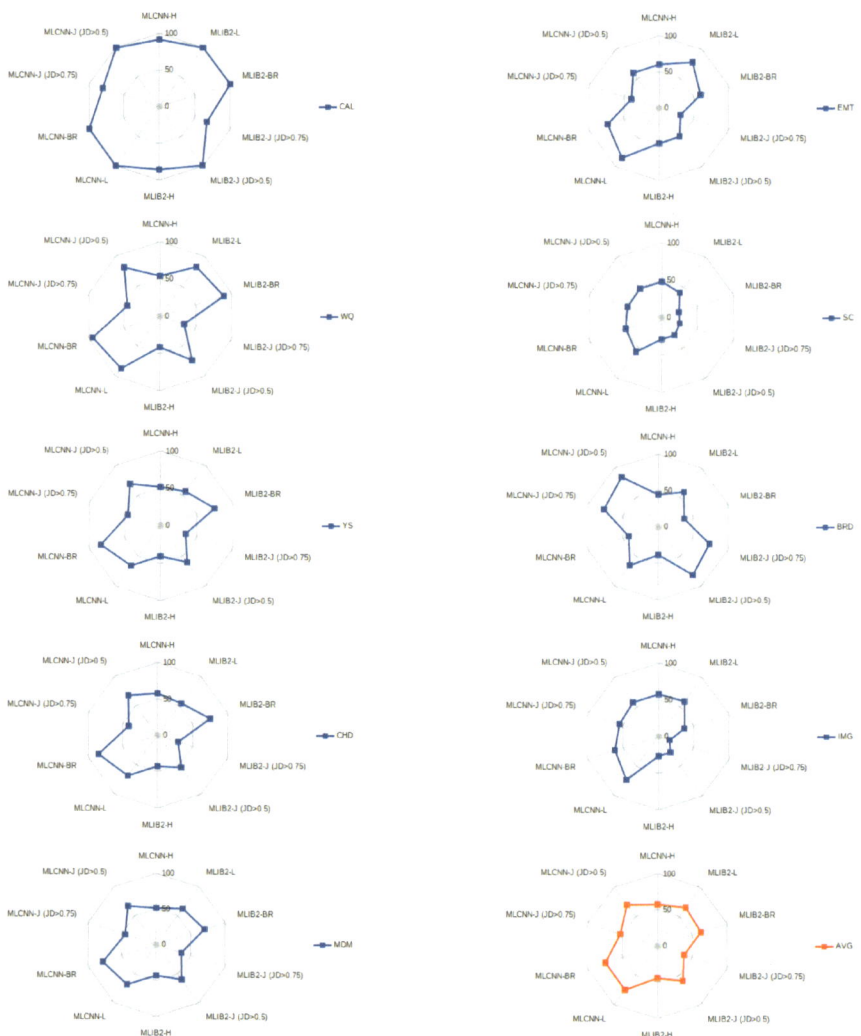

Figure 2. Percentage of retained instances per dataset (CAL, EMT, WQ, SC, YS, BRD, CHD, IMG, MDM) and algorithm. Last figure reports the average percentage of retained instances over all datasets and for each algorithm.

Table 5. Comparison table of the reduction rate (RR (%)) and the Hamming loss (HL (%)).

Dataset		BRkNN	MLCNN-H	MLCNN-J ($JD > 0.5$)	MLCNN-J ($JD > 0.75$)	MLCNN-BR	MLCNN-L	MLIB2-H	MLIB2-J ($JD > 0.5$)	MLIB2-J ($JD > 0.75$)	MLIB2-BR	MLIB2-L
CAL	HL:	0.19	0.19	0.19	0.19	0.19	0.19	0.19	0.19	0.19	0.18	0.19
	RR:	-	8.28	0.75	19.45	0.0	0.10	14.26	1.20	32.92	0.0	0.35
EMT	HL:	0.24	0.26	0.26	0.25	0.29	0.24	0.26	0.26	0.26	0.30	0.24
	RR:	-	39.88	40.26	60.62	26.64	14.67	51.05	51.64	69.06	39.97	21.67
WQ	HL:	0.33	0.34	0.33	0.34	0.36	0.33	0.35	0.33	0.34	0.37	0.33
	RR:	-	45.71	18.99	54.65	5.33	12.74	58.16	26.68	66.20	10.85	17.73
SC	HL:	0.11	0.12	0.12	0.12	0.13	0.12	0.14	0.14	0.14	0.17	0.14
	RR:	-	51.93	51.93	53.03	50.44	43.05	70.75	70.75	74.50	75.69	58.85
YS	HL:	0.24	0.27	0.26	0.26	0.30	0.26	0.27	0.26	0.27	0.30	0.26
	RR:	-	48.26	31.27	54.72	16.61	33.95	59.09	39.80	65.07	24.90	43.61
BRD	HL:	0.05	0.06	0.05	0.05	0.08	0.05	0.06	0.05	0.05	0.08	0.05
	RR:	-	55.70	15.50	22.71	58.18	34.81	61.59	18.80	26.28	62.87	40.62
CHD	HL:	0.35	0.36	0.36	0.38	0.40	0.37	0.37	0.37	0.38	0.40	0.38
	RR:	-	42.18	32.20	58.64	14.74	30.98	57.20	44.97	70.82	24.89	45.33
IMG	HL:	0.20	0.22	0.22	0.21	0.23	0.21	0.25	0.25	0.25	0.26	0.22
	RR:	-	42.11	42.11	44.82	38.43	26.61	73.23	73.23	84.67	63.43	40.47
MDM	HL:	0.031	0.035	0.032	0.036	0.042	0.032	0.037	0.033	0.039	0.043	0.033
	RR:	-	48.70	33.19	54.58	21.76	31.30	57.08	39.90	63.43	30.06	37.61

We have also utilized the non-parametric Friedman test to rank the algorithms individually for each dataset. This test assigned a rank to each algorithm, with the best performer receiving rank 1, the second best receiving rank 2 and so on. The PSPP statistical software was employed to conduct the Friedman test that was executed twice, once for each measured criterion.

6.3.1. Wilcoxon Signed Rank Test Results

The results of the Wilcoxon signed rank test for the Hamming loss (ACC) and Reduction Rate (RR) measurements are shown in Table 6. The column labeled "w/l/t" presents the number of wins, losses and ties for each comparison test. The last column, labeled "Wilc.", indicates a numerical value that quantifies the significance of the difference between the two compared algorithms. If this value is less than 0.05, it can be concluded that the difference is statistically significant. In Table 6, the Wilc. values that is less than 0.05 are in bold face.

The results indicate that there is no statistical difference in accuracy between the BRkNN classifier and the proposed MLCNN-L and MLIB2-L. Therefore, the test proves that the BRkNN classifier that operates on the CS generated by the proposed MLCNN-L and MLIB2-L algorithms achieves a comparable level of accuracy to the conventional BRkNN classifier that operates on the original training set.

Moreover, the test shows that there is a significant statistical difference in accuracy between BRkNN and the remaining multilabel variations in CNN. Nevertheless, in certain datasets, MLCNN-J ($JD > 0.5$) and MLIB2-J ($JD > 0.5$) achieve a comparable level of accuracy to the "conventional" BRkNN classifier that operates on the original training set. Moreover, MLCNN-L achieved the best Hamming loss in comparison to all other algorithms and close to the one of BRkNN.

Furthermore, the test affirms that there is statistical difference in terms of accuracy between the pairs MLCNN-L versus MLIB2-BR, MLIB2-J ($JD > 0.5$), MLIB2-L, MLCNN-H, MLCNN-BR, MLIB2-H, MLIB2-J ($JD > 0.75$) and MLCNN-J ($JD > 0.75$). The version of MLCNN-J ($JD > 0.5$) presents discrepancy in statistical terms of accuracy with the versions MLCNN-H, MLCNN-BR, MLIB2-H, MLIB2-J ($JD > 0.75$) and MLIB2-BR.

According to the Wilcoxon test, we have statistical difference in terms of reduction rate between the pairs MLIB2-H and the versions MLCNN-BR, MLCNN-H, MLIB2-J ($JD > 0.5$), MLIB2-BR, MLCNN-L, MLIB2-L and MLCNN-J ($JD > 0.5$). Moreover, the pairs between MLIB2-J ($JD > 0.75$) and the versions MLIB2-J ($JD > 0.5$), MLIB2-BR, MLCNN-L and MLIB2-L have statistical difference in terms of reduction rate.

Table 6. Results of Wilcoxon signed rank test on ACC and RR measurements.

Methods	Accuracy		Reduction Rate	
	w/l/t	Wilc.	w/l/t	Wilc.
BRkNN vs. MLCNN-H	8/0/1	**0.012**	-	-
BRkNN vs. MLCNN-J ($JD > 0.5$)	6/0/3	**0.027**	-	-
BRkNN vs. MLCNN-J ($JD > 0.75$)	7/0/2	**0.018**	-	-
BRkNN vs. MLCNN-BR	8/0/1	**0.012**	-	-
BRkNN vs. MLIB2-H	8/0/1	**0.012**	-	-
BRkNN vs. MLIB2-J ($JD > 0.5$)	6/0/3	**0.026**	-	-
BRkNN vs. MLIB2-J ($JD > 0.75$)	7/0/2	**0.018**	-	-
BRkNN vs. MLIB2-BR	8/1/0	**0.011**	-	-
BRkNN vs. MLCNN-L	5/1/3	**0.046**	-	-
BRkNN vs. MLIB2-L	5/1/3	**0.046**	-	-
MLCNN-H vs. MLCNN-J ($JD > 0.5$)	1/6/2	**0.028**	6/1/2	**0.028**
MLCNN-H vs. MLCNN-J ($JD > 0.75$)	3/6/0	0.260	1/8/0	0.110
MLCNN-H vs. MLCNN-BR	1/8/0	**0.011**	8/1/0	**0.015**

Table 6. *Cont.*

Methods	Accuracy		Reduction Rate	
	w/l/t	Wilc.	w/l/t	Wilc.
MLCNN-H vs. MLIB2-H	8/1/0	**0.011**	0/9/0	**0.008**
MLCNN-H vs. MLIB2-J ($JD > 0.5$)	4/5/0	0.859	5/4/0	0.767
MLCNN-H vs. MLIB2-J ($JD > 0.75$)	7/2/0	0.066	1/8/0	0.086
MLCNN-H vs. MLIB2-BR	8/1/0	**0.015**	5/4/0	0.515
MLCNN-H vs. MLCNN-L	1/8/0	**0.021**	9/0/0	**0.008**
MLCNN-H vs. MLIB2-L	2/7/0	0.260	7/2/0	0.051
MLCNN-J ($JD > 0.5$) vs. MLCNN-J ($JD > 0.75$)	6/3/0	0.214	0/9/0	**0.008**
MLCNN-J ($JD > 0.5$) vs. MLCNN-BR	8/1/0	**0.011**	8/1/0	0.110
MLCNN-J ($JD > 0.5$) vs. MLIB2-H	9/0/0	**0.008**	0/9/0	**0.008**
MLCNN-J ($JD > 0.5$) vs. MLIB2-J ($JD > 0.5$)	8/1/0	**0.011**	0/9/0	**0.008**
MLCNN-J ($JD > 0.5$) vs. MLIB2-J ($JD > 0.75$)	9/0/0	**0.008**	0/9/0	**0.008**
MLCNN-J ($JD > 0.5$) vs. MLIB2-BR	8/1/0	**0.011**	6/3/0	0.859
MLCNN-J ($JD > 0.5$) vs. MLCNN-L	1/8/0	0.051	7/2/0	0.214
MLCNN-J ($JD > 0.5$) vs. MLIB2-L	3/6/0	0.678	4/5/0	0.314
MLCNN-J ($JD > 0.75$) vs. MLCNN-BR	8/1/0	**0.011**	8/1/0	0.051
MLCNN-J ($JD > 0.75$) vs. MLIB2-H	7/2/0	0.066	3/6/0	0.214
MLCNN-J ($JD > 0.75$) vs. MLIB2-J ($JD > 0.5$)	3/6/0	0.953	7/2/0	0.374
MLCNN-J ($JD > 0.75$) vs. MLIB2-J ($JD > 0.75$)	7/2/0	**0.021**	0/9/0	**0.008**
MLCNN-J ($JD > 0.75$) vs. MLIB2-BR	8/1/0	**0.015**	6/3/0	0.260
MLCNN-J ($JD > 0.75$) vs. MLCNN-L	0/9/0	**0.008**	8/1/0	**0.015**
MLCNN-J ($JD > 0.75$) vs. MLIB2-L	2/7/0	0.214	7/2/0	0.086
MLCNN-BR vs. MLIB2-H	3/6/0	0.139	0/9/0	**0.008**
MLCNN-BR vs. MLIB2-J ($JD > 0.5$)	3/6/0	0.066	1/8/0	0.110
MLCNN-BR vs. MLIB2-J ($JD > 0.75$)	3/6/0	0.173	1/8/0	**0.015**
MLCNN-BR vs. MLIB2-BR	8/1/0	**0.038**	0/8/1	**0.012**
MLCNN-BR vs. MLCNN-L	1/8/0	**0.011**	4/5/0	0.953
MLCNN-BR vs. MLIB2-L	2/7/0	**0.021**	2/7/0	0.139
MLIB2-H vs. MLIB2-J ($JD > 0.5$)	1/6/2	0.056	6/1/2	**0.028**
MLIB2-H vs. MLIB2-J ($JD > 0.75$)	5/4/0	1.000	1/8/0	0.110
MLIB2-H vs. MLIB2-BR	8/1/0	**0.015**	7/2/0	**0.021**
MLIB2-H vs. MLCNN-L	0/9/0	**0.008**	9/0/0	**0.008**
MLIB2-H vs. MLIB2-L	1/8/0	**0.028**	9/0/0	**0.008**
MLIB2-J ($JD > 0.5$) vs. MLIB2-J ($JD > 0.75$)	7/2/0	0.051	0/9/0	**0.008**
MLIB2-J ($JD > 0.5$) vs. MLIB2-BR	8/1/0	**0.011**	7/2/0	0.173
MLIB2-J ($JD > 0.5$) vs. MLCNN-L	1/8/0	**0.011**	8/1/0	0.051
MLIB2-J ($JD > 0.5$) vs. MLIB2-L	1/7/1	0.093	6/3/0	0.214
MLIB2-J ($JD > 0.75$) vs. MLIB2-BR	8/1/0	**0.015**	7/2/0	**0.066**
MLIB2-J ($JD > 0.75$) vs. MLCNN-L	0/9/0	**0.008**	8/1/0	**0.011**
MLIB2-J ($JD > 0.75$) vs. MLIB2-L	0/9/0	**0.008**	8/1/0	**0.011**
MLIB2-BR vs. MLCNN-L	1/8/0	**0.011**	4/5/0	0.374
MLIB2-BR vs. MLIB2-L	1/8/0	**0.011**	4/5/0	0.678
MLCNN-L vs. MLIB2-L	4/1/4	0.078	0/9/0	**0.008**

6.3.2. Friedman Test Results

The results of the Friedman test for the ACC and RR measurements are displayed in Table 7. Notice that the RR ranks are inverted; i.e., the larger the number, the higher the rank of the algorithm is. The Friedman test shows that

Table 7. Results of Friedman test on ACC and RR measurements.

Algorithm	Mean Rank	
	ACC	RR
MLCNN-H	6.00	5.78
MLCNN-J ($JD > 0.5$)	4.22	4.00
MLCNN-J ($JD > 0.75$)	5.17	7.22
MLCNN-BR	9.00	2.28
MLCNN-L	3.67	2.67
MLIB2-H	7.72	8.33
MLIB2-J ($JD > 0.5$)	5.72	6.11
MLIB2-J ($JD > 0.75$)	7.44	9.22
MLIB2-BR	9.72	4.61
MLIB2-L	5.06	4.78
BRkNN	2.28	-

- MLCNN-L is the most accurate approach. MLCNN-J ($JD > 0.5$), MLCNN-J ($JD > 0.75$) and MLIB2-L are the runners up.
- MLIB2-J ($JD > 0.75$) and MLIB2-H achieve the highest reduction rates. MLCNN-J ($JD > 0.75$) and MLIB2-J ($JD > 0.5$) are the runners up.

7. Conclusions

The main objective of this paper is to address the issue of data reduction techniques specifically tailored for multilabel datasets. Here, the focus is on reducing instances rather than features. This type of reduction is crucial in the context of instance-based classification as it helps mitigate the computational burden associated with large datasets. However, it is important to note that most existing data reduction techniques are primarily designed for single-label classification problems and are not well-suited for multilabel classification. Additionally, these techniques cannot seamlessly integrate with problem transformation methods such as binary relevance or label powerset, which are commonly used in multilabel classification scenarios.

This paper presents novel algorithms focused on accelerating the instance-based classifiers in the context of multilabel classification. The study introduces four variations of the well-known CNN-rule and four variations of IB2 specifically designed for multilabel classification. The proposed MLCNN-H, MLCNN-J, MLCNN-L and MLCNN-BR algorithms and the corresponding MLIB2-H, MLIB2-J, MLIB2-L and MLIB2-BR algorithms can be considered as the first prototype selection algorithms for multilabel data condensing.

The proposed algorithms do not require any specific parameters. MLCNN-H and MLIB2-H consider two neighboring instances to be different if their Hamming distance exceeds the dataset density. MLCNN-J and MLIB2-J identify two neighboring instances as distinct if their Jaccard distance surpasses a predefined threshold. MLCNN-BR and MLIB2-BR construct separate prototypes for each label using the conventional CNN method and subsequently merge them by combining different labels to form the final condensing sets. Finally, in MLCNN-L and MLIB2-L, if the Levenshtein distance between two neighboring instances exceeds half the cardinality of the dataset, they are considered as different.

Consequently, the proposed algorithms generate a multilabel condensing set. This condensing set can be utilized by BRkNN to conduct multilabel prediction.

The experimental study demonstrated that switching from the initial training set to the condensing sets produced by the proposed algorithms did not greatly affect the accuracy achieved by BRkNN. However, it reduced the computational cost required for the classification process. The new variations achieved a reduction of more than 50% in computational costs.

Looking in more detail at the results of the experiments, it appears that, in terms of the overall classification performance of the algorithms compared to BRkNN, the MLCNN-L and MLCNN-J ($JD > 0.5$) outperformed the other variations. Further, regarding the algorithms' overall reduction rate performance, the MLIB2-J ($JD > 0.75$) and MLIB2-H achieved superior results.

This study highlights the ongoing significance of data reduction in multilabel problems within the domains of data mining and machine learning. Our goal is to extend popular data reduction techniques, typically applied to single-label datasets, to the realm of multilabel datasets. Additionally, our future work involves the development of novel parameter-free data reduction methods, as well as scalable approaches for training set classification in the context of multilabel problems.

Author Contributions: Conceptualisation, P.F., S.O. and G.E.; methodology, P.F., S.O. and G.E.; software, P.F., S.O. and G.E.; validation, P.F., S.O. and G.E.; formal analysis, P.F., S.O. and G.E.; investigation, P.F., S.O. and G.E.; resources, P.F., S.O. and G.E.; data curation, P.F., S.O. and G.E.; writing—original draft preparation, P.F., S.O. and G.E.; writing—review and editing, P.F., S.O. and G.E.; visualisation, P.F., S.O. and G.E.; supervision, S.O. and G.E.; project administration, P.F., S.O. and G.E. All authors have read and agreed to the published version of the manuscript.

Funding: This research received no external funding.

Institutional Review Board Statement: Not applicable.

Data Availability Statement: Publicly available datasets were analyzed in this study. These data can be found here: https://mulan.sourceforge.net/datasets-mlc.html (accessed on 17 October 2023) and https://www.uco.es/kdis/mllresources/ (accessed on 17 October 2023).

Conflicts of Interest: The authors declare no conflicts of interest.

Abbreviations

The following abbreviations are used in this manuscript:

DRT	Data Reduction Technique
PS	Prototype Selection
CNN	Condensed Nearest Neighbor
IB2	Instance-Based Learning 2
CS	Condensing set
TS	Training set
MLCNN-H	Multilabel Condensed Nearest Neighbor with Hamming Distance
MLCNN-J	Multilabel Condensed Nearest Neighbor with Jaccard Distance
MLCNN-L	Multilabel Condensed Nearest Neighbor with Levenshtein Distance
MLCNN-BR	Multilabel Condensed Nearest Neighbor with Binary Relevance
MLIB2-H	Multilabel Instance-Based Learning 2 with Hamming Distance
MLIB2-J	Multilabel Instance-Based Learning 2 with Jaccard Distance
MLIB2-L	Multilabel Instance-Based Learning 2 with Levenshtein Distance
MLIB2-BR	Multilabel Instance-Based Learning 2 with Binary Relevance

References

1. Tsoumakas, G.; Katakis, I. Multi-label classification: An overview. *Int. J. Data Warehous. Min.* **2007**, *3*, 1–13. [CrossRef]
2. Cover, T.; Hart, P. Nearest neighbor pattern classification. *IEEE Trans. Inf. Theory* **1967**, *13*, 21–27. [CrossRef]
3. Liu, H.; Motoda, H. *Feature Selection for Knowledge Discovery and Data Mining*; Kluwer Academic Publishers: New York, NY, USA, 1998.
4. Garcia, S.; Derrac, J.; Cano, J.; Herrera, F. Prototype Selection for Nearest Neighbor Classification: Taxonomy and Empirical Study. *IEEE Trans. Pattern Anal. Mach. Intell.* **2012**, *34*, 417–435. [CrossRef] [PubMed]
5. Triguero, I.; Derrac, J.; Garcia, S.; Herrera, F. A Taxonomy and Experimental Study on Prototype Generation for Nearest Neighbor Classification. *Trans. Systems Man Cyber Part C* **2012**, *42*, 86–100. [CrossRef]

6. Spyromitros, E.; Tsoumakas, G.; Vlahavas, I. An Empirical Study of Lazy Multilabel Classification Algorithms. In *Proceedings of the Artificial Intelligence: Theories, Models and Applications*; Darzentas, J.; Vouros, G.A.; Vosinakis, S.; Arnellos, A., Eds.; Springer: Berlin/Heidelberg, Germany, 2008; pp. 401–406. [CrossRef]
7. Hart, P.E. The condensed nearest neighbor rule. *IEEE Trans. Inf. Theory* **1967**, *18*, 515–516.
8. Aha, D.W.; Kibler, D.; Albert, M.K. Instance-based learning algorithms. *Mach. Learn.* **1991**, *6*, 37–66. [CrossRef]
9. Filippakis, P.; Ougiaroglou, S.; Evangelidis, G. Condensed Nearest Neighbour Rules for Multi-Label Datasets. In Proceedings of the International Database Engineered Applications Symposium Conference, Heraklion, Greece, 5–7 May 2023; pp. 43–50. [CrossRef]
10. Levenshtein, V.I. Binary codes capable of correcting deletions, insertions, and reversals. *Sov. Phys. Dokl.* **1965**, *10*, 707–710.
11. Tsoumakas, G.; Spyromitros-Xioufis, E.; Vilcek, J.; Vlahavas, I. Mulan: A Java Library for Multi-Label Learning. *J. Mach. Learn. Res.* **2011**, *12*, 2411–2414.
12. Read, J.; Reutemann, P.; Pfahringer, B.; Holmes, G. MEKA: A Multi-label/Multi-target Extension to WEKA. *J. Mach. Learn. Res.* **2016**, *17*, 1–5.
13. Charte, F.; Rivera, A.J.; del Jesus, M.J.; Herrera, F. MLeNN: A First Approach to Heuristic Multilabel Undersampling. In *Intelligent Data Engineering and Automated Learning–IDEAL 2014*; Springer: New York, NY, USA, 2014; pp. 1–9. [CrossRef]
14. Wilson, D.L. Asymptotic Properties of Nearest Neighbor Rules Using Edited Data. *IEEE Trans. Syst. Man Cybern.* **1972**, *SMC-2*, 408–421. [CrossRef]
15. Kanj, S.; Abdallah, F.; Denœux, T.; Tout, K. Editing training data for multi-label classification with the k-nearest neighbor rule. *Pattern Anal. Appl.* **2015**, *19*, 145–161. [CrossRef]
16. Arnaiz-González, Á.; Díez-Pastor, J.F.; Rodríguez, J.J.; García-Osorio, C. Local sets for multi-label instance selection. *Appl. Soft Comput.* **2018**, *68*, 651–666. [CrossRef]
17. Leyva, E.; González, A.; Pérez, R. Three new instance selection methods based on local sets: A comparative study with several approaches from a bi-objective perspective. *Pattern Recognit.* **2015**, *48*, 1523–1537. [CrossRef]
18. Li, H.; Fang, M.; Li, H.; Wang, P. Prototype selection for multi-label data based on label correlation. *Neural Comput. Appl.* **2023**. [CrossRef]
19. Chou, C.H.; Kuo, B.H.; Chang, F. The Generalized Condensed Nearest Neighbor Rule as A Data Reduction Method. In Proceedings of the 18th International Conference on Pattern Recognition (ICPR'06), Hong Kong, China, 20–24 August 2006; Volume 2, pp. 556–559. [CrossRef]
20. Suyal, H.; Singh, A. Improving Multi-Label Classification in Prototype Selection Scenario. In *Computational Intelligence and Healthcare Informatics*; Wiley: Hoboken, NJ, USA, 2021; pp. 103–119. [CrossRef]
21. Zhang, M.L.; Zhou, Z.H. ML-KNN: A lazy learning approach to multi-label learning. *Pattern Recognit.* **2007**, *40*, 2038–2048. [CrossRef]
22. Arnaiz-González, Á.; Díez-Pastor, J.F.; Rodríguez, J.J.; García-Osorio, C. Study of data transformation techniques for adapting single-label prototype selection algorithms to multi-label learning. *Expert Syst. Appl.* **2018**, *109*, 114–130. [CrossRef]
23. Calvo-Zaragoza, J.; Valero-Mas, J.J.; Rico-Juan, J.R. Improving kNN multi-label classification in Prototype Selection scenarios using class proposals. *Pattern Recognit.* **2015**, *48*, 1608–1622. [CrossRef]
24. González, M.; Cano, J.R.; García, S. ProLSFEO-LDL: Prototype Selection and Label- Specific Feature Evolutionary Optimization for Label Distribution Learning. *Appl. Sci.* **2020**, *10*, 3089. [CrossRef]
25. Geng, X. Label Distribution Learning. *IEEE Trans. Knowl. Data Eng.* **2016**, *28*, 1734–1748. [CrossRef]
26. Ougiaroglou, S.; Filippakis, P.; Evangelidis, G. Prototype Generation for Multi-label Nearest Neighbours Classification. In *Proceedings of the Hybrid Artificial Intelligent Systems*; Sanjurjo González, H., Pastor López, I., García Bringas, P., Quintián, H., Corchado, E., Eds.; Springer: Cham, Germany, 2021; pp. 172–183.
27. Ougiaroglou, S.; Filippakis, P.; Fotiadou, G.; Evangelidis, G. Data reduction via multi-label prototype generation. *Neurocomputing* **2023**, *526*, 1–8. [CrossRef]
28. Sánchez, J. High training set size reduction by space partitioning and prototype abstraction. *Pattern Recognit.* **2004**, *37*, 1561–1564. [CrossRef]
29. Valero-Mas, J.J.; Gallego, A.J.; Alonso-Jiménez, P.; Serra, X. Multilabel Prototype Generation for data reduction in K-Nearest Neighbour classification. *Pattern Recognit.* **2023**, *135*, 109190. [CrossRef]
30. Chen, C.; Jóźwik, A. A sample set condensation algorithm for the class sensitive artificial neural network. *Pattern Recognit. Lett.* **1996**, *17*, 819–823. [CrossRef]
31. Sun, L.; Ji, S.; Ye, J. Hypergraph Spectral Learning for Multi-Label Classification. In Proceedings of the Proceedings of the 14th ACM SIGKDD International Conference on Knowledge Discovery and Data Mining, Las Vegas, NV, USA, 24–27 August 2008; pp. 668–676. [CrossRef]
32. Byerly, A.; Kalganova, T. Class Density and Dataset Quality in High-Dimensional, Unstructured Data. *arXiv* **2022**. [CrossRef]
33. Zhang, S.; Hu, Y.; Bian, G. Research on string similarity algorithm based on Levenshtein Distance. In Proceedings of the 2017 IEEE 2nd Advanced Information Technology, Electronic and Automation Control Conference (IAEAC), Chongqing, China, 25–26 March 2017; pp. 2247–2251. [CrossRef]
34. Pedregosa, F.; Varoquaux, G.; Gramfort, A.; Michel, V.; Thirion, B.; Grisel, O.; Blondel, M.; Prettenhofer, P.; Weiss, R.; Dubourg, V.; et al. Scikit-learn: Machine Learning in Python. *J. Mach. Learn. Res.* **2011**, *12*, 2825–2830.

35. Sechidis, K.; Tsoumakas, G.; Vlahavas, I. On the Stratification of Multi-label Data. In *Proceedings of the Machine Learning and Knowledge Discovery in Databases*; Gunopulos, D., Hofmann, T., Malerba, D., Vazirgiannis, M., Eds.; Springer: Berlin/Heidelberg, Germany, 2011; pp. 145–158. [CrossRef]
36. Czarnowski, I.; Jędrzejowicz, P. An Approach to Data Reduction for Learning from Big Datasets: Integrating Stacking, Rotation, and Agent Population Learning Techniques. *Complexity* **2018**, *2018*, 7404627. [CrossRef]
37. Gallego, A.J.; Calvo-Zaragoza, J.; Valero-Mas, J.J.; Rico-Juan, J.R. Clustering-Based k-Nearest Neighbor Classification for Large-Scale Data with Neural Codes Representation. *Pattern Recogn.* **2018**, *74*, 531–543. [CrossRef]
38. Ougiaroglou, S.; Evangelidis, G. RHC: Non-Parametric Cluster-Based Data Reduction for Efficient k-NN Classification. *Pattern Anal. Appl.* **2016**, *19*, 93–109. [CrossRef]
39. Escalante, H.J.; Graff, M.; Morales-Reyes, A. PGGP: Prototype Generation via Genetic Programming. *Appl. Soft Comput.* **2016**, *40*, 569–580. [CrossRef]
40. Escalante, H.J.; Marin-Castro, M.; Morales-Reyes, A.; Graff, M.; Rosales-Pérez, A.; Montes-Y-Gómez, M.; Reyes, C.A.; Gonzalez, J.A. MOPG: A Multi-Objective Evolutionary Algorithm for Prototype Generation. *Pattern Anal. Appl.* **2017**, *20*, 33–47. [CrossRef]
41. Calvo-Zaragoza, J.; Valero-Mas, J.J.; Rico-Juan, J.R. Prototype Generation on Structural Data Using Dissimilarity Space Representation. *Neural Comput. Appl.* **2017**, *28*, 2415–2424. [CrossRef]
42. Sheskin, D. *Handbook of Parametric and Nonparametric Statistical Procedures*; A Chapman & Hall Book; Chapman & Hall/CRC: Boca Raton, FL, USA, 2011.

Disclaimer/Publisher's Note: The statements, opinions and data contained in all publications are solely those of the individual author(s) and contributor(s) and not of MDPI and/or the editor(s). MDPI and/or the editor(s) disclaim responsibility for any injury to people or property resulting from any ideas, methods, instructions or products referred to in the content.

Article

Convolutional Neural Networks Analysis Reveals Three Possible Sources of Bronze Age Writings between Greece and India †

Shruti Daggumati and Peter Z. Revesz *

School of Computing, College of Engineering, University of Nebraska-Lincoln, Lincoln, NE 68588, USA; sdaggumati@unl.edu (S.D.); revesz@cse.unl.edu (P.Z.R.)
* Correspondence: revesz@cse.unl.edu; Tel.: +1-402-421-6990
† This paper is an extended version of our paper published in the 23rd International Database Engineering and Applications Symposium, IDEAS 2019, Athens, Greece, 10–12 June 2019.

Abstract: This paper analyzes the relationships among eight ancient scripts from between Greece and India. We used convolutional neural networks combined with support vector machines to give a numerical rating of the similarity between pairs of signs (one sign from each of two different scripts). Two scripts that had a one-to-one matching of their signs were determined to be related. The result of the analysis is the finding of the following three groups, which are listed in chronological order: (1) Sumerian pictograms, the Indus Valley script, and the proto-Elamite script; (2) Cretan hieroglyphs and Linear B; and (3) the Phoenician, Greek, and Brahmi alphabets. Based on their geographic locations and times of appearance, Group (1) may originate from Mesopotamia in the early Bronze Age, Group (2) may originate from Europe in the middle Bronze Age, and Group (3) may originate from the Sinai Peninsula in the late Bronze Age.

Keywords: classification; epigraphy; neural networks; script family; support vector machine

Citation: Daggumati, S.; Revesz, P.Z. Convolutional Neural Networks Analysis Reveals Three Possible Sources of Bronze Age Writings between Greece and India. *Information* **2023**, *14*, 227. https://doi.org/10.3390/info14040227

Academic Editor: Xin Ning

Received: 7 February 2023
Revised: 4 April 2023
Accepted: 4 April 2023
Published: 7 April 2023

Copyright: © 2023 by the authors. Licensee MDPI, Basel, Switzerland. This article is an open access article distributed under the terms and conditions of the Creative Commons Attribution (CC BY) license (https://creativecommons.org/licenses/by/4.0/).

1. Introduction

In this paper, we use data mining methods to analyze the relationships among eight Bronze Age scripts from between Greece and India, namely the Brahmi script [1], Cretan hieroglyphs [2], the Greek alphabet [3], the Indus Valley script [4–8], the Linear B syllabary [9], the Phoenician alphabet [10–12], the proto-Elamite script [13,14], and Sumerian pictographs [15].

We are interested in testing the hypothesis that these eight scripts had a single origin. This is probable given that the eight scripts originate from geographic locations along an east–west line between India and Greece, as is shown in Figure 1.

We are going to test this hypothesis by applying data mining to the scripts. The data mining method that we have chosen for this study is a convolutional neural networks analysis. Convolutional neural networks have previously been applied to the recognition of various signs, including alphabets, but they have not been used in a multiscript analysis.

The novel idea in our approach is to first train separate convolutional neural networks to recognize various scripts (see Section 5.1 for a review of works that are related to this first phase). Then, in the second phase, we pass one script's signs into another's convolutional neural network. The sign 'recognized' by the convolutional neural network can be considered the closest to the input sign. If the two scripts are related to each other, then a one-to-one mapping may be found between the signs of the two scripts. If the two scripts are not related to each other, then there will be no one-to-one mapping.

Our study is motivated by a desire to contribute to the decipherment of ancient, Bronze Age scripts, especially the Indus Valley script [8]. Decipherment can be greatly facilitated by understanding the precise relationships among these ancient scripts. A one-to-one mapping of the signs of an undeciphered and a deciphered script would suggest phonetic values for the signs of the undeciphered script because the visual forms and the phonetic values of the signs tend to change simultaneously and gradually.

The outline of the paper is as follows. Section 2 introduces the eight ancient scripts that are to be compared and classified. Section 3 presents the machine learning software algorithms used to learn and group together the various signs. Section 4 describes the major findings of our study. Section 5 analyses the results and compares them with related work. Finally, Section 6 presents some open problems.

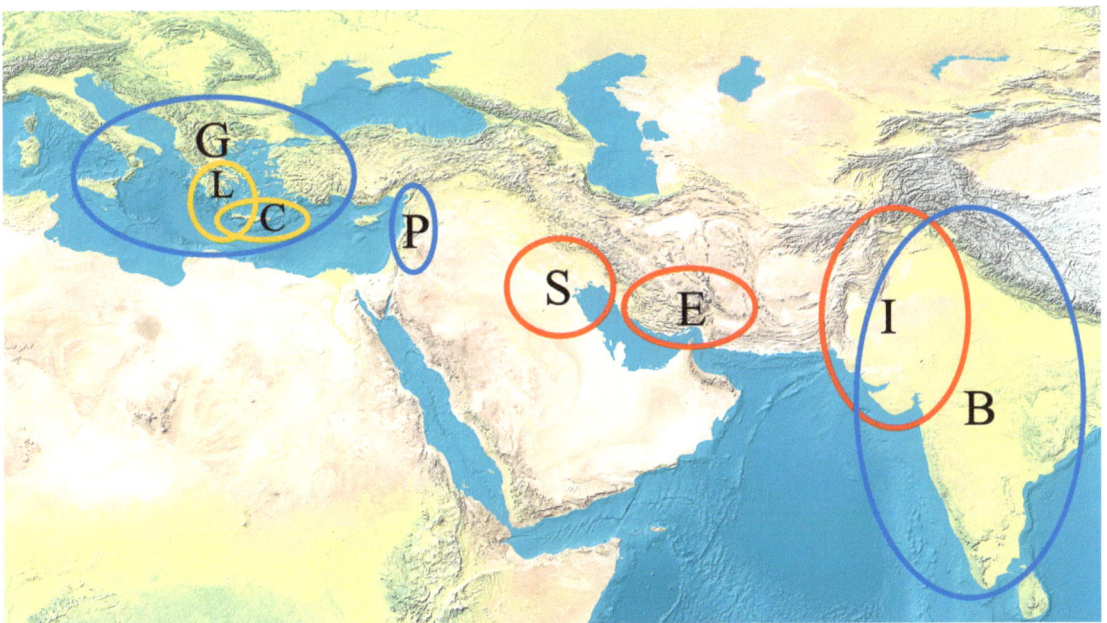

Figure 1. The approximate locations of the eight ancient scripts. The legend is as follows: B—Brahmi, C—Cretan hieroglyphs, E—Elamite, G—Greek, I—Indus Valley, L -Linear B, P—Phoenician, and S—Sumerian. Red indicates the three earliest scripts, orange the middle two scripts, and blue the three most recent of the eight ancient scripts. Source of background map: https://upload.wikimedia.org/wikipedia/commons/f/f3/Map_of_Eurasia.png (accessed on 2 February 2023).

2. Data Source

We used the following ancient scripts as data sources.

1. Brahmi, which has an unknown origin, was an abugida script in India and was written left-to-right [1]. We used 34 of the signs from the Brahmi script, as shown in Figure 2.

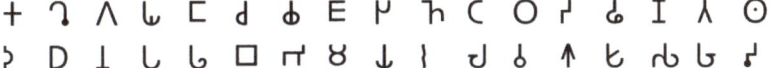

Figure 2. The 34 Brahmi signs used in this study.

2. Cretan hieroglyphs also have an unknown origin. Cretan hieroglyphs were used between 2100 to 1700 BCE [2], that is, mainly contemporaneously with Linear A, but both were superseded by Linear B. We used 22 signs from the Cretan hieroglyphs, as shown in Figure 3.

Figure 3. The 22 Cretan hieroglyphs used in this study.

3. Starting around 800 BCE, the ancient *Greek alphabet* had several variants according to various local Greek dialects [3]. We used all 27 letters of the Greek alphabet, as shown in Figure 4.

Figure 4. The 27 ancient Greek alphabet letters.

4. The Indus Valley script was in use in what is today Pakistan and India from around 2400 BCE to 1900 BCE [4–8]. Its writing direction was mainly right-to-left, although there are some left-to-right and boustrophedon writing examples, too. Remarkably, the Indus Valley script has over 700 different signs. Since only those signs that occur at least three times seem significant, we used only the 23 most frequent Indus Valley script signs, as shown in Figure 5.

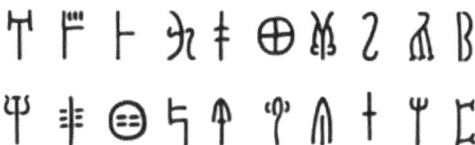

Figure 5. The 23 Indus Valley script signs used in this study.

5. The Mycenaean Greeks used the Linear B script, which is an adaptation of the earlier Linear A that was used by the Minoans. In 1952, Michael Ventris succeeded in determining that Linear B was the older written form of the Greek language that was written using syllabic signs [9]. We used 20 signs from Linear B, as shown in Figure 6.

Figure 6. The 20 Linear B signs used in this study.

6. Beginning around 1200 BCE, the Phoenician alphabet was written on clay tablets [10]. According to some proposals, the Phoenician alphabet may be derived from Egyptian hieroglyphs [11], but its development may also have been influenced by Linear B [12]. Since the 22 Phoenician alphabet letters originally denoted only consonants, it is classified as an abjad. Phoenician texts also usually run right-to-left. We used all 22 Phoenician alphabet letters, as shown in Figure 7.

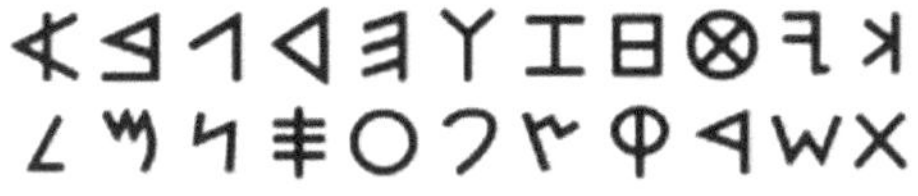

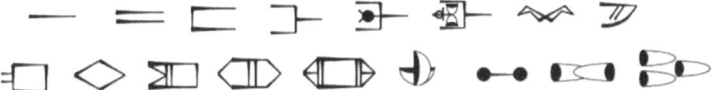

Figure 7. The 22 Phoenician alphabet letters used in this study.

7. The proto-Elamite script existed primarily in the region that today is Iran during the fourth millennium BCE [13]. The proto-Elamite script had almost two thousand signs, but most of those signs were used infrequently [14]. Currently, the proto-Elamite script is currently undeciphered. We used 17 signs from the proto-Elamite script, as shown in Figure 8.

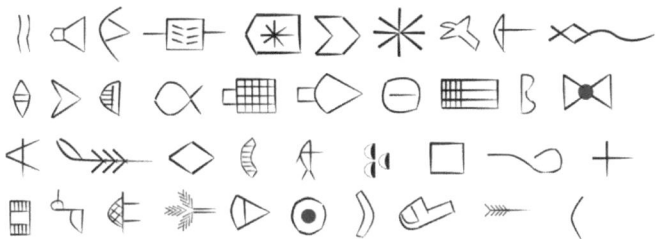

Figure 8. The 17 proto-Elamite signs used in this study.

8. Sumerian pictograms were a novel development and were mostly logographic, according to researchers. They were formed in the fourth millennium BCE, but they developed into cuneiform signs, which were used over several millennia until the first century [15]. The Sumerian language is distantly related to the Dravidian and Uralic languages [16,17]. We used 34 signs from the Sumerian pictograms, as shown in Figure 9.

Figure 9. The 34 Sumerian Pictograms used in this study.

Figure 10 gives a timeline of the eight scripts mentioned above. The Sumerian pictograms were used only for a few hundred years and then gradually developed into the cuneiform script that was used by later cultures over three thousand years. It is important to consider this timeline and the locations together with the similarity of the scripts in order to identify ancestor–successor relationships. Figure 1 shows a map of the approximate locations of the scripts compared in this paper. Neural networks can identify similarities in the scripts, but they are unaware of the timeline or the locations in which the various scripts were used.

Another consideration is the orientation of the signs. For example, in the Sumerian pictograms, the signs initially stood upright, while in later times they had been rotated 90 degrees. Figure 9 shows this later stage, after the signs had been rotated. This rotation is obvious for some signs, such as the bird sign, which is the second from the right in the last row of Figure 9. One of the advantages of neural networks is that they can learn to recognize signs regardless of their orientation. However, to achieve this rotation independence, the training examples need to include several rotated versions of the same sign.

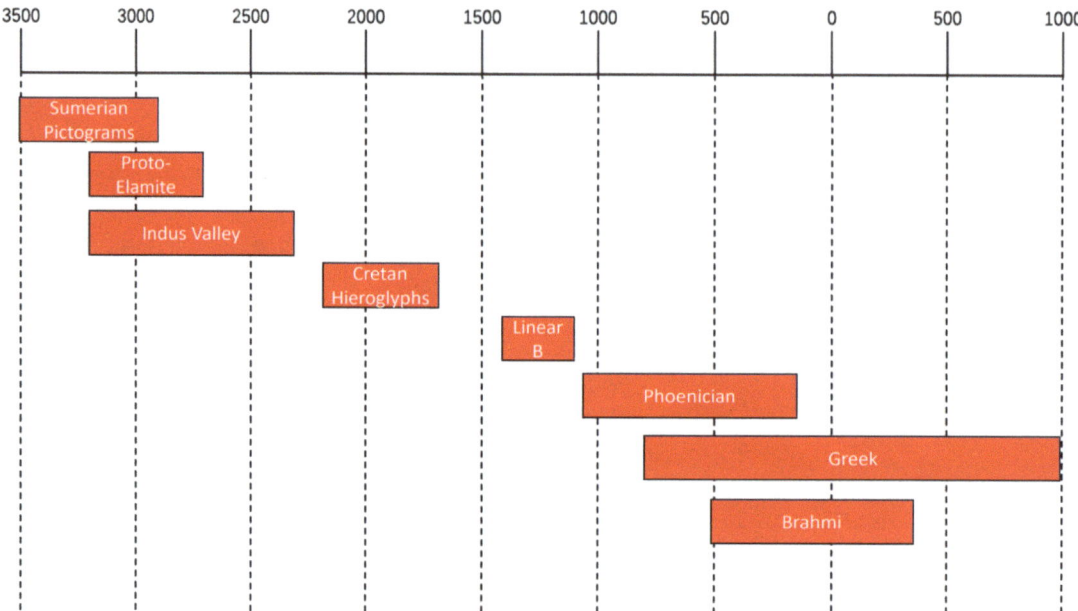

Figure 10. A timeline of the eight ancient scripts analyzed in this study.

Another consideration is the vertical mirror symmetry of the signs. Many ancient scripts were written in a boustrophedon style. This meant that the writer wrote the first line from left-to-write, then reversed the direction to right-to-left in the second line, and then kept switching the direction for each successive line of the inscription. As an aid to the reader, the boustrophedon inscriptions often used a vertical mirror-symmetric version of the usual sign. For example, instead of an E, they might use an Ǝ. These two forms of the letter E are considered allographs of each other and should be treated as a single letter during script comparisons. Neural networks can also learn to recognize mirror-symmetric signs if the training examples include mirror-symmetric examples of the signs.

We took MNIST as a model for preprocessing the data and built a database [18]. We used 780 training images and 120 validation images, a total of 900 hand-drawn or computer-distorted images for each sign. The size of each grayscale image was 50 × 50 pixels.

3. Experimental Design

3.1. Design of the CNN

Python and TensorFlow together with a Keras wrapper were used to build neural networks with different accuracy levels depending on the learned script. We used a convolutional neural network (CNN) architecture similar to the architecture of LeNet [19], though with some changes that are illustrated below in Figure 11. The primary difference between LeNet and our network is that a support vector machine (SVM) was added at the end. The addition of an SVM was also effectively used in [20].

We first reduced the input images to 46 × 46 pixels by applying 5 × 5 filters. Second, we further cut the size of the images to 23 × 23 pixels using a pooling layer. Third, by another set of convolution filters, the images were reduced to 20 × 20 pixels. Fourth, another pooling produced 10 × 10 pixels. Fifth, the images passed a layer that had 1024 fully connected neurons. The output of the neurons was fed into the SVM, which we further detail in the next section.

We added to the convolution layers rectified linear unit (ReLU) activation functions, which produced a linear value with a slope of one when $x > 0$. The 2 × 2 filters picked for the feature map the maximum of the four quadrants' values. Max pooling was applied by the pooling layers.

Overfitting was avoided by a 0.4 drop rate. A small 0.001 learning rate was used by the Adam optimizers [21] within the convolutional neural networks.

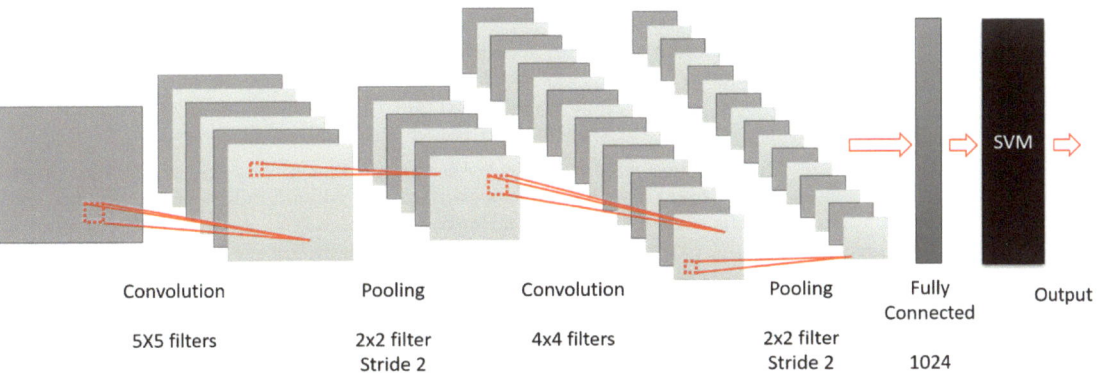

Figure 11. The classifier architecture.

3.2. Design of the SVM

We used a Python library package and Python for the development of the support vector machine in the software architecture described in Figure 11. Within the last layer of Figure 11, we used L2-SVM for multiclass classification, which is considered better than Softmax, which is a common alternative [22]. The L2-SVM optimized the sum of the squared errors using the following function, where the vector variable w has the dimension N, ξ_i are the slack variables, and C is the penalty parameter.

Minimize:

$$\frac{1}{2}||w||^2 + \frac{C}{2}\sum_{i=1}^{N}\xi_i^2$$

Subject to:

$$y_i(x_i \cdot w + b) \geq 1 - \xi_i \quad i = 1, \ldots, N$$

where b is a bias term.

3.3. The Sign Classifier

Figure 12 shows the scheme according to which the trained and validated classifiers for the eight scripts were used to test the similarity of any pair of scripts. Figure 12 specifically shows how the N signs of any one of the seven other scripts (called the 'unknown script' in the diagram) can be compared with the 22 letters of the Phoenician alphabet.

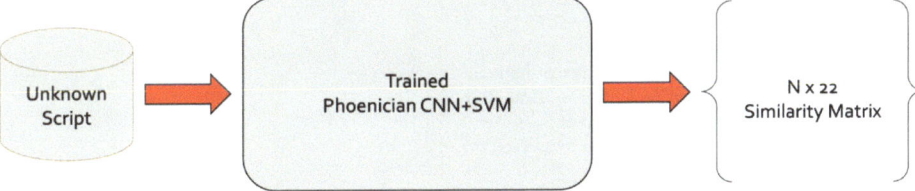

Figure 12. The scheme of comparing any 'unknown script' (which can be any of the other scripts with N signs) with the letters of the Phoenician alphabet.

After passing in the 'unknown scripts' to each of the trained and validated script classifiers, the scheme in Figure 12 yielded 56 N × M dimensional similarity matrices, where N and M are the number of different signs in the two scripts. A strength measure between a pair of scripts can be defined in either of the following two ways.

1. The average of all considers all the of signs by averaging the maximum probability matches between the input signs and the trained script signs. If an input sign had a low correlation with all of the trained signs, then the average of all value would be lowered.
2. The selective average takes the average of only those pairs of signs which have a higher than 75 percent (or other chosen threshold) similarity match.

There are two advantages to the second approach. The first advantage is that the selective average yields a higher measure compared with the average of all. The second advantage is that we can, if we want, also simultaneously obtain the number of input signs which have a pair in the trained script with a similarity threshold above 75 percent.

3.4. Generation of Classification Dendrograms

As was described in Section 3.3, there are two different ways to obtain a strength measure between a pair of scripts. Furthermore, it is convenient to consider the number of input signs for which there is a trained sign with an above 75 percent similarity. The different measures lead to two different algorithms for the generation of classification trees or dendrograms.

1. Similarity classification dendrograms: The weighted pair group method with arithmetic mean (WPGMA) algorithm was used to create a dendrogram as follows. We repeatedly merged those sets of scripts that were most similar according to the similarity matrix. The similarity matrix was updated after each merge. The update requires that the most similar script sets, x and y, are merged into the union $x \cup y$ of the two sets. This means merging the corresponding two rows into one row and the corresponding two columns into one column in the similarity matrix. In addition, the distance to another set, z and $x \cup y$, is updated using the following equation:

$$d_{(x \cup y),z} = \frac{d_{x,z} + d_{y,z}}{2} \quad (1)$$

The similarity is taken to be the negative of the distance.

2. Hierarchical classification dendrograms: In generating a hierarchal tree, it is assumed that some scripts have an ancestor–descendant relationship. This requires a modification of the WPGMA algorithm, but must also consider the periods during which the scripts were used. If x and y are the most similar to each other, that is, they can be considered to be closest script pair, and x's period of use preceded y's period of use, or vice versa, then we consider x to be an ancestor or parent of y. Algorithm 1 was built on this idea.

Algorithm 1 Time-Based Descendant Tree

1: Create parent node P
2: Create a node for each script
3: **for all** Closest Script Pairs S_x and S_y **do**
4:    **if** S_x.Time > S_y.Time **then**
5:       Parent of S_x is P
6:       Parent of S_y is S_x
7:    **else**
8:       Parent of S_y is P
9:       Parent of S_x is S_y
10: **for all** Singleton Scripts S_z **do**
11:    Parent of S_z is P
12: **return** Tree

4. Experimental Results

The three main ideas that we have presented above are the creation of the dataset, the design of the classifiers for each script and their use in a scheme to generate a script similarity matrix, and the algorithm for the generation of the hierarchical dendrograms. These three components must all work smoothly together to create a satisfying result. Table 1 shows the accuracy of the individual classifiers for each script. The classifiers of each script reached over 97 percent accuracy at 100 epochs.

Table 1. Validation accuracy for the eight scripts after 25, 50, 75, and 100 epochs of training.

Script	25	50	75	100
Brahmi Script	95.09	98.15	98.24	99.35
Cretan Hieroglyphs	91.09	92.84	94.47	97.53
Greek Alphabet	93.49	96.26	97.23	98.63
Indus Valley Script	93.50	95.70	96.85	98.23
Linear B Script	91.19	93.15	96.42	99.48
Phoenician Alphabet	93.18	94.77	95.36	97.52
Proto-Elamite Script	91.93	94.55	97.05	99.09
Sumerian Pictograms	90.79	93.21	96.94	97.40

To validate the automatic identification of ancestor–descendant relationships, we conducted an experiment in which scripts were grouped as follows: Known Origin, i.e., scripts that are used for validation by showing that we can reproduce established results, and Unknown Origin. Below are some specific examples that fall within these two categorizations.

1. Known Origin: It is well-known that the Phoenician alphabet was adopted by the ancient Greeks, who extended it by four letters that are specific to the Greek alphabet, as is shown in Table 2. It is also known to be an ancestor of Aramaic, which is an ancestor of Brahmi. By transitivity, Phoenician is an ancestor of Brahmi too. In addition, Cretan hieroglyphics are often said to be an ancestor of the Linear B script.
2. Unknown Origin: Sumerian pictographs have no known ancestors. A similar situation holds for the proto-Elamite and Indus Valley scripts.

We can validate Phoenician as the ancestor of Greek by passing the letter of the Greek alphabet into the classifier that was trained to recognize the Phoenician alphabet, or vice versa. Figures 13 and 14 show the heatmaps for the Phoenician and the Greek alphabets. The heatmaps were generated from the similarity matrices and show high similarities along the main diagonal. This proves that there is an almost perfect one-to-one function between the letters of the Phoenician and Greek alphabets. Moreover, this mapping matches our original expectations.

After this validation step, we were able to continue with confidence to test the relationship between other pairs of scripts with an unknown relationship. Whenever our CNN+SVM finds an almost one-to-one mapping between two scripts, we can be confident that the two scripts have an ancestor–descendant relationship such as that between the Phoenician and Greek alphabets. Table 3 records the number of signs which have over 75 percent correlation from among the pairs of the eight scripts.

Table 2. Adaptation of the Phoenician alphabet to the Greek alphabet, including four extra letters.

Phoenician Letter	Phoenician Name	Greek Letter	Greek Name
𐤀	aleph	A	alpha
𐤁	beth	B	beta
𐤂	giml	Γ	gamma
𐤃	daleth	Δ	delta
𐤄	he	E	epsilon
Y	waw	F or Y	digamma or upsilon
𐤆	zayin	I	zeta
𐤇	heth	H	eta
⊗	teth	⊗	theta
𐤉	yodh	I	iota
𐤊	kaph	K	kappa
𐤋	lamedh	Λ	lambda
𐤌	mem	M	mu
𐤍	nun	N	nu
╪	samekh	Ξ	xi
O	ayin	O	omicron
𐤐	pe	Π	pi
𐤑	sade	M	san
Φ	qoph	Ϙ	koppa
𐤓	res	P	rho
W	sin	Σ	sigma
X	taw	T	tau
		Φ	phi
		X	chi
		Ψ	psi
		Ω	omega

Table 3. The number of signs with over 75 percent correlation between pairs of various scripts.

	Brahmi	Cretan Hieroglyphs	Greek	Indus Valley	Linear B	Phoenician	Proto-Elam.	Sumerian Pictograms
Brahmi	34	2	9	8	3	9	2	6
Cretan Hieroglyphs	2	22	4	5	20	6	2	6
Greek	9	4	26	9	7	22	2	7
Indus Valley	8	5	9	23	4	9	4	20
Linear B	3	20	7	4	20	9	0	5
Phoenician	9	6	22	9	9	22	3	7
Proto-Elamite	2	2	2	4	0	3	17	3
Sumerian Pictograms	6	6	7	20	5	7	3	39

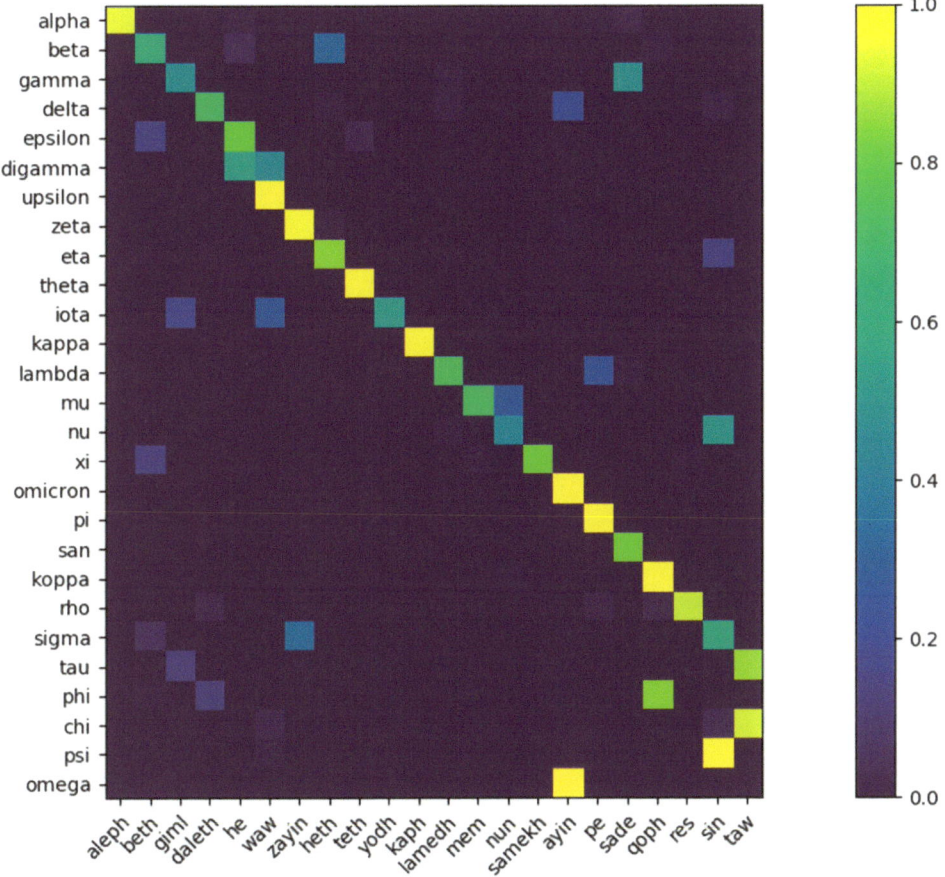

Figure 13. The heatmap generated when Greek letters were passed into the Phoenician letter classifier.

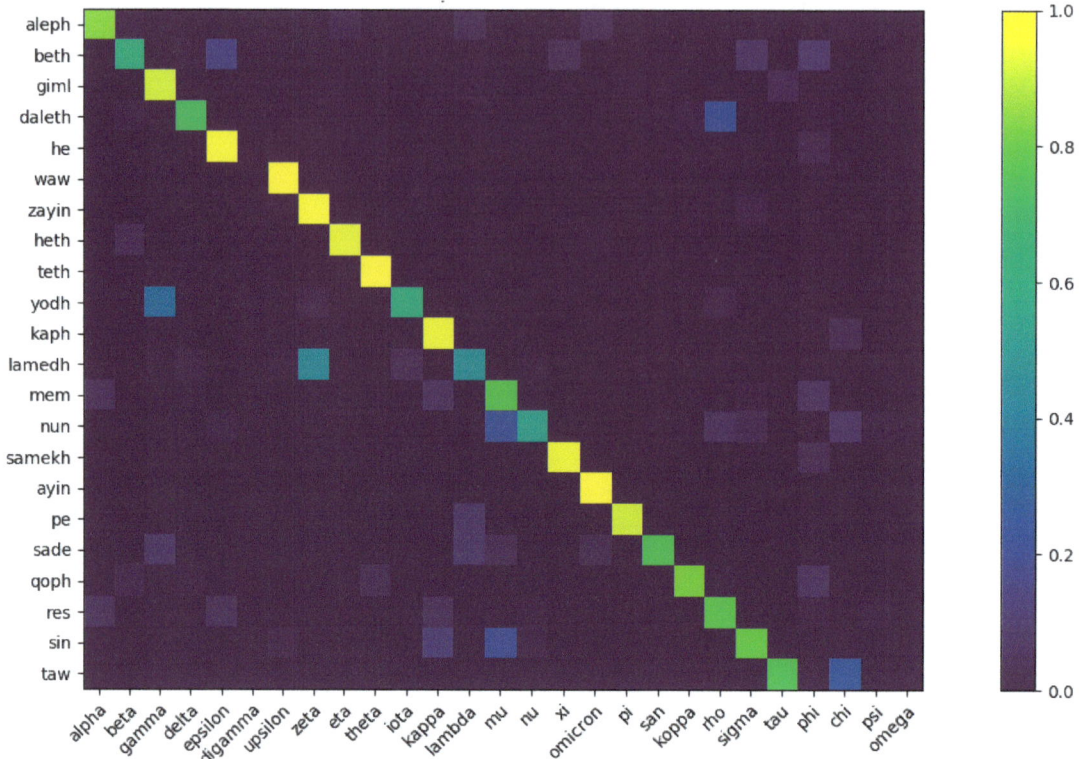

Figure 14. The heatmap generated when Phoenician letters were passed into the Greek letter classifier.

Our CNN+SVM predictor method discovered some previously unrecognized ancestor–descendant relationships. The heatmap in Figure 15 illustrates that there is also an almost perfect one-to-one function between Sumerian pictograms and Indus Valley signs.

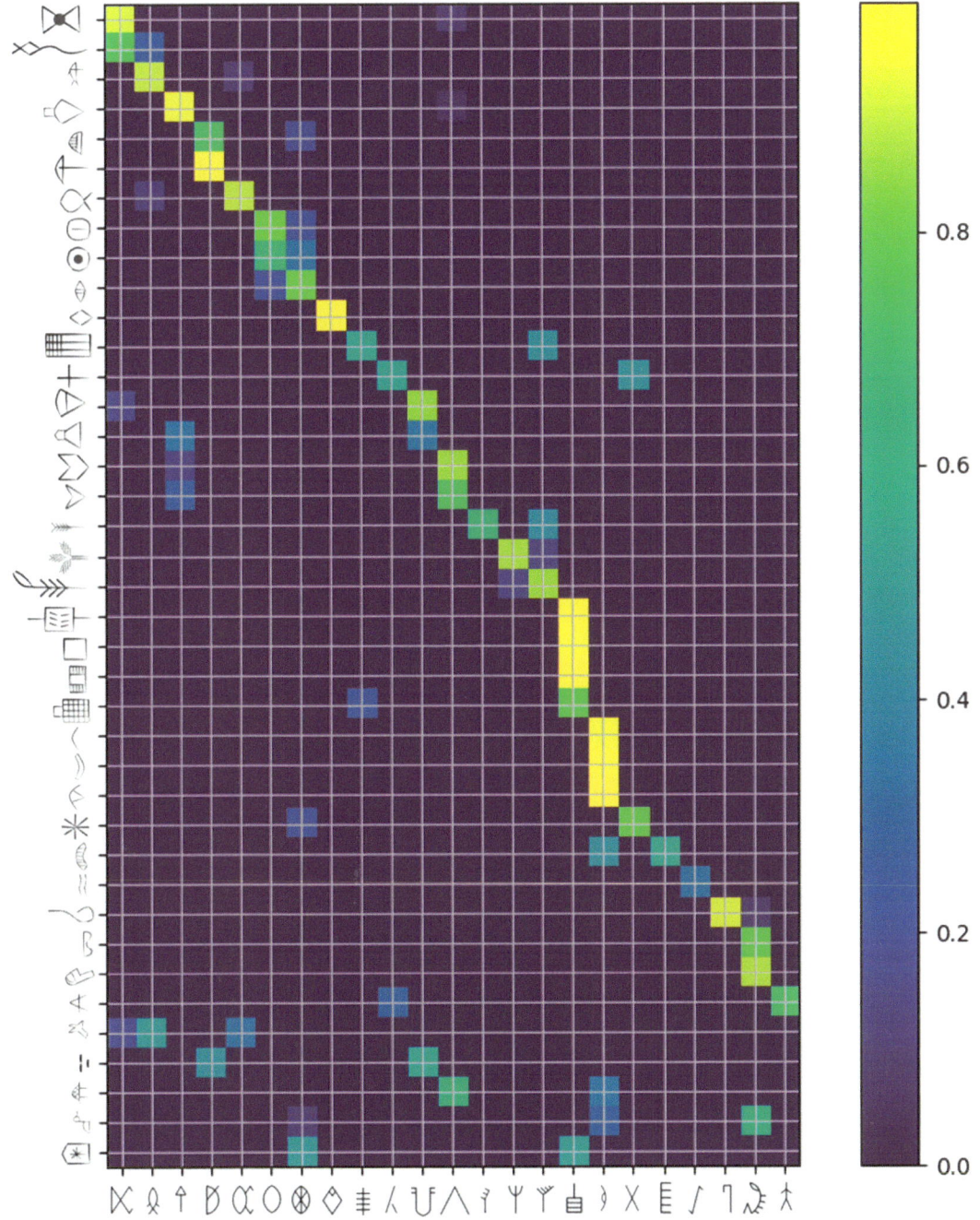

Figure 15. The heatmap generated when Sumerian pictograms were passed into the trained Indus Valley script classifier.

5. Discussion of the Results

Figures 16 and 17 show the classifications and the hierarchical dendrograms that were generated from among the scripts using the similarity matrices.

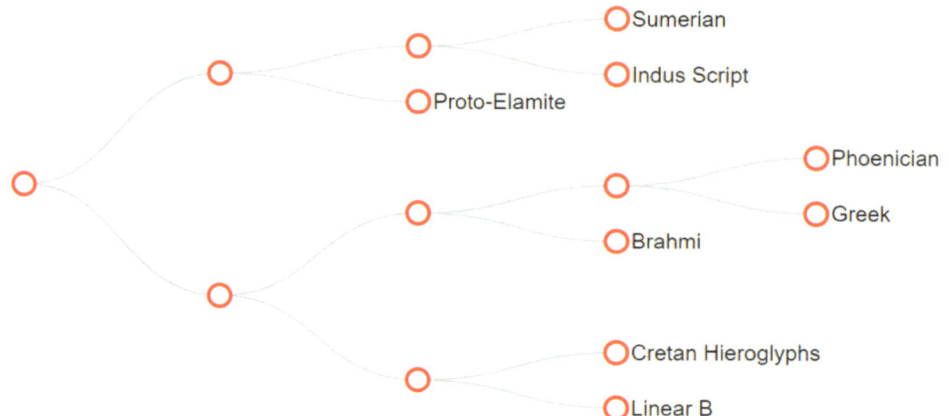

Figure 16. Classification dendrogram generated using the WPGMA algorithm.

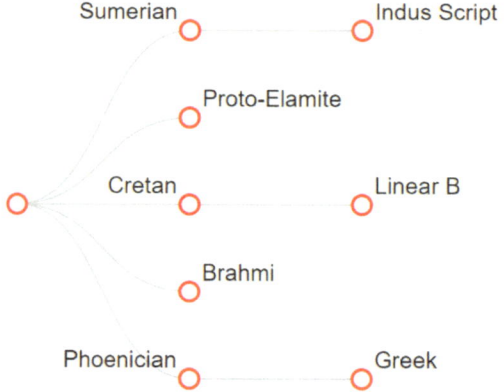

Figure 17. Hierarchical dendrogram generated by considering ancestor–descendant relationships.

In addition to verifying the known origins, the classification dendrogram reveals some new information. The most interesting seems to be that shown by the two main branches in Figure 16. The first branch is composed of proto-Elamite, Sumerian pictographs, and the Indus Valley script, while the second major branch is composed of the remaining scripts.

The hierarchical tree of Figure 17 takes into consideration the time intervals during which the scripts were used. Figure 17 considers Greek to be a Phoenician descendant, while Linear B is a descendant of Cretan hieroglyphs. Interestingly, Sumerian pictographs are identified as ancestors of the Indus Valley script signs. Figure 17 shows no known ancestor for proto-Elamite or Brahmi. The most tentative aspect of Figures 16 and 17 is the assumption of a common origin of all the scripts. This is only because the algorithm is designed to draw the best tree to explain the development of these eight scripts from a single source. The existence of a single source is only a hypothesis built into the algorithms that generated the classification trees. In fact, it is rather unlikely that a single unidentified source would spread independently in five different directions, as is shown in Figure 17. It is more plausible that the unknown source spread in two separate directions, as indicated by the two main branches of Figure 16.

A possible criticism of the above methodology is that the scripts are assumed to be related a priori. Of course, that may not necessarily be the case. There could have been independent developments in writing taking place in various regions of the world. By dropping the built-in assumption that there must be a single source for all eight ancient scripts, it is possible to obtain an alternative classification that is consistent with the timeline of use of these scripts and all the script similarity information that we obtained, but which allows for three different inventions and the spreading of ancient writing as shown in Figure 18. Figure 18 shows a classification forest with three roots instead of a classification tree with only one root.

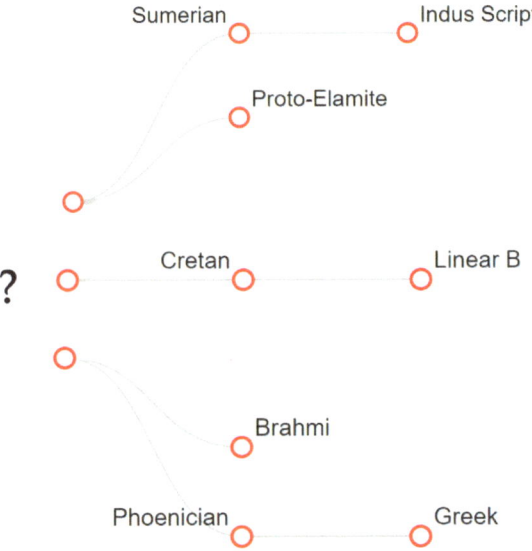

Figure 18. A modified classification that does not insist on a single source for all eight scripts.

The three groups that we obtained correspond to the red, orange, and blue set of scripts shown in Figure 10. The red scripts are the earliest scripts and are located near each other, as is shown in Figure 10. These correspond to the first group in Figure 18. It is possible that ancient traders spread the Sumerian script to the Indus Valley via a sea route. There are also many similarities between the two locations, such as in architecture and food production [23]. Moreover, Sumerian inscriptions called the Indus Valley Civilization Meluhha. This name may be related to the present-day region of Baluchistan [8].

The orange scripts are the middle two scripts in Figure 10, and these correspond to the second group in Figure 18. The location of Cretan hieroglyphs and Linear B overlap. It is possible that Cretan hieroglyphs developed into Linear A, which was also used by the Minoan civilization, and then Linear A developed into Linear B after the Mycenaean conquest of Crete. The Linear B script may then have spread to other Mycenaean areas.

Finally, the blue scripts are the three most recent scripts in Figure 10, and these correspond to the third group in Figure 18. The locations of the ancient Greek alphabet, the Phoenician alphabet, and the Brahmi script are farther apart. Of these three, the Phoenician alphabet is the oldest and may have originated in the Bronze Age as a descendant of proto-Sinaitic, which was invented in the Sinai Peninsula under the influence of Egyptian hieroglyphs [11]. The Phoenician alphabet could also have been spread by traders [24]. The analysis of ancient weight measures shows that there was an ancient version of the Silk Road between Greece and India in the Bronze Age [25].

Clearly, the smaller values in Table 3 mean that there are fewer pairs of signs with a one-to-one match between the two scripts. Since there is no established threshold value for saying that two scripts are related, we presented different possible solutions in Figures 16–18. Nevertheless, if script S_1 is an ancestor of script S_n, then there has to be a chain of temporally overlapping scripts S_i and S_{i+1} for $1 <= i <= n$, such that S_i is an ancestor of S_{i+1}. The red scripts have this overlap. The orange scripts do not have this overlap, but we obtain an overlap if we add the missing Linear A script. Finally, the blue scripts also have an overlap.

However, there appears to be a time gap between the latest red and the earliest orange script, and no additional script is known to have existed that can bridge this time gap. Hence, the red and the orange scripts seem to be independent developments. Similarly, there is a time gap between the latest orange and the earliest blue scripts in Figure 1. Therefore, these also appear to be independent developments. Hence, Figure 18 seems to be the best solution.

To better analyze the relationships among the eight scripts, we introduced a new assumption. We assumed that if script x's maximum connection is to another script, y, then x and y are related scripts. The relationship could be either because of an ancestor–descendant relationship or because of a common ancestor. The reason for this assumption is that while some scripts change significantly over time, they always stay most like those scripts with which they share a recent common ancestor. Implicitly, being closest to a particular script matters more than the actual number of similarities between the scripts.

As an example, suppose that Table 4 is a simple estimate of vocabulary similarities among six languages ranging from 10 (most similar) to 1 (least similar). To aid the analysis, the maximum value in each column and row is highlighted by a color. Our assumption allows the grouping together of Arabic and Hebrew, which are Semitic languages, English and German, which belong of the Germanic branch of the Indo-European languages, and Finnish and Hungarian, which are Uralic languages that diverged thousands of years ago. Hence, our assumption led to a grouping that corresponds to the usual classification of language families. The clustering algorithm ignored some values that may have arisen due to word borrowings such as the value 4 between Hebrew and German.

Table 4. Grouping of six languages highlighting maximal non-diagonal values in each column and row. Semitic languages are highlighted in green, Indo-European languages in blue, and Uralic languages in yellow.

	Arabic	Hebrew	English	German	Finnish	Hungarian
Arabic		8	2	1	1	1
Hebrew	8		2	4	1	1
English	2	2		9	2	1
German	1	4	9		2	3
Finish	1	1	2	2		4
Hungarian	1	1	1	3	4	

Similarly, the existence of three separate sources for the eight scripts is implied by a rearrangement of Table 3, as is shown in Table 5. In Table 5, the maximum non-diagonal values in each column and row are highlighted. The columns and rows highlighted in red form the cluster that is associated with the first branch in Figure 18, the columns and rows highlighted in orange form the cluster that is associated with the second branch in Figure 18, and the columns and rows highlighted in blue form the cluster that is associated with the third branch in Figure 18. The red scripts may have originated in Mesopotamia and were likely logographic, the orange scripts may have originated in Europe, and the blue scripts may be traced back to the Egyptian hieroglyphs, which is likely to have influenced the development of the proto-Sinaitic script that is an ancestor of the Phoenician alphabet. Figure 18 seems to present a logical explanation of the development of writing originating from three different locations, although some cross-influence among the three groups cannot be ruled out. These cross-influences would include such things as word borrowings.

Table 5. Highlighting of the maximal non-diagonal values in each column and row of Table 3. The Phoenician alphabet and its descendants are highlighted in blue, the Cretan-origin scripts are highlighted in orange, and the Mesopotamian-origin scripts are highlighted in red.

	Brahmi	Greek	Phoenician	Cretan Hieroglyphs	Linear B	Indus Valley	Proto-Elam.	Sumerian Pictograms
Brahmi		9	9	2	3	8	2	6
Greek	9		22	4	7	9	2	7
Phoenician	9	22		6	9	9	3	7
Cretan Hieroglyphs	2	4	6		20	5	2	6
Linear B	3	7	9	20		4	0	5
Indus Valley	8	9	9	5	4		4	20
Proto-Elamite	2	2	3	2	0	4		3
Sumerian Pictograms	6	7	7	6	5	20	3	

The grouping of the scripts is not intended to imply that the languages are related. For example, the Latin alphabet is used to write many different modern languages that belong to complete different language families. Therefore, similarity between languages is independent from the similarity of the scripts. Only after a decipherment can we say whether the languages are related to each other.

There are some even more advanced character-recognition algorithms beyond our convolutional neural networks. However, the convolutional neural network we used was already able to accurate perform character recognition. The main problem addressed in this paper was the comparison of characters from different scripts, rather than character recognition within a single script. The character comparison also gave high similarity measures for related signs, as is shown in the heat maps of Figures 14–16. Low character comparison values were obtained only when the signs were not related to each other.

5.1. Related Work

Sir Alexander Cunningham assumed that the Indus Valley seals were imports. He and other scholars also thought that Brahmi may have been a descendant of the Indus Valley Script [26,27] and that it may have expressed a Dravidian language [8,28,29].

A weakness of these proposals is the large time gap between the latest Indus Valley and the earliest Brahmi script inscriptions, which are likely from the time of Ashoka's Empire in the 3rd century BCE, although some authors assume around 500 BCE for the beginning of this script. Salomon [30] has proposed a Phoenician alphabet origin of the Brahmi script. This latter proposal agrees more closely with our proposal in Figure 16, where Brahmi and Phoenician are placed in the same branch of the script evolution tree.

The proto-Elamite script also reflects a Dravidian language, according to the Elamo–Dravidian hypothesis. McAlpin [31] thinks that the Elamo–Dravidian language family also includes the underlying language of the Indus Valley script. McAlpin's theory agrees with our proposal in Figure 16, which places the proto-Elamite script, the Indus Valley script, and Sumerian pictograms in the same branch of the script evolution tree. Relationships among these three are also suggested by archaeological evidence of connections among the Elamite, Indus Valley, and Sumerian civilizations [8,24].

Farmer et al. [32] question whether the Indus Valley Script reflects a language. They propose that it is more like the clan names/signs on heraldic coats of arms or the symbols of various gods. Regardless of whether it is a language, its apparent similarity to the Sumerian pictographs suggests that it is a descendant of the latter. Since the Indus Valley script inscriptions are rather short, they may not represent a full language. That is also the case for the proto-cuneiform writing in Mesopotamia dating from about 3300 BCE. In these proto-cuneiform inscriptions, the signs record calculations concerning products such as beer, and various occupations.

5.2. Machine Learning

Convolutional neural networks have been used for a long time for optical character recognition [33]. Convolutional neural networks and support vector machines have been applied to an increasing number of scripts. For example, Elleuch et al. [20] applied another combination of CNN and SVM to the recognition of Arabic letters. He et al. [34] and Yang et al. [35] used CNNs for handwritten Chinese character recognition. Arora et al. [36] compared neural networks and support vector machines using the Devanagari script, which is a Brahmi descendant. However, these earlier works did not apply CNNs to generate various script classification dendrograms, as in the current paper, and in the preliminary conference papers of the authors [37,38]. The authors also applied non-neural network-based techniques to identify allographs within the Indus Valley script [39].

Some recent works have used a feature vector-based analysis instead of neural networks to decipher Cretan hieroglyphic [40], Linear A [41], and Old Hungarian inscriptions [42], and to investigate the reading direction of the Phaistos Disk [43]. Other promising computer-based methods for the analysis of scripts are described in [44,45]. It remains to be seen whether neural networks can also be used for the decipherment of scripts.

6. Conclusions and Open Problems

The invention of writing was a major milestone, although the exact time and circumstances, as well as the details of its early spread, remain mostly a mystery. In this paper, we have presented some strong arguments for several surprising ancestor–descendant relationships among some of the oldest known scripts. We plan to expand this work to many more scripts to explore ancient script families. Future work will go beyond the region of the Near East and the Mediterranean Sea to other likely independent script families in America, East Asia, and other regions.

Author Contributions: Conceptualization, S.D. and P.Z.R.; methodology, S.D. and P.Z.R.; investigation, S.D. and P.Z.R.; writing—original draft preparation, S.D. and P.Z.R.; writing—review and editing, S.D. and P.Z.R. All authors have read and agreed to the published version of the manuscript.

Funding: This research received no external funding.

Institutional Review Board Statement: Not applicable.

Informed Consent Statement: Not applicable.

Data Availability Statement: Not applicable.

Conflicts of Interest: The authors declare no conflict of interest.

References

1. Salomon, R. *Indian Epigraphy: A Guide to the Study of Inscriptions in Sanskrit, Prakrit, and the other Indo-Aryan Languages*; Oxford University Press: Oxford, UK, 1998.
2. Olivier, J.-P. Cretan writing in the second millennium BCE. *World Archaeol.* **1986**, *17*, 377–389. [CrossRef]
3. Cook, B.F. *Greek Inscriptions*; University of California Press: Berkeley, CA, USA, 1987; Volume 5.
4. Mahadevan, I. *The Indus Script: Texts, Concordance and Tables, Memoirs*; Archaeological Survey of India: Delhi, India, 1977; Volume 77.
5. Joshi, J.P.; Parpola, A. Corpus of Indus Seals and Inscriptions. vol. 1, Collections in India. In *Annales Academiae Scientiarum Fennicae*; Series B; Suomalainen Tiedeakatemia: Helsinki, Finland, 1987; Volume 239.
6. Shah, S.G.M.; Parpola, A. Corpus of Indus Seals and Inscriptions, vol 2. Collections in Pakistan. In *Annales Academiae Scientiarum Fennicae*; Series B; Suomalainen Tiedeakatemia: Helsinki, Finland, 1991; Volume 240.
7. Parpola, A.; Pande, B.M.; Koskikallio, P. *Corpus of Indus Seals and Inscriptions, Vol. 3. New Material, Untraced Objects, and Collections Outside India and Pakistan*; Suomalainen Tiedeakatemia: Helsinki, Finland, 2010.
8. Parpola, A. *Deciphering the Indus Script*; Cambridge University Press: Cambridge, UK, 2009.
9. Chadwick, J. *The Decipherment of Linear B*; Cambridge University Press: Cambridge, UK, 1958.
10. Fischer, S.R. *History of Writing*; Reaktion Books: London, UK, 2004.
11. Colless, B.E. The origin of the alphabet: An examination of the Goldwasser hypothesis. *Antig. Oriente* **2014**, *12*, 71–104.
12. Revesz, P.Z. Bioinformatics evolutionary tree algorithms reveal the history of the Cretan Script Family. *Int. J. Appl. Math. Inform.* **2016**, *10*, 67–76.

13. Englund, R.K. The Proto-Elamite script. In *The World's Writing Systems*; Daniels, P.T., Bright, W., Eds.; Oxford University Press: Oxford, UK, 1996; pp. 160–164.
14. Dahl, J.L. Complex graphemes in Proto-Elamite. *Cuneif. Digit. Libr. J.* **2005**, *4*. Available online: https://cdli.mpiwg-berlin.mpg.de/articles/cdlj/2005-3 (accessed on 3 April 2023).
15. Labat, R.; Malbran-Labat, F. Manuel D'épigraphie Akkadienne: Signes, Syllabaire, Idéogrammes, Librairie Orientaliste Paul Geuthner; Enlarged édition (1 avril 2002). Available online: https://www.amazon.fr/Manuel-dépigraphie-akkadienne-Syllabaire-Idéogrammes/dp/2705335838 (accessed on 3 April 2023).
16. Parpola, S. Etymological Dictionary of the Sumerian Language. *J. Indo-Eur. Stud.* **2022**, *3*, 247–252.
17. Revesz, P.Z. Sumerian contains Dravidian and Uralic substrates associated with the Emegir and Emesal dialects. *WSEAS Trans. Inf. Sci. Appl.* **2019**, *16*, 8–30.
18. LeCun, Y.; Cortes, C.; Burges, C. MNIST Handwritten Digit Database. Available online: http://yann.lecun.com/exdb/mnist/ (accessed on 5 April 2019).
19. LeCun, Y.; Bottou, L.; Bengio, Y.; Haffner, P. Gradient-based learning applied to document recognition. *Proc. IEEE* **1998**, *86*, 2278–2324. [CrossRef]
20. Elleuch, M.; Tagougui, N.; Kherallah, M. A novel architecture of CNN based on SVM classifier for recognizing Arabic handwritten script. *Int. J. Intell. Syst. Technol. Appl.* **2016**, *15*, 323–340.
21. Kingma, D.P.; Ba, J. Adam: A method for stochastic optimization. In Proceedings of the 3rd International Conference on Learning Representations, San Diego, CA, USA, 7–9 May 2015.
22. Yann, M.L.; Tang, Y. Learning deep convolutional neural networks for X-ray protein crystallization image analysis. In Proceedings of the Thirtieth AAAI Conference on Artificial Intelligence, Phoenix, AZ, USA, 12–17 February 2016; AAAI Press: Palo Alto, CA, USA, 2016; pp. 1373–1379.
23. Collon, D. Mesopotamia and the Indus: The evidence of the seals. In *The Indian Ocean in Antiquity*; The British Museum and Kegan Paul International: London, UK; New York, NY, USA, 1996; pp. 209–225.
24. Howard, M.C. *Transnationalism in Ancient and Medieval Societies: The Role of Cross-Border Trade and Travel*; McFarland: Jefferson, NC, USA, 2014.
25. Revesz, P.Z. Data science applied to discover ancient Minoan-Indus Valley trade routes implied by common weight measures. In Proceedings of the 26th International Database Engineered Applications Symposium (IDEAS), Budapest, Hungary, 22–24 August 2022; ACM Press: New York, NY, USA, 2022; pp. 150–155.
26. Rao, R.P.; Yadav, N.; Vahia, M.N.; Joglekar, H.; Adhikari, R.; Mahadevan, I. Entropic evidence for linguistic structure in the Indus script. *Science* **2009**, *324*, 5931. [CrossRef] [PubMed]
27. Rao, R.P.; Yadav, N.; Vahia, M.N.; Joglekar, H.; Adhikari, R.; Mahadevan, I. A Markov model of the Indus Script. *Proc. Natl. Acad. Sci. USA* **2009**, *106*, 13685–13690. [CrossRef] [PubMed]
28. Wells, B.K. *Epigraphic Approaches to Indus Writing*; Oxbow Books: Oxford, UK, 2011.
29. Zide, A.R.; Zvelebil, K.V. (Eds.) The Soviet Decipherment of the Indus Valley Script: Translation and Critique. In *Janua Linguarum. Series Practica*; de Gruyter Mouton: Berlin, Germany, 1976; Volume 156. [CrossRef]
30. Salomon, R. On the origin of the early Indian scripts. *J. Am. Orient. Soc.* **1995**, *115*, 271–279. [CrossRef]
31. McAlpin, D.W. Proto-Elamo-Dravidian: The evidence and its implications. *Trans. Am. Philos. Soc.* **1981**, *71*, 1–155. [CrossRef]
32. Farmer, S.; Sproat, R.; Witzel, M. The collapse of the Indus-script thesis: The myth of a literate Harappan civilization. *Electron. J. Vedic Stud.* **2016**, *11*, 19–57.
33. Jaderberg, M.; Simonyan, K.; Vedaldi, A.; Zisserman, A. Reading text in the wild with convolutional neural networks. *Int. J. Comput. Vis.* **2016**, *116*, 1–20. [CrossRef]
34. He, M.; Zhang, S.; Mao, H.; Jin, L. Recognition confidence analysis of hand-written Chinese character with CNN. In Proceedings of the 13th International Conference on Document Analysis and Recognition, Nancy, France, 23–26 August 2015; IEEE: Piscataway, NJ, USA, 2015; pp. 61–65.
35. Yang, W.; Jin, L.; Liu, M. Chinese character-level writer identification using path signature feature, DropStroke and deep CNN. In Proceedings of the 13th International Conference on Document Analysis and Recognition, Nancy, France, 23–26 August 2015; IEEE: Piscataway, NJ, USA, 2015; pp. 546–550.
36. Arora, S.; Bhattacharjee, D.; Nasipuri, M.; Malik, L.; Kundu, M.; Basu, D.K. Performance comparison of SVM and ANN for handwritten Devnagari character recognition. *arXiv* **2010**, arXiv:1006.5902.
37. Daggumati, S.; Revesz, P.Z. Data mining ancient script image data using convolutional neural networks. In Proceedings of the 22nd International Database Engineering and Applications Symposium, Villa San Giovanni, Italy, 18–20 June 2018; ACM Press: New York, NY, USA, 2018; pp. 267–272.
38. Daggumati, S.; Revesz, P.Z. Data mining ancient scripts to investigate their relationships and origins. In Proceedings of the 23rd International Database Engineering and Applications Symposium, Athens, Greece, 10–12 June 2019; ACM Press: New York, NY, USA, 2019; pp. 209–218.
39. Daggumati, S.; Revesz, P.Z. A method of identifying allographs in undeciphered scripts and its application to the Indus Valley Script. *Humanit. Soc. Sci. Commun.* **2021**, *8*, 50. [CrossRef]
40. Revesz, P.Z. A translation of the Arkalochori Axe and the Malia Altar Stone. *WSEAS Trans. Inf. Sci. Appl.* **2017**, *14*, 124–133.

41. Revesz, P.Z. Establishing the West-Ugric language family with Minoan, Hattic and Hungarian by a decipherment of Linear, A. *WSEAS Trans. Inf. Sci. Appl.* **2017**, *14*, 306–335.
42. Revesz, P.Z. Decipherment challenges due to tamga and letter mix-ups in an Old Hungarian runic inscription from the Altai Mountains. *Information* **2022**, *13*, 422. [CrossRef]
43. Revesz, P.Z. Experimental evidence for a left-to-right reading direction of the Phaistos Disk. *Mediterr. Archaeol. Archaeom.* **2022**, *22*, 79–96.
44. Hosszú, G. *Scriptinformatics: Extended Phenetic Approach to Script Evolution*; Nap Kiadó: Budapest, Hungary, 2021.
45. Tóth, L.; Hosszú, G.; Kovács, F. Deciphering Historical Inscriptions Using Machine Learning Methods. In Proceedings of the 10th International Conference on Logistics, Informatics and Service Sciences, Beijing, China, 23 February 2020; Liu, S., Bohács, G., Shi, X., Shang, X., Huang, A., Eds.; Springer: Singapore, 2020; pp. 419–435. [CrossRef]

Disclaimer/Publisher's Note: The statements, opinions and data contained in all publications are solely those of the individual author(s) and contributor(s) and not of MDPI and/or the editor(s). MDPI and/or the editor(s) disclaim responsibility for any injury to people or property resulting from any ideas, methods, instructions or products referred to in the content.

Article

Archaeogenetic Data Mining Supports a Uralic–Minoan Homeland in the Danube Basin [†]

Peter Z. Revesz

School of Computing, College of Engineering, University of Nebraska-Lincoln, Lincoln, NE 68588, USA; peter.revesz@unl.edu; Tel.: +1-402-421-6990

[†] This paper is an extended version of a paper published in the 25th International Database Engineering and Applications Symposium, IDEAS 2021, Montreal, QC, Canada, 14–19 July 2021.

Abstract: Four types of archaeogenetic data mining are used to investigate the origin of the Minoans and the Uralic peoples: (1) six SNP mutations related to eye, hair, and skin phenotypes; (2) whole-genome admixture analysis using the G25 system; (3) an analysis of the history of the U5 mitochondrial DNA haplogroup; and (4) an analysis of the origin of each currently known Minoan mitochondrial and y-DNA haplotypes. The uniform result of these analyses is that the Minoans and the Uralic peoples had a common homeland in the lower and middle Danube Basin, as well as the Black Sea coastal regions. This new result helps to reconcile archaeogenetics with linguistics, which have shown that the Minoan language belongs to the Uralic language family.

Keywords: admixture; archaeogenetics; data mining; haplogroup; Minoan; mitochondria; Uralic

1. Introduction

Archaeogenetic and linguistic studies give contradictory results regarding the origins of the Minoan civilization, which flourished on the island of Crete in the Bronze Age. An influential whole-genome archaeogenetic study by Lazaridis et al. [1] concluded that

"Minoans and Mycenaeans were genetically similar, having at least three-quarters of their ancestry from the first Neolithic farmers of western Anatolia and the Aegean, and most of the remainder from ancient populations related to those of the Caucasus and Iran."

In contrast, recent linguistic studies by Revesz [2–4] have indicated that the Minoan language belongs to the Uralic language family, which had a homeland near the Ural Mountains [5], the Northern Black Sea region [6], or the Carpathian Basin [7]—or more generally, the Danube Basin [8]. According to the traditional view, the Uralic language family originated about 7000 to 10,000 years ago [9]. It contains both Finno-Ugric and Samoyedic languages, and the Finno-Ugric languages have two main branches: the Finno-Permic—which includes Finnish, Estonian, Saami, and other languages—and the Ugric branch—which includes Hungarian, Khanty, and Mansi [10].

Revesz [2–4] identified Minoan as an extinct member of the Ugric branch based on translating thirty-one Minoan inscriptions as Proto-Ugric language documents. This linguistic classification was strengthened recently [11] (pp. 208–212) by showing regular sound changes between Pre-Greek origin Greek words identified by Beekes [12] and Proto-Uralic, Proto-Finno-Ugric, and Proto-Ugric words reconstructed by Rédei [13]. The overwhelming number of Pre-Greek words shows a uniformity that implies borrowings from a single source [14] (p. 45). Since the Minoan culture preceded the Greek-speaking Mycenaean culture on the islands of Crete and Santorini [15], the Pre-Greek words are likely to be borrowings from the Minoan language. Furthermore, demonstrating regular sound changes is the primary way to prove linguistic relationships among languages and is also used in Indo-European linguistics [16]. Bernal [17], Best [18], Campbell-Dunn [19],

---

Citation: Revesz, P.Z. Archaeogenetic Data Mining Supports a Uralic–Minoan Homeland in the Danube Basin. *Information* **2024**, *15*, 646. https://doi.org/10.3390/info15100646

Academic Editor: Haridimos Kondylakis

Received: 26 August 2024
Revised: 7 October 2024
Accepted: 10 October 2024
Published: 16 October 2024

Copyright: © 2024 by the author. Licensee MDPI, Basel, Switzerland. This article is an open access article distributed under the terms and conditions of the Creative Commons Attribution (CC BY) license (https://creativecommons.org/licenses/by/4.0/).

Gordon [20], Kvashilava [21], La Marle [22], and other authors who proposed a different linguistic affiliation of the Minoan language did not show regular sound changes.

The aim of the present paper is to reconcile the archaeogenetic and linguistic data and to show that they are compatible with a Danube Basin homeland of the Uralic languages. The Danube Basin includes the Danube Delta area from which the Minoans could have sailed south to the Aegean Sea via the Bosporus strait, while the rest of the Uralic language speakers could have migrated eastward along the Northern Black Sea coast and then northward along the major rivers as described by Wiik [6]. Figure 1 shows the hypothetical dispersal of the Uralic language family based on Krantz [7] and extended by an Ugric-Minoan link by Revesz [8]. Hence, the primary focus of the reconciliation proposed in this paper is to show that the archaeogenetic data support a Uralic-Minoan homeland in the Danube Basin.

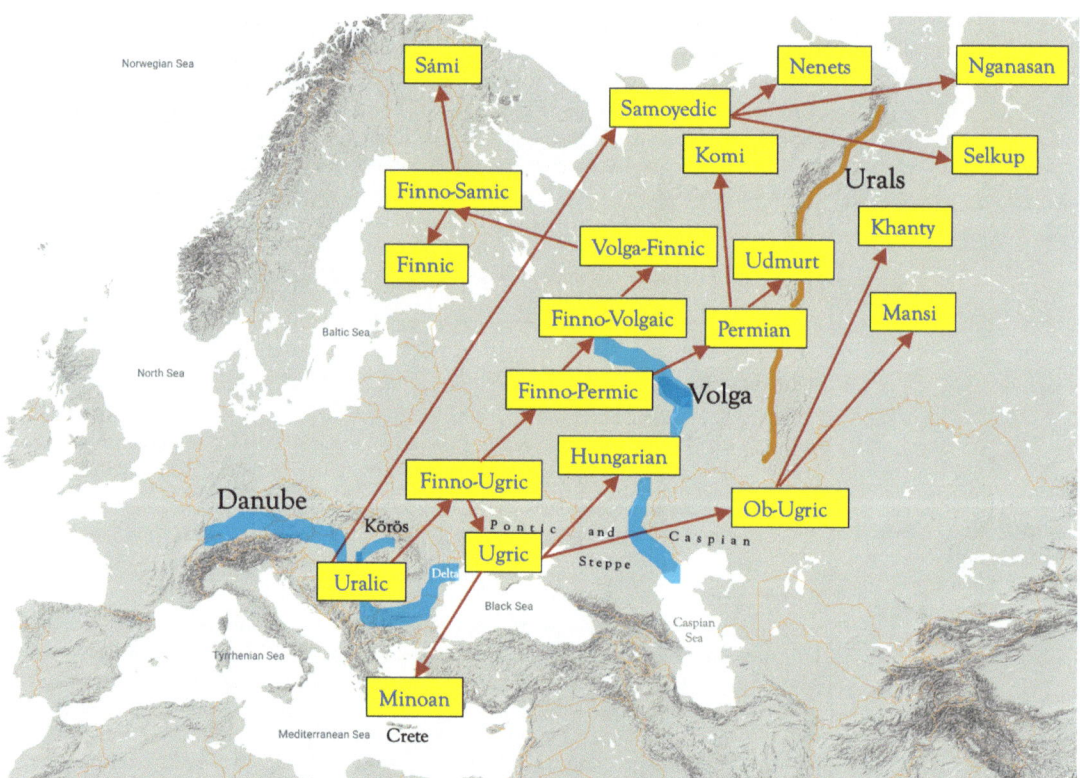

Figure 1. A hypothetical dispersal of the Uralic languages from a Danube Basin homeland based on Krantz [7] with the Ugric to Minoan link added by Revesz [8].

This rest of this paper is organized as follows. Section 2 discusses SNP mutations related to eye, hair, and skin phenotypes. Section 3 discusses whole-genome G25 admixture analysis. Section 4 presents an analysis of U5 mtDNA haplotypes. Section 5 presents an analysis of the origin of all Minoan mtDNA and y-DNA haplotypes. Section 6 summarizes the results of the analyses given in Sections 2–4 and provides a discussion of the results. Finally, Section 7 gives some conclusions and directions for further work.

2. Method and Experiment 1: SNP Mutations Related to Eye, Hair and Skin Phenotypes

2.1. Method of Analyzing Archaeogenetic Phenotype Data

Human eye, hair, and skin phenotypes are genetically determined by various alleles. The lighter eye, hair, and skin phenotypes have some selective advantages at higher latitudes. Hence, they are spread widely among Eurasian populations. Table 1 shows some data regarding six alleles that affect eye, hair, or skin pigmentation.

Table 1. Genetic loci associated with lighter eye, hair, or skin color.

Gene	Loci	Allele Mutation	Phenotype
HERC2	rs12913832	A > G	Blue Eye
SLC24A4	rs2402130	G > A	Light Hair
SLC24A5	rs1426654	G > A	Light Skin
SLC45A2	rs28777	C > A	Low Melanin
SLC45A2	rs16891982	C > G	Light Skin
TYR	rs1042602	C > A	Light Skin

Table 2 lists the eleven archaeological cultures that we considered in this paper. The first nine archaeological cultures preceded the Minoan civilization. Hence, these nine archaeological cultures were considered possible ancestors of the Minoan civilization. The Mycenaean civilization largely followed the Minoan civilization and was included as a comparison with the Minoan civilization. The comparison would reveal whether the Minoans and Mycenaeans had different ancestors.

Table 2. Archaeological cultures with their periods based on the estimates or sample dates in the references.

Name	Abbreviation	Period	References
Caucasian Hunter-Gatherers	CHG	15,000–8000 BP	[23]
Eastern-European Hunter-Gatherers	EHG	10,000–7000 BP	[23,24]
Western-European Hunter-Gatherers	WHG	15,000~5000 BP	[23]
Lower-Danube Mesolithic Hunter-Gatherers	L_Danube_ME	9075–8435 BP	[25]
Fertile Crescent Neolithic Farmers	FertileC_NE	8300–7800 BP	[23,25–27]
Aegean Early Neolithic Farmers	Aegean_NE	8438–8030 BP	[28]
Körös Neolithic	Körös_NE	7800–7300 BP	[29]
Hungarian Middle Neolithic Farmers	Hungary_MN	7310–6950 BP	[29]
Hungarian Bronze Age	Hungary_BA	3900–3450 BP	[30]
Minoan Civilization (Early and Middle)	Minoan	5100–3450 BP	[1,31]
Mycenaean Civilization	Mycenaean	3750–3050 BP	[1,31]

After collecting loci mutation and allele data from samples from these archaeological cultures, we computed the percentage of the various alleles. Then, we computed the root mean square error (RMSE) between every pair of archaeological cultures x and y, where $x \neq y$ using the following formula:

$$RMSE(x, y) = \sqrt{\frac{\sum_{k=1}^{k=n} (p_{x,k} - p_{y,k})^2}{n}} \qquad (1)$$

where n is the number of genetic loci considered, and $p_{x,k}$ is the percentage of the kth allele associated with lighter eye, hair, or skin pigmentation among the samples from x. A lower RMSE value indicates a greater overall genotypic similarity between the populations of two cultures.

2.2. Experiment with Archaeogenetic Phenotype Data

Table 3 records the SNP variations data we could collect from 48 samples from the eleven archaeological cultures in Table 2. The 48 samples are listed in the second column of Table 3, and their locations are shown in Figure 2. For each culture, the boldface row gives the percentages for each of the six alleles of Table 1 that were associated with a lighter eye, hair, or skin pigmentation.

Table 3. Six eye, hair, skin alleles found in the samples from eleven ancient cultures.

Culture	Sample	Location	Ref.	rs12913832	rs2402130	rs1426654	rs28777	rs16891982	rs1042602
CHG	Kotias	Kotias, Georgia	[23]	AA	AA	AA	CC	CC	CC
CHG	Satsurblia	Satsurblia, Georgia	[23]	AG	AA	AA	CC	CC	CC
CHG				**0.25 G**	**1.00 A**	**1.00 A**	**0.00 A**	**0.00 G**	**0.00 A**
EHG	Ukr_HG1	Vasil'evka, Ukraine	[24]	AA	-	AA	-	GG	-
EHG	SVP44	Samara Oblast, Russia	[23]	GG	AA	AA	-	GG	CC
EHG	UzOO77	Yuzhnyy Oleniy, Russia	[23]	AA	AA	AA	AC	CG	CC
EHG				**0.33 G**	**1.00 A**	**1.00 A**	**0.50 A**	**0.83 G**	**0.00 A**
WHG	Braña	Braña, Spain	[23]	GG	AG	GG	CC	CC	CC
WHG	Loschbour	Loschbour, Luxemburg	[23]	GG	AG	GG	CA	CC	CC
WHG	Bichon	Bichon, France	[23]	AG	AA	GG	CC	CC	CC
WHG	Villabruna	Villabruna, Italy	[23]	GG	AG	GG	CC	CC	CC
WHG				**0.88 G**	**0.63 A**	**0.00 A**	**0.13 A**	**0.00 G**	**0.00 A**
L_Danube_ME	SC1	Schela Cladovei, Romania	[25]	AG	AA	GG	CA	CC	CC
L_Danube_ME	SC2	Schela Cladovei, Romania	[25]	AA	AA	GG	AA	CC	CC
L_Danube_ME	OC1	Ostrovul Corbului, Romania	[25]	AG	GA	GG	CC	CC	CC
L_Danube_ME				**0.33 G**	**0.83 A**	**0.00 A**	**0.50 A**	**0.00 G**	**0.00 A**
FertileC_NE	WC1	Wezmeh cave, Iran	[32]	AG	AA	GA	CC	CC	CC
FertileC_NE	AH1	Tepe Abdul Hosein, Iran	[32]	-	AA	GA	-	-	CC
FertileC_NE	AH2	Tepe Abdul Hosein, Iran	[32]	AA	AA	GG	CC	-	-
FertileC_NE	AH4	Tepe Abdul Hosein, Iran	[32]	-	AA	AA	-	-	-
FertileC_NE	GD13a	Ganj Dareh, Iran	[26]	AA	GG	AA	CC	CC	CC
FertileC_NE	Bon004	Boncuklu, Turkey	[27]	AA	-	AA	-	CC	CC
FertileC_NE	Bon014	Boncuklu, Turkey	[27]	AG	-	AA	-	CC	-
FertileC_NE	Bon001	Boncuklu, Turkey	[27]	GG	-	AA	-	CC	CC
FertileC_NE	Bon002	Boncuklu, Turkey	[23]	AG	-	AG	-	-	-
FertileC_NE				**0.36 G**	**0.80 A**	**0.72 A**	**0.00 A**	**0.00 G**	**0.00 A**
Aegean_NE	Bar8	Barcin, Turkey	[28]	AA	GA	AA	CA	CG	CC
Aegean_NE	Bar31	Barcin, Turkey	[28]	AA	AA	AA	AA	CG	CC
Aegean_NE	Rev5	Revenia, Greece	[28]	AA	-	AA	-	CC	CC
Aegean_NE				**0.00 G**	**0.25 A**	**1.00 A**	**0.75 A**	**0.33 G**	**0.00 A**
Körös_NE	KO1	Tiszaszőlős, Hungary	[29]	GG	GA	AG	CC	CC	CC
Körös_NE	KO2	Berettyóújfalu, Hungary	[29]	AG	AA	AG	CA	CG	CC
Körös_NE				**0.75 G**	**0.75 A**	**0.50 A**	**0.25 A**	**0.25 G**	**0.00 C**
Hungary_MN	NE1	Polgár-Ferenci-hát, Hungary	[29]	AG	AA	AG	CA	CC	CC
Hungary_MN	NE2	Debrecen Tócópart, Hungary	[29]	AA	GA	AA	CC	CC	CA
Hungary_MN	NE3	Garadna, Hungary	[29]	AG	GG	AA	CA	CG	CA
Hungary_MN	NE4	Polgár-Ferenci-hát, Hungary	[29]	GG	GA	AA	CA	CG	CC
Hungary_MN	NE5	Kompolt-Kigyósér, Hungary	[29]	AG	AA	AA	CA	CC	CA
Hungary_MN	NE6	Apc-Berekalja I., Hungary	[29]	GG	GA	AA	CA	CC	CC
Hungary_MN				**0.58 G**	**0.58 A**	**0.92 A**	**0.42 A**	**0.17 G**	**0.25 A**
Hungary_BA	BR1	Kompolt-Kigyósér, Hungary	[29]	AG	GA	AA	CA	CG	CC
Hungary_BA	BR2	Ludas-Varjú-dűlő, Hungary	[29]	AG	GA	AA	AA	GG	CC
Hungary_BA	S11	Balatonkeresztúr, Hungary	[30]	GG	GG	-	-	CC	CC
Hungary_BA	S14	Balatonkeresztúr, Hungary	[30]	AG	GG	-	AA	-	CC
Hungary_BA	S21	Balatonkeresztúr, Hungary	[30]	GG	-	AA	-	-	AA
Hungary_BA				**0.70 G**	**0.25 A**	**1.00 A**	**0.84 A**	**0.50 G**	**0.20 A**
Minoan	Pta08	Petras Siteia, Greece	[31]	AA	GG	AA	-	CG	-
Minoan	I0070	Charalambos Cave, Greece	[1]	AA	AA	AA	-	GG	AA
Minoan	I0071	Charalambos Cave, Greece	[1]	AA	AA	AA	CC	CC	CA
Minoan	I0073	Charalambos Cave, Greece	[1]	AA	AA	AA	AA	GG	-
Minoan	I0074	Charalambos Cave, Greece	[1]	AA	AG	AA	AA	GG	-
Minoan	I9005	Charalambos Cave, Greece	[1]	AA	AA	AA	-	GG	CA
Minoan	I9130	Odigitria, Greece	[1]	-	-	-	-	GG	CC
Minoan				**0.00 G**	**0.75 A**	**1.00 A**	**0.67 A**	**0.79 G**	**0.50 A**
Mycenaean	I9006	A. Kyriaki, Salamis, Greece	[1]	AA	GG	AA	-	GG	AA
Mycenaean	I9010	Galatas Apatheia, Greece	[1]	-	GG	AA	-	GG	AA
Mycenaean	I9033	Peristeria Tryfilia, Greece	[1]	AA	GG	AA	-	CG	AA
Mycenaean	I9041	Galatas Apatheia, Greece	[1]	AG	AG	AA	AA	CG	CA
Mycenaean	Log02	Logkas Elati, Greece	[31]	AA	GA	AA	CA	CC	-
Mycenaean				**0.13 G**	**0.20 A**	**1.00 A**	**0.75 A**	**0.60 G**	**0.88 A**

Figure 2. Locations of the archaeological samples listed in Table 3.

We calculated the RMSE between each pair of archaeological cultures using Equation (1), with $n = 6$. A lower RMSE value indicates a greater overall genotypic similarity between the populations of two cultures. For example, Figure 3 shows an RMSE value of 0.15 between the CHG and the FertileC_NE cultures. This indicates that the two cultures had similar genotypes and presumably also had similar phenotypes regarding eye, hair, and skin pigmentation. For more discussion of the results, see Section 6.

	CHG	EHG	WHG	L_Danube_ME	FertileC_NE	Aegean_NE	Körös_NE	Hungary_MN	Hungary_BA	Minoan	Mycenaean
CHG		0.40	0.51	0.46	0.15	0.46	0.34	0.31	0.54	0.49	0.63
EHG	0.40		0.61	0.54	0.42	0.40	0.38	0.35	0.40	0.27	0.51
WHG	0.51	0.61		0.28	0.37	0.63	0.24	0.43	0.57	0.70	0.74
L_Danube_ME	0.46	0.54	0.28		0.36	0.52	0.30	0.42	0.56	0.58	0.66
FertileC_NE	0.15	0.42	0.37	0.36		0.44	0.23	0.26	0.50	0.51	0.60
Aegean_NE	0.46	0.40	0.63	0.52	0.44		0.47	0.33	0.31	0.35	0.38
Körös_NE	0.34	0.38	0.24	0.30	0.23	0.47		0.24	0.40	0.51	0.59
Hungary_MN	0.31	0.35	0.43	0.42	0.26	0.33	0.24		0.26	0.38	0.42
Hungary_BA	0.54	0.40	0.57	0.56	0.50	0.31	0.40	0.26		0.40	0.37
Minoan	0.49	0.27	0.70	0.58	0.51	0.35	0.51	0.38	0.40		0.29
Mycenaean	0.63	0.51	0.74	0.66	0.60	0.38	0.59	0.42	0.37	0.29	

Figure 3. The root means square error values between each pair of archaeological cultures. The lower values (red) indicate a stronger genotypic connection, while the higher values (blue) indicate a weaker genotypic connection.

3. Method and Experiment 2: A G25 Admixture Analysis of Archaeogenetic Data

3.1. Method of Identifying the Ancestors of the Minoans Using the G25 System

Given a set of archaeological cultures $S_1, S_2, \ldots, S_n$, and T, an admixture analysis finds an apportionment among the $S_1, S_2, \ldots, S_n$ cultures, which are called the source cultures, that seems to best explain the archaeological culture T, which is called the test culture. For example, Lazaridis et al. [1] used the popular qpADM admixture analysis system with S_1 = CHG, S_2 = Anatolia_N, and T = Minoan_Odigitria, that is, all the samples from the Minoan site of Moni Odigitria. The qpADM admixture analysis system returned the result that the Minoan_Odigitria culture is composed of 14.4 percent CHG and 85.6 percent Anatolia_N.

Unfortunately, the qpADM system is limited to 2 or 3 possible sources in most archaeogenetic publications. This creates a severe limitation, because hundreds of archaeological cultures located on the coastal areas of the Mediterranean Sea, the Black Sea, and the Atlantic Ocean could be possible genetic sources of the Minoans to some extent. Hence, we would need to simultaneously compare hundreds of possible sources for a completely fair apportionment among all those archaeological cultures. Luckily, the G25 genome admixture analysis system, which is available at https://www.dnagenics.com/products/g25studio (accessed on 30 July 2021), can compare hundreds of possible sources. The G25 system describes each archaeological culture by a numerical vector of length 25, which summarizes thousands of SNPs.

3.2. Experiment with the G25 System

The G25 system listed 271 different Neolithic, Mesolithic, and Paleolithic cultures in Africa, Asia, and Europe. We considered these 271 cultures as potential sources. While the Minoan civilization flourished on the island of Crete, many of the other Aegean islands were part of the Cycladic culture (c. 5100–3000 BP) [15], which we included in this experiment for comparison. We separately tested the Cycladic samples from Koufonisia island (Kou01 and Kou03) and the Minoan samples from the Charalambos Cave (I0070, I0071, I0073, I0074, I9005), Moni Odigitria (I9129, I9130, I9131), and Petras (Pta08).

Figure 4 shows the G25 admixture analysis results based on [33] with a listing of only those rows that had some non-zero value for at least one of the eleven tested samples. Figure 4 presents the data by grouping together the archaeological culture sources into five main regions: (1) Africa, (2) Greece and Macedonia, (3) Danube Basin, (4) Caucasus, Russia and Ukraine, and (5) Fertile Crescent and Iran.

The G25 analysis found a previously completely overlooked genetic connection between the Odigitria I9129 sample and a hunter-gatherer from the Shum Laka rock shelter in Cameroon about 8000 years ago. One hypothesis to explain the connection is that a common source population once lived in the Sahara. When the Sahara dried up, some people moved north into Europe and reached the island of Crete, while others from the same group moved south to Cameroon. These hypothetical movements could explain the linguistic connections between African and European mountain names [34].

The G25 analysis also reveals large differences among the various Cycladic and Minoan groups that were overlooked by previous admixture analysis publications. Figure 4 shows that there are great differences between the Charalambos Cave and the Moni Odigitria samples. The Charalambos Cave's primary source is the Greek Neolithic (62.2 percent), and the secondary sources are the Danube Basin (19.4 percent), the Caucasus (15.1 percent), and the Fertile Crescent (3.2 percent). In contrast, the Moni Odigitria's primary source is the Danube Basin (72.1 percent), and the secondary sources are the Fertile Crescent (15.1 percent), the Greek Neolithic (9.5 percent), the Caucasus (3.1 percent), and Africa (0.2 percent) on average. Hence, the two cultures are very different in origin and possibly came to Crete at different times as well. The Cyclades' primary source is the Greek Neolithic (42 percent). Hence, the Cyclades and the Charalambos Cave samples form a natural cluster. The Petras' primary source is the Danube Basin. Hence, the Moni Odigitria and the Petras samples also form a natural cluster.

Figure 5 shows some hypothetical population movements based on the above analysis. The map suggests that while the Anatolian and Fertile Crescent Neolithic culture reached Crete, major population movements to the island only happened later.

Regions (below)	Archaeological Culture (below)	Cycladic samples Koufonisia		Minoan samples Charalambos					Odigitria			Petras	Averages Cycladic Average	Charalambos Average	Odigitria Average
		Kou01	Kou03	I0070	I0071	I0073	I0074	I9005	I9129	I9130	I9131	Pta08			
Africa	CMR_Shum_Laka_8000BP	0	0	0	0	0	0	0	0.6	0	0	0	0	0.0	0.2
Greece and Macedonia	GRC_N	0	13.4	29.6	18	13	0	13	0	0	0	0	6.7	14.7	0.0
	GRC_Peloponnese_N	21.4	49.2	51.8	49.6	28.8	35.4	40.8	0	19.6	9	0	35.3	41.3	9.5
	MKD_N	0	0	0	0	0	0	31.2	0	0	0	0	0	6.2	0.0
	total Greece	21.4	62.6	81.4	67.6	41.8	35.4	85	0	19.6	9	0	42	62.2	9.5
Danube Basin	BGR_Krepost_N	0	0	0	18.2	0	7.6	0	0	0	27.4	0	0	5.2	9.1
	Corded_Ware_DEU	0.6	0	0	0	0	0	0	0	0	0	0	0.3	0.0	0.0
	Corded_Ware_POL_early	0	0	0	0	0	0	0	0	0	1.6	0	0	0.0	0.5
	DEU_LBK_KD	1	0	0	3.6	0	3.2	0	0	33.4	6	7.6	0.5	1.4	13.1
	DEU_LBK_SCH	0	0	0	0	0	0	0	2.8	0	0	0	0	0.0	0.9
	DEU_LBK_UW	0	4.6	0	0	0	0	0	0	0	0	0	2.3	0.0	0.0
	FRA_Grand_Est_EN	0	0	0	0	0	0	0	4.2	0	0	0	0	0.0	1.4
	HUN_ALPc_I_MN	0	0	0	0	0	0	0	10.6	0	0	0	0	0.0	3.5
	HUN_ALPc_Tiszadob_MN	0	0	0	0	0	0	0	0	13.8	0	5.2	0	0.0	4.6
	HUN_Koros_N	0	0	0	0	0	0	0	0	0	45.6	0	0	0.0	15.2
	HUN_LBK_MN	0	0	0	0	32.4	0	0	0	0	0	0	0	6.5	0.0
	HUN_Sopot_LN	0	0	0	0	0	0	0	0	13.8	0	0	0	0.0	4.6
	HUN_Starcevo_N	13.4	0	0	0	0	32.2	0	0	0	0	49.4	6.7	6.4	0.0
	ROU_N	0	10.4	0	0	0	0	0	47.4	0	0	0	5.2	0.0	15.8
	SRB_N	3.4	0	0	0	0	0	0	2.4	0	0	0	1.7	0.0	0.8
	SRB_Starcevo_N	0	0	0	0	0	0	0	7.2	0	0	0	0	0.0	2.4
	total Danube Basin	18.4	15	0	21.8	32.4	43	0	74.6	61	80.6	62.2	16.7	19.4	72.1
Caucasus, Russia and Ukraine	AZE_Caucasian_lowlands_LN	24.2	0	11.8	2.8	6.6	20.4	6.4	0	0	0	33.6	12.1	9.6	0.0
	GEO_CHG	0	8.6	6.6	7.8	3	0	8.6	0	2.6	6.6	0	4.3	5.2	3.1
	RUS_AfantovaGora3	0	2	0	0	0	0	0	0	0	0	0	1	0.0	0.0
	RUS_Yakutia_N	0	0	0	0	0.2	0.8	0	0	0	0	0	0	0.2	0.0
	RUS_Yakutia_Ymyiakhatkh_LN	0	0	0	0	0	0.4	0	0	0	0	0	0	0.1	0.0
	UKR_N	14	0	0	0	0	0	0	0	0	0	0	7	0.0	0.0
	total Caucasus	38.2	10.6	18.4	10.6	9.8	21.6	15	0	2.6	6.6	33.6	24.4	15.1	3.1
Fertile Crescent and Iran	IRN_Seh_Gabi_LN	0	11.8	0	0	0	0	0	0	10	0	0	5.9	0.0	3.3
	IRN_Wezmeh_N	0	0	0	3.8	0	0	0	10.2	0	0	0	0	0.8	3.4
	Levant_PPNB	0	0	0.2	0	0	0	0	0	0	1.4	0	0	0.0	0.0
	Levant_PPNC	0	0	0	0	0	0	0	0	3.8	0	0	0	0.0	1.3
	MYS_LN	0	0	0	0	0	0	0	0	0	0.2	0	0	0.0	0.0
	TUR_Tepecik_Ciftlik_N	0.8	0	0	0	0	0	0	6.8	0	2.6	0	0.4	0.0	2.3
	TUR_Kumtepe_N	21.2	0	0	0	12.2	0	0	14.6	0	0	0	10.6	2.4	4.9
	total Fertile Crescent	22	11.8	0.2	0	16	0	0	24.8	16.8	3.8	4.2	16.9	3.2	15.1
		100	100	100	100	100	100	100	100	100	100	100	100	100.0	100.0

Figure 4. The sources (rows) for various Cycladic samples (columns 2–3) and Minoan samples (columns 4–12), as well as some averages (columns 13–15) according to the G25 admixture analysis system. The sources have been grouped into five regions (column 1).

Figure 5. The Greek and Macedonian Neolithic cultures are the primary sources of the Cycladic and the Minoan Charalambos samples (red), while the Danube Basin Neolithic cultures are the primary sources of the Minoan Odigitria and Petras samples (blue) according to the G25 admixture analysis. The red and dark blue lines show hypothetical migrations.

One of the major population movements was from the Peloponnese Peninsula. This movement reached both the Cyclades and Central Crete, including the Lassithi Plateau, where the Charalambos Cave is located. Another major movement was from the Danube Basin. This movement reached eastern Crete, including the town of Petras and southern Crete, where Moni Odigitria can be found near Phaistos.

Figure 6 shows on the x and y axes the 1st and 2nd principal components of the principal component analysis generated by the G25 system. The principal component analysis shows that the Moni Odigitria samples (the red triangle) are directly below the Middle Neolithic Linear Pottery culture samples from Hungary (HUN_LBK_MN). The Petras sample (GRC_Minoan_EBA) is located directly below the Moni Odigitria samples. In contrast, the Hagia Charalambos samples from the Lassithi Plateau (the green pentagon), the two Cycladic samples (the dashed brown line), and the Mycenean samples (purple quadrangle) are located below the Moni Odigitria samples and most to its left or right. Hence, the principal component analysis in Figure 6 supports the analysis in Figure 4.

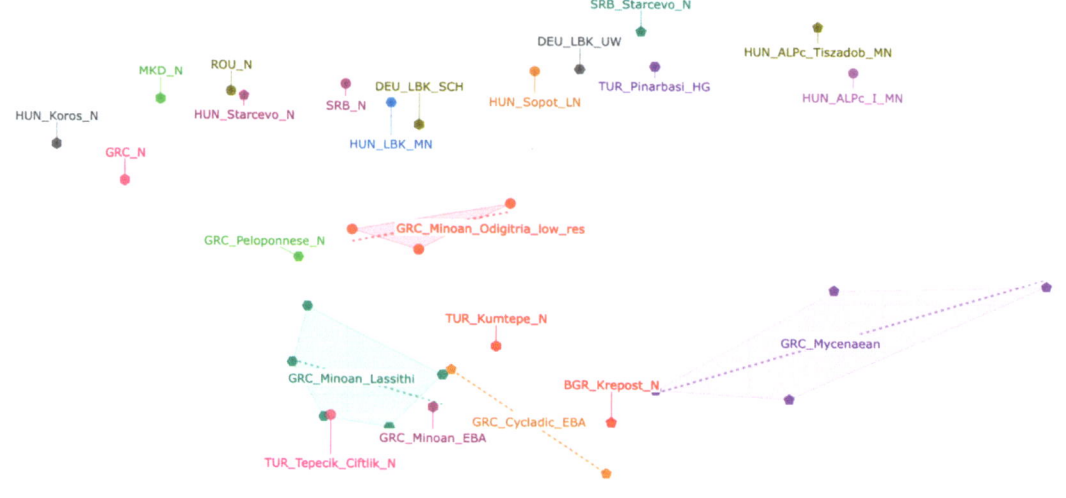

Figure 6. Principal component analysis of archaeogenetic samples, including Minoan samples from the Charalambos Cave on the Lassithi Plateau (green pentagon), Moni Odigitria (red triangle), and Mycenaean samples (purple quadrangle).

4. Method and Experiment 3: Analysis of the Origin of the Minoan U5a1 Haplogroup

4.1. Method of Analyzing the Origin of the Minoan U5a1 Haplogroup

The method of analyzing the origin of the Minoan U5a1 Haplogroup relies on the available data from the Ancient DNA Database [35]. We found where U5, U5a1 and U5a1d2b samples are found at times before the arrival of Indo-Europeans to Europe. If these are found exclusively in Europe, then these haplogroups cannot originate from other continents and could not have been brought to Europe by Indo-Europeans.

The motivation to focus on the U5 haplogroup is that it is known to be associated with Uralic speakers. Table 4 shows that the U5 mtDNA haplogroup percentages are almost always higher among the Uralic speakers than among their Indo-European neighbors according to Simoni et al. [36], who examined more than 2600 mtDNA sequences from current European populations. For example, the U5 mtDNA haplogroup percentage is 48 percent among the Saami, while it is only 11.4 percent among the Norwegians, their Indo-European speaking neighbors.

While some caution is warranted because there are still relatively few ancient mtDNA samples from Greece, they show the same trend, with the U5a1 haplotype reported for 2 out of 11 of Minoan samples by Lazaridis et al. [1] (sample I0071 from the Charalambos Cave and sample I9123 from the Late Minoan cemetery at Armenoi, Crete), and the U5 haplotype reported for 1 out of 40 Mycenaean samples by Skourtanioti et al. [37]. This suggests that Proto-Uralic speakers had a high percentage of the U5 mtDNA haplogroup, and Proto-Indo-European speakers had no U5 mtDNA haplogroup initially. However, the percentage decreased among the Uralic speakers and increased among their Indo-European neighbors due to millennia of genetic admixture.

Table 4. mtDNA haplogroup U5 within Uralic language speaking and neighboring populations.

Uralic Speakers	U5 Percent	Indo-European Neighbors	U5 Percent	Source
Saami	48	Norway	11.4	Simoni et al. [36]
Finland	20.7	Sweden	12.1	Simoni et al. [36]
Moksha	18.9	Russia	10.4	Bramanti et al. [38]
Minoan	18.2	Mycenaean	2.5	[1], Skourtanioti et al. [37]
Mordovians	15.7	Russia	10.4	Simoni et al. [36]
Mari	14	Russia	10.4	Simoni et al. [36]
Estonia	13.3	Latvia	10	Simoni et al. [36]
Basques	11.7	Spain	8.1	Simoni et al. [36]
Udmurt	8.9	Russia	10.4	Simoni et al. [36]
Hungary	7.4	Romania	7.2	Simoni et al. [36]

4.2. Experiment with U5, U5a1, and U5a1d2b Haplogroup Data

We used the Ancient DNA Database [35] to map all the U5 haplogroup samples before 8000 BP (Figure 7 top), the U5a1 haplogroup samples before 6000 BP (Figure 7 middle), and the U5a1d2b haplogroup samples before 5500 BP (Figure 7 bottom).

Figure 7 (top) shows (purple symbols) that the U5 haplogroup appeared first among Gravettian hunter-gatherers in present-day Dolní Věstonice, Czechia, around 30,800 BP [39] spread to other areas of Europe but not to other continents by 8000 BP.

Figure 7 (middle) shows (dark blue symbol) that the U5a1 mtDNA haplogroup appeared first in the Iron Gates gorges area in the lower Danube Basin around 10,530 BP. The U5a1 haplogroup was concentrated in the Danube Basin, as well as some areas that are considered to have been long inhabited by Uralic speakers such as the Baltic Sea region and the middle Volga region.

Figure 7 (bottom) shows the distribution of the even more specific haplogroup U5a1d2b before 5500 BP. The Baltic Sample, Tamula22, is from the Combed Ware culture in Estonia, and the other sample, MUR019, from Murzikhinsky II, Russia, is associated with the Eneolithic Volga-Kama culture. These cultures are commonly associated with early Uralic speakers. In addition, three 'early Hungarian' U5a1d2b samples, BAL23.6B, NTHper1, and VPBoer51, from 1000 to 1000 CE are also known [35].

The U5, U5a1, and U5a1d2b haplogroups do not have an Indo-European origin because these haplogroups are native to Europe, while the Indo-Europeans came to Europe only around 5300 BP according to the Kurgan Hypothesis of Indo-European origin [16]. However, these haplogroups were absorbed by the Mycenaeans and other Indo-Europeans after their arrival to Europe, as shown by the Mycenaean U5a1d2b sample from Aidonia, northern Greece, that was found by Skourtanioti et al. [37].

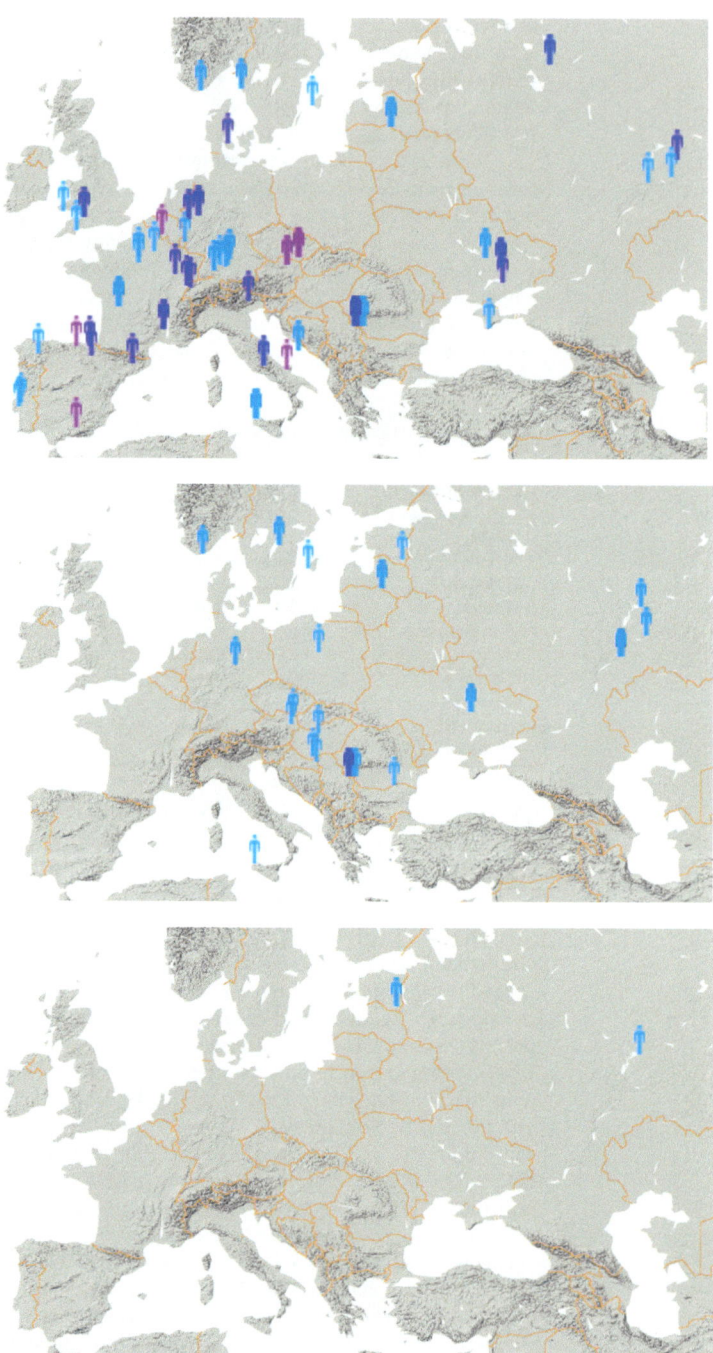

Figure 7. Location of samples based on [35], accessed on 20 September 2024: (**top**) U5 before 15,000 BP (purple), 15,000–10,000 BP (dark blue), and 10,000–8000 BP (light blue); (**middle**) U5a1 before 10,000 BP (dark blue), and 10,000–6000 BP (light blue); and (**bottom**) U5a1d2b before 5500 BP.

5. Method and Experiment 4: Analysis of Maximal Minoan Haplogroups

The previous experiment showed that the U5a1 haplogroup, which was found in two Minoan samples, had to come from the Danube Basin from where it originated. The next experiment investigates the origin of all the mtDNA and y-DNA haplogroups that were observed in the Minoan samples.

5.1. Method of Analyzing Maximal Minoan Haplogroups

We start by giving a definition to better describe the experiment in this section.

Definition 1. *Given a set $S = \{S_1, \ldots, S_n\}$ of mtDNA (or y-DNA) haplogroups, any S_i is a maximal mtDNA (or y-DNA) haplogroup in S if there is not another $S_j \in S$ such that S_i is a prefix or beginning of S_j.*

For example, if S = {U5, U5a1, U5a1d2b} is a set of mtDNA haplogroups, then U5 is not a maximal haplogroup in S, because U5 is the prefix or beginning of U5a1, which is also in S. However, U5a2d2b is a maximal mtDNA haplogroup.

Clearly, the maximal haplogroups carry the most valuable information, because they can be found in fewer places than the non-maximal haplogroups. Hence, to make our search efficient, it is enough to focus on identifying the origin of those Minoan mtDNA and y-DNA haplogroups that are maximal.

Hence, the method is to find for each maximal Minoan mtDNA and y-DNA from which the following three regions it could possibly come from: (1) the western Mediterranean coastal regions, (2) the Black Sea coastal regions, or (3) the Fertile Crescent. We searched for samples with the same haplogroup or even a more specific haplogroup from these three regions from a time before 3700 BP. If we found several samples, then we picked the one that was closest to Crete. If we found no samples, then we wrote down N/A for 'not available'. At the end of the process, we found the total number of N/As for each of these three regions. The higher the number of N/As for a region, the less likely that the Minoans came from that region.

5.2. Experiment with Maximal Minoans Haplogroups

We only analyzed the Early Minoan and Middle Minoan samples from Aposelemis, Charalambos, Odigitria, and Petras because some later samples from the other sites could be Mycenaean samples, meaning that due to the Mycenaeans' occupation of the island Crete during the Late Minoan period [15]. All the Minoan samples that we considered are dated to 3700 BP or earlier.

Table 5 lists and Figure 8 shows the Minoan mtDNA and y-DNA samples and their closest matches to Mediterranean, Black Sea region, and Fertile Crescent samples.

The second column of Table 5 gives the sample ID and reference to the source for each y-DNA and mtDNA haplogroup sample. If there are no 3700 BP samples in the database, then a 'not available' (N/A) is shown.

We revised some of the reported mtDNA haplogroups of Hughey et al. [40], because their earlier classifications were sometimes not as specific as possible. The revisions are indicated by '(rev)' in the first column of Table 5. Some revisions were already given by Revesz [41], but the ones listed in Table 6 are further improvements.

The revisions use Hughey et al. [40]'s reported mutations with respect to the rCRS reference sequence. The reported mutations are shown in black in the last column of Table 6. Unfortunately, the technique used by Hughey et al. [40] can miss many mutations, because only fragments of the mtDNA are scanned. It appears that the mutations shown in red were missed, because they were needed on a path from the root to the new haplogroup in the PhyloTree [42] used for mtDNA haplogroup classification. For example, the reported mutation 14055T implies that the haplogroup classification should be H41a1a, even though the red ones are missing in the second row.

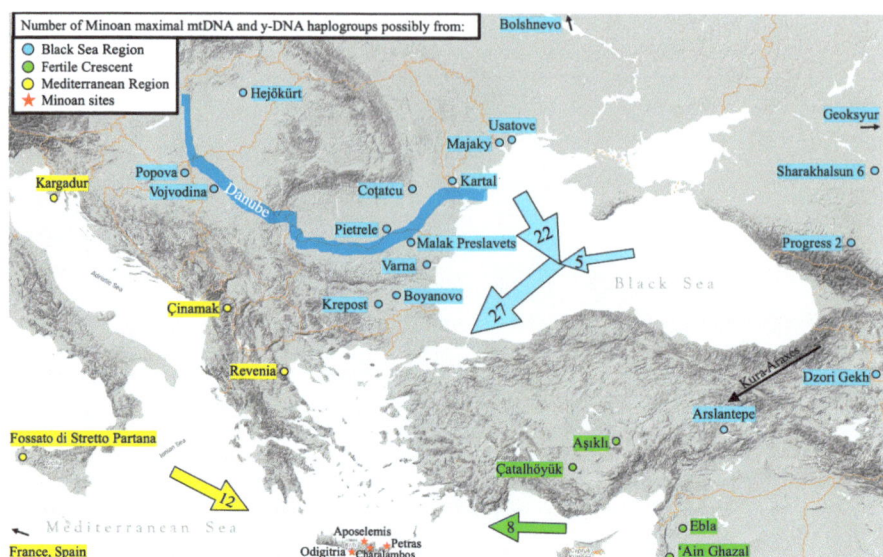

Figure 8. Location of the archaeological samples listed in Table 5.

Table 5. mtDNA and y-DNA (blue) haplotypes shared by DNA samples (1st–3rd columns) from Minoan, Mediterranean, Black Sea, and Fertile Crescent locations (4th–7th columns). N/A means there are no non-Minoan samples from at least 3700 BP.

mtDNA/y-DNA	Sample ID	BP	Minoan	Mediterranean Region	Black Sea Region	Fertile Crescent
G2a2b2a1a1	Pta08 [43]	4685	Petras	N/A		
G2a2b2a1a1c2	PIE015 [44]	6534			Pietrele, Romania	N/A
H1bm	I0073 [35]	4000	Charalambos	N/A		
H1bm	I8531 [35]	5050			Geoksyur, Turkmenistan	N/A
H2a2a1d (rev)	8H [40]	3700	Charalambos	N/A	N/A	
H2a2a1d	CCH290 [35]	8590				Çatalhöyük, Turkey
H4a1	HGC005 [37]	4178	Charalambos			
H4a1a	CRE14 [35]	6302		Béziers, France		
H4a1	PIE048 [44]	6586			Pietrele, Romania	N/A
H5	HGC017 [37]	EMBA	Charalambos			
H5	I4565 [35]	4915		Galls Carboners, Cat., Spain		
H5	I0679 [35]	7617			Krepost, Bulgaria	N/A
H7	12AH [40]	3700	Charalambos			
H7c	I5072 [35]	7551		Kargadur, Croatia		
H7	PIE014 [44]	6455			Pietrele, Romania	N/A
H13a1a	I0070 [1]	4000	Charalambos	N/A		
H13a1a1	BOY009 [44]	4799			Boyanovo, Bulgaria	N/A
H41a1a (rev)	6AH [40]	3700	Charalambos	N/A		
H41a	BOL003 [35]	4408			Bolshnevo, Tver, Russia	N/A
H102	HGC041 [37]	EMBA	Charalambos			
H102	I14689 [45]	4568		Çinamak, Albania	N/A	N/A
HV-b	HGC018 [37]	EMBA	Charalambos	N/A		
HV-b	PIE057 [44]	6421			Pietrele, Romania	N/A

Table 5. Cont.

mtDNA/y-DNA	Sample ID	BP	Minoan	Mediterranean Region	Black Sea Region	Fertile Crescent
I1	HGC040 [37]	4134	Charalambos			
I1	Neolithic 5 [35]	5200		Camí de Can Grau, Spain		
I1a1	MAJ008 [44]	6110			Majaky, Ukraine	N/A
I5a	HGC024 [37]	3700	Charalambos	N/A		
I5a	PIE063 [44]	6460			Pietrele, Romania	N/A
J1a2a1a2−	HGC001 [37]	EMBA	Charalambos	N/A		
J1a2a1a2d2b2b2−	I16120 [45]	3390			Dzori Gekh, Armenia	
J1a2a1a2d2b2	ETM012 [35]	4470				Ebla, Syria
J2a1a1a2b1b2	HGC006 [37]	EMBA	Charalambos	N/A		
J2a1a1a2b1b	ART020 [45]	5177			Arslantepe, Turkey	N/A
J2b1a1	ERS1770867 [1]	3895	Odigitria			
J2b1a1	I8153 [35]	4650		Sima del Ángel, Luc., Spain		
J2b1a1b	I23210 [43]	3900			Vojvodina, Serbia	N/A
K1a2	ERS1770871 [1]	3895	Odigitria			
K1a2a	CB13 [35]	7345		Cova Bonica, Cat., Spain		
K1a2	I2532 [35]	7614			Coțatcu, Romania	N/A
K1a4	HGC027 [37]	EMBA	Charalambos	N/A		
K1a4	PIE065 [44]	6568			Pietrele, Romania	
K1a4	Ash129 [35]	10093				Aşıklı, Turkey
K2b1	APO023 [37]	3558	Aposelemis			
K2b1	I4065 [35]	6815		Fossato di Stretto Partana, IT		
K2b1c	POP06 [43]	6450			Popova, Croatia	N/A
T1	9H [40]	3700	Charalambos	N/A		
T1a	VAR016 [44]	6452			Varna, Bulgaria	
T1a2	I1727 [35]	10050				'Ain Ghazal, Jordan
T2b25	HGC008 [37]	4219	Charalambos			
T2b	584 [35]	4950		Treilles cave, France		
T2b	PIE008 [44]	6422			Pietrele, Romania	N/A
T2c1d	HGC020 [37]	EMBA	Charalambos			
T2c1d1	I15946 [35]	5968		Anghelu Ruju, Sardinia		
T2c1d1	PIE030 [44]	6259			Pietrele, Romania	N/A
T2e6 (rev)	21H [40]	3700	Charalambos			
T2e	Bar10 [35]	4710		Barranc d'en Rifà, Spain		
T2e	I0700 [35]	7912			Malak Preslavets, Bulgaria	
T2e	CCH311 [35]	8520				Çatalhöyük, Turkey
U1a1a-a	HGC010 [37]	EMBA	Charalambos	N/A		
U1a1a3a*	PG2002 [43]	4361			Progress 2, Russia	N/A
U3b3	I9130 [35]	3895	Odigitria	N/A		
U3b3	KTL005 [44]	4905			Kartal, Ukraine	N/A
U5a1f1 (rev)	4H [40]	3700	Charalambos	N/A		
U5a1f1	MAJ020 [44]	5871			Majaky, Ukraine	N/A
U7b	HGC053 [37]	EMBA	Charalambos	N/A		
U7b*	SA6001 [43]	5444			Sharakhalsun 6, Russia	N/A
U8b1b4 (rev)	M4 [40]	3700	Charalambos	N/A		
U8b1b4	I2378 [35]	7050			Hejőkürt, Hungary	N/A
W	6H [40]	3700	Charalambos	N/A		
W3b	PIE022 [44]	6392			Pietrele, Romania	
W1c4*	MK308703.1 [43]	8365				Çatalhöyük, Turkey
X2b (rev)	M8 [40]	3700	Charalambos			
X2b	Rev5 [35]	8316		Revenia, Greece		
X2b	USV005 [44]	5588			Usatove, Ukraine	
X2b4*	MK308702.2 [43]	8365				Çatalhöyük, Turkey
--	Total N/As	--		17	2	21
--	Percent N/As	--		58.6	6.9	72.4

Another problem with Hughey et al. classifications [40] is that since they reported the mutations with respect to rCRS, which belongs to haplogroup H2a2a1, some mutations with respect to RSRS cannot be expected to be reported if there is an agreement on these rCRS mutations between the analyzed sample and the rCRS. These not-expected-to-be-reported mutations could be assumed to be present in the third sample because it also has the mutation 16172C, which indicates that its haplogroup classification is most likely H2a2a1d, although H66a is also a possible classification based on the 2706A and 16172C mutations.

Table 6. Updating the mtDNA haplogroup classifications of Hughey et al. [40]. The sample IDs are from the European Nucleotide Archive database, which is available online: https://www.ebi.ac.uk/ena/browser/home (accessed on).

Sample ID	mtDNA Old	New	Mutations with Respect to RSRS
HM022275	H	H41a1a	15617A (H41), 262T, 5460A, 10124C, 14118G (H41a), 16362T! (H41a1), 14055T (H41a1a)
HM022291	U5a	U5a1f1	16192T, 16270T (U5), 3197C, 9477A, 13617C (U5a'b), 14793G, 16256T (U5a), 15218G, 16399G (U5a1), 6023A (U5a1f), 5585A, 7569C, 16311C! (U5a1f1)
HM022294	H	H2a2a1d	2706A, 7028C (H), 1438A (H2), 4769A (H2a), 750A (H2a2), 8860A, 15326A (H2a2a), 263A (H2a2a1), 16172C (H2a2a1d)
HM022303	T	T2e6	11812G, 14233G, 16296T (T2), 150T (T2-a), 16153A (T2e), 16240C (T2e6)
HM022308	U	U8b1b4	9698C (U8), 3480G (U8b'c), 9055A, 14167T (U8b), 195C!, 16189C!, 16234T (U8b1), 1811A!, 5165T, 16324C (U8b1b), 16290T (U8b1b4)
HM022312	X	X2b	6221C, 6371T, 13966G, 14470C, 16189C!, 16278T! (X), 153G (X1'2'3), 195C!, 1719A (X2), 225A (X2-a), 13708A (X2b'd), 8393T, 15927A (X2b)

We explain some of the regional classifications as follows. The rare mtDNA H1bm sample from Geoksyur, Turkmenistan was counted as 'Black Sea region' for regional classification in Table 5 because of likely ancient migration from the Caspian Sea area via the Volga and the Don to the Black Sea area. These regions were connected by trade routes since at least the Bronze Age [46].

The rare mtDNA H41a sample from Bolshnevo, Tver, Russia, which is part of the Fatyanovo culture, was also counted as 'Black Sea region' because the Fatyanovo culture resulted from a migration from the Middle-Dnieper culture [47].

Finally, the rare y-DNA J2a1a1a2b1b sample from Arslantepe, Turkey was counted as the 'Black Sea region' because the Kura-Araxes culture moved south from the north of the Caucasus to Arslantepe around 3000 BCE, when there was widespread burning and destruction, after which Kura–Araxes culture pottery appeared in the area [48]. Moreover, this J2a1a1a2b1b sample, ART20, had blue eyes according to Lazaridis et al. [45] (supplement Table 5), further implying that this individual was part of the Kura–Araxes migration to the south.

6. Results and Discussion

6.1. Summary of the Results

Section 2 showed that six SNP mutations associated with light eyes, hair, and skin phenotypes and originating in Europe or Eurasia have a very high presence among the Minoan samples. The SNP mutations analysis showed that the Eastern-European Hunter-Gatherer culture (EHG) was closest to the Minoans, because Minoans had the lowest distance, 0.27, to the EHGs according to the root mean square error measure. Since the area of EHGs included present day Ukraine, the proximal genetic source of the Minoans was likely the northern Black Sea coastal region.

Section 3 presented a G25 admixture analysis that showed that the Minoan samples from Odigitria and Petras likely have a Danube Basin origin, while the Minoan Charalambos samples likely have a Greek mainland origin.

Section 4 showed that the U5, U5a1, and U5a1d2b mtDNA haplogroups are native to Europe. The U5 mtDNA haplogroup is frequent among Uralic speakers and could be found among several Minoans, whereas the U5 haplogroup was absent in Neolithic Anatolia.

Section 5 traced back the origins of each known Minoan mtDNA and y-DNA haplogroup. The analysis involved looking at the most specific haplogroups that can be identified based on the current PhyloTree [42] classification. Table 5 listed 29 different Minoan mtDNA and y-DNA haplotypes. The bottom row showed that out of these 29 Minoan haplotypes 17, or 58.6 percent, could not have come from the Mediterranean region, and 21, or 72.4 percent, could not have come from the Fertile Crescent, while only 2, or 6.9 percent, could not have come from the Black Sea region. This also seems to suggest that the proximal genetic source of the Minoans was overwhelmingly the Black Sea region.

6.2. Discussion on the Phenotypes

Blue eyes, which are associated with the HERC2 rs12913832-A allele, likely originated about 42,000 years ago among the WHGs, where it has the highest frequency [49]. This allele spread widely and can be found among hunter-gatherers, as well as farmers, except the Aegean early Neolithic farmers and the Bronze Age Minoans.

Light skin color is associated with the SLC45A2 rs28777-A and the SLC45A2 rs16891982-G alleles among other alleles. Both alleles can be found among the EHGs with a high percentage, and the second allele was also found in a Paleolithic hunter-gatherer (Kostenki14) in the Don River area [23]. Hence, these alleles seem ancient in the area where EHGs also lived. These alleles apparently spread from the EHGs to the WHGs at Loschbour, to the Lower Danube Mesolithic hunter-gatherers, and to European Neolithic and Bronze Age cultures, except the Fertile Crescent Neolithic culture [23].

The SLC24A4 rs2402130-A allele is associated with light hair, and the SLC24A5 rs1426654-A allele is associated with light skin. These alleles are present in all CHG and EHG samples. Hence, these alleles likely originated in a common ancestor of these hunter-gatherers around the Caucasus area and spread to other regions, except the rs1426654-A allele did not reach the Mesolithic Lower Danube and the WHG cultures.

The TYR rs1042602-A allele is also associated with light skin and a lower occurrence of freckles. This allele is absent from both hunter-gatherers and the early farmers of the Fertile Crescent, the Aegean, and the Körös River area. It seems to first have occurred in the Hungarian Middle Neolithic culture around 5000 BC in three samples (NE2, NE3, and NE5) [29]. This allele continued in the Hungarian Bronze Age and has been found among the Minoans and the Mycenaeans. This allele apparently spread from the Danube Basin southward to the Aegean area.

Hence, the SLC45A2 rs28777-A, SLC45A2 rs16891982-G, SLC24A4 rs2402130-A, and SLC24A5 rs1426654-A alleles suggest a genetic connection between the EHGs and the Minoans. The TYR rs1042602-A allele suggests another genetic connection between the Hungarian Middle Neolithic farmers and the Minoans. Furthermore, the lack of the HERC2 rs12913832-A allele makes it unlikely that either WHGs or FertileC_NE farmers reached Crete in significant numbers. While farming spread to Crete during the early phase of the Neolithic, the lack of the HERC2 rs12913832-A allele suggests that farming reached Crete from the Aegean_NE culture, which also lacks this allele, rather than directly from the FertileC_NE, where the allele is present in a significant percentage of the samples. While the Aegean_NE culture learned farming from the FertileC_NE culture, a likely genetic admixture with local Aegean hunter-gatherers who lacked the HERC2 rs12913832-A allele may have diluted this allele to an insignificant percentage before reaching Crete.

The genetic admixture between hunter-gatherers and early farmers is most noticeable in the Körös_NE culture. In fact, KO1 did not cluster together with early European farmers according to a study by Gamba et al. [29]. The genetic admixture between hunter-gatherers and farmers is also well-documented at the Iron Gates gorges area [50], where the Danube crosses the Carpathian Mountains. Most early farmers likely passed through the Iron Gates before entering the Carpathian Basin, where the early Neolithic Körös_NE and the Middle Neolithic Hungary_MN cultures also flourished.

A common problem in archaeogenetics is the low sample size, which may cause statistical errors. For example, we have only twelve Minoan allele samples (two allele samples from six individuals) regarding the rs12913832 genetic locus. While none of these allele samples had the mutation that causes blue eye color, there is still a certain probability that a blue-eyed Minoan sample will be found later.

6.3. Discussion on the G25 System

The results of the G25 admixture analysis system need to be handled with caution, because the reliability of the system is not yet well-tested. However, the main result of a movement from the Danube Basin to Crete is also supported by some studies on climate change and the spread of agriculture.

The exact time and reason for these population movements shown in Figure 5 is unknown currently. However, the second movement may be related to the 4.2 kiloyear BP aridification event [51] that dried out the Danube Basin and made agriculture infeasible there. That may have caused the Danube Basin farmers to move to the Messara Plain in southern Crete, which may have provided better agricultural and fishing opportunities. The distinguishing of these two major population movements has major implications regarding the languages spoken in different areas of Crete and the decipherment of the Minoan scripts.

These hypothetical migrations are also supported by the presence of millet grains at the Minoan sites such as Chania, Knossos and Zominthos starting from the Neopalatial period [52]. Livarda and Kotzamani [52] speculate that millet had reached Crete from Bronze Age Central Europe, where it was commonly cultivated.

6.4. Discussion on U5, U5a1, and U5a1d2b Haplogroups

Some of the Gravettian hunter-gatherers found refuge in the lower Danube Basin and the northern Pontic coastal areas during the subsequent Ice Age. The Proto-Uralic language likely developed in this refuge area during the Ice Age and broke up sometime during the Mesolithic period when some of the Uralic speakers went northeast. These early Uralic speakers may have followed the mammoth herds, which also moved from this area north to the Baltic and southern Scandinavia, where remains were found between 17 and 12 thousand years ago [53].

The U5 mtDNA haplogroup is strongly associated with Uralic language speakers, because the Uralic language speakers had matrilineal cultures in the past. Since the husband moves to the village of the wife in a matrilineal society, their children will speak the mother's language. Hence, in matrilineal societies the mother's language is passed on in parallel to the mother's mtDNA. In contrast, the Yamnaya and other early Indo-European cultures were patrilineal. Since the wife moves to the village of the husband in a patrilineal culture, their children will speak the father's language. Hence, Indo-European language speakers are more commonly associated with the R1 y-DNA haplogroup [54].

6.5. Discussion on the Minoan Maximal Haplogroups

While the U haplogroup was the dominant European hunter-gatherer haplogroup, other haplogroups arrived from the Fertile Crescent and the Caucasus. Despite the new haplogroups, the Neolithic Old European Civilization (Gimbutas [55,56]) or the Danube Civilization (Haarmann [57]) was likely Uralic speaking, because the neolithization of the Danube Basin was a slow process taking place over thousands of years. Hence, those who came earlier from the Anatolia may have learned the local Uralic language, and they and their descendants may have taught it to those who came later from Anatolia. Hence, while the incoming Anatolian famers' total genetic effect on the local population was considerable after several millennia, their linguistic effect may have been small.

The geography of the Carpathian Basin may have helped in the process of unifying the spoken language there. If one follows the Danube River, then entering the Carpathian Basin requires going through the Iron Gates gorge, which is a natural geographic barrier. The Iron Gates barrier likely slowed down the inward movement of any wave of newcomers. Since it is a defensible barrier, passing through it may have required cultivating friendly relations with the locals, and that likely resulted in intermingling between the local population and the newcomers.

Brami et al. [50] found evidence of this intermingling studied at the site of Lepenski Vir, near the Iron Gates gorge. Brami at al. [50] found one individual with only hunter-gatherer genes, three individuals with some genetic admixture between hunter-gatherers and Aegean farmers, and two individuals with only Aegean farmer genes between 6100 and 6000 BCE. In addition, two individuals had only hunter-gatherer genes, and three individuals had only Aegean farmer genes before this transition period. Furthermore, out of the three admixed individuals, two belonged to the U5 and one to the H mtDNA haplogroup. These two haplogroups were already present in the hunting-gathering period before 7400 BCE. This suggests that Aegean farmers moving into the Lepenski Vir community married local hunter-gatherer women. Moreover, if such an intermingling happened at the Iron Gates area between hunter-gatherers and Aegean farmers, then it likely happened with even greater ease later between the already neolithic local Iron Gates population and later Aegean farmer newcomers. This suggests that the local hunter-gatherer language did not change with the arrival of Aegean farmers. Therefore, the Old European mtDNA and y-DNA haplogroups can be associated with Uralic languages.

7. Conclusions and Further Work

Four different experiments of Sections 2–5 suggest that the proximal sources of the Minoans were the Danube Basin and the Black Sea coastal area, which overlap in the Danube Delta area, providing easy migration opportunities between them. Future work needs to look at the rapidly increasing ancient DNA data to be able to make a statistically stronger conclusion and to further narrow down the proximal source of the Minoans.

Lazaridis et al. [1] presented the first whole-genome sequences for Minoan samples. That is a lasting contribution to archaeogenetics, but their data analysis is flawed, because they overlooked the Danube Basin and the Black Sea coastal area as a possible proximal source. Hence, their claim that the proximal source of the Minoans was Anatolia or the Fertile Crescent has to be abandoned. This correction regarding the origin of the Minoans helps to reconcile archaeogenetics with linguistic work that links the Minoan language to the Uralic language family. The reconciliation of archaeogenetics and linguistics would follow, because the Danube Basin and the northern Black Sea coastal areas were identified by some researchers as potential Uralic speaking areas before the arrival of the Yamnaya people, who are believed to have spoken an Indo-European language [6,7]. The arrival of the Yamnaya may have prompted the Minoans to sail to Crete. It also may have prompted other Uralic peoples to move away from the Steppe, although the precise route and timing of their migrations remains an open problem.

Unfortunately, furthering the incorrect view that the proximal source of the Minoans was Anatolia or the Fertile Crescent would continue to lead linguists to suspect that the Minoan language is related to Near Eastern or African languages [17–22]. A search in those regions for language connections with Minoan did not yield any result for over a century. Both the sequencing and the data mining of archaeogenetic data have to be correct to aid instead of hinder linguistic discoveries.

Funding: This research received no external funding.

Institutional Review Board Statement: Not applicable.

Informed Consent Statement: Not applicable.

Data Availability Statement: The original contributions presented in the study are included in the article, further inquiries can be directed to the author.

Conflicts of Interest: The author declares no conflicts of interest.

References

1. Lazaridis, I.; Mittnik, A.; Patterson, N.; Mallick, S.; Rohland, N.; Pfrengle, S.; Furtwängler, A.; Peltzer, A.; Posth, C.; Vasilakis, A.; et al. Genetic origins of the Minoans and Mycenaeans. *Nature* **2017**, *548*, 214–218. [CrossRef] [PubMed]
2. Revesz, P.Z. Establishing the West-Ugric language family with Minoan, Hattic and Hungarian by a decipherment of Linear A. *WSEAS Trans. Inf. Sci. Appl.* **2017**, *14*, 306–335.
3. Revesz, P.Z. A translation of the Arkalochori Axe and the Malia Altar Stone. *WSEAS Trans. Inf. Sci. Appl.* **2017**, *14*, 124–133.
4. Revesz, P.Z. A computer-aided translation of the Phaistos Disk. *Int. J. Comput.* **2016**, *10*, 94–100.
5. Hajdú, P. Über die alten Siedlungsraume der uralischen Sprachfamilie. *Acta Linguist. Hung.* **1964**, *14*, 47–83.
6. Wiik, K. The Uralic and Finno-Ugric phonetic substratum in Proto-Germanic. *Linguist. Ural.* **1997**, *33*, 258–280. [CrossRef]
7. Krantz, G. *Geographical Development of European Languages*; P. Lang: New York, NY, USA, 1988.
8. Revesz, P.Z. *Was the Uralic Homeland in the Danube Basin?* Magyarok Világszövetsége: Budapest, Hungary, 2021.
9. Harms, R.T. Uralic Languages, In: *Encyclopaedia Britannica*; Oxford University Press: Oxford, UK, 2024.
10. Syrjänen, K.; Honkola, T.; Korhonen, K.; Lehtinen, J.; Vesakoski, O.; Wahlberg, N. Shedding more light on language classification using basic vocabularies and phylogenetic methods: A case study of Uralic. *Diachronica* **2013**, *30*, 323–352. [CrossRef]
11. Revesz, P.Z. A tale of two sphinxes: Proof that the Potaissa Sphinx is authentic and other Aegean influences on early Hungarian inscriptions. *Mediterr. Archaeol. Archaeom.* **2024**, *24*, 79–96.
12. Beekes, R.S.P. *Etymological Dictionary of Greek*; Brill NV: Leiden, Netherlands, 2009.
13. Rédei, K. (Ed.) *Uralisches Etymologisches Wörterbuch*; Akadémiai Kiadó: Budapest, Hungary, 1988.
14. Beekes, R.S.P. *Pre-Greek Phonology, Morphology, Lexicon*; Brill NV: Leiden, The Netherlands, 2014.
15. Manning, S. Chronology and Terminology. In *The Oxford Handbook of the Bronze Age Aegean*; Cline, E., Ed.; Oxford University Press: Oxford, UK, 2012; pp. 11–28.
16. Clackson, J.P.T. *Indo-European Linguistics: An Introduction*; Cambridge University Press: Cambridge, UK, 2007.
17. Bernal, M. *Black Athena: The Afroasiatic Roots of Classical Civilization*; Rutgers University Press: New Brunswick, NJ, USA, 1991; Volume 2.
18. Best, J.G.P. *Some Preliminary Remarks on the Decipherment of Linear A*; A.M. Hakkert Publishing: Las Palmas, Spain, 1972.
19. Campbell-Dunn, G. *Who were the Minoans? An African Answer*; Author House: Bloomington, IN, USA, 2006.
20. Gordon, C.H. *Evidence for the Minoan Language*; Ventnor Publishing: Ventnor, NJ, USA, 1966.
21. Kvashilava, G. *On Reading Pictorial Signs of the Phaistos Disk and Related Scripts*; Ivane Javakhishvili Institute of History and Ethnology: Tbilisi, Georgia, 2010.
22. La Marle, H. *Reading Linear A: Script, Morphology, and Glossary of the Minoan Language*; Geuthner: Paris, France, 2010.
23. Günther, T. Population genomics of Mesolithic Scandinavia: Investigating early postglacial migration routes and high-latitude adaptation. *PLoS Biol.* **2018**, *16*, e2003703. [CrossRef]
24. Paraskevi, K. Meta-Analysis of Phenotypic Traits in Prehistoric European Populations from Paleogenetic Analysis Data. Ph.D. Dissertation, Aristotle University of Thessaloniki, Thessaloniki, Greece, 2022. Appendix data table.
25. González-Fortes, G.; Jones, E.R.; Lightfoot, E.; Bonsall, C.; Lazar, C.; Grandal-d'Anglade, A.; Garralda, M.D.; Drak, L.; Siska, V.; Simalcsik, A.; et al. Paleogenomic evidence for multi-generational mixing between Neolithic farmers and Mesolithic hunter-gatherers in the Lower Danube Basin. *Curr. Biol.* **2017**, *27*, 1801–1810. [CrossRef]
26. Llorente, M.G. The Origins and Spread of the Neolithic in the Old World Using Ancient Genomes. Ph.D. Dissertation, Cambridge University, Cambridge, UK, 2017.
27. Feldman, M.; Fernández-Domínguez, E.; Reynolds, L.; Baird, D.; Pearson, J.; Hershkovitz, I.; May, H.; Goring-Morris, N.; Benz, M.; Gresky, J.; et al. Late Pleistocene human genome suggests a local origin for the first farmers of central Anatolia. *Nat. Commun.* **2019**, *10*, 1218. [CrossRef] [PubMed]
28. Hofmanová, Z.; Kreutzer, S.; Hellenthal, G.; Sell, C.; Diekmann, Y.; Díez-del-Molino, D.; Van Dorp, L.; López, S.; Kousathanas, A.; Link, V.; et al. Early farmers from across Europe directly descended from Neolithic Aegeans. *Proc. Natl. Acad. Sci. USA* **2016**, *113*, 6886–6891. [CrossRef] [PubMed]
29. Gamba, C.; Jones, E.R.; Teasdale, M.D.; McLaughlin, R.L.; Gonzalez-Fortes, G.; Mattiangeli, V.; Domboróczki, L.; Kővári, I.; Pap, I.; Anders, A.; et al. Genome flux and stasis in a five millennium transect of European prehistory. *Nat. Commun.* **2014**, *5*, 5257. [CrossRef] [PubMed]
30. Gerber, D.; Szeifert, B.; Székely, O.; Egyed, B.; Gyuris, B.; Giblin, J.I.; Horváth, A.; Palcsu, L.; Köhler, K.; Kulcsár, G.; et al. Interdisciplinary Analyses of Bronze Age Communities from Western Hungary Reveal Complex Population Histories. *Mol. Biol. Evol.* **2023**, *40*, msad182. [CrossRef] [PubMed]
31. Clemente, F.; Unterländer, M.; Dolgova, O.; Amorim, C.E.G.; Coroado-Santos, F.; Neuenschwander, S.; Ganiatsou, E.; Dávalos, D.I.C.; Anchieri, L.; Michaud, F.; et al. The genomic history of the Aegean palatial civilizations. *Cell* **2021**, *184*, 2565–2586. [CrossRef]

32. Broushaki, F.; Thomas, M.G.; Link, V.; López, S.; Van Dorp, L.; Kirsanow, K.; Hofmanová, Z.; Diekmann, Y.; Cassidy, L.M.; Díez-del-Molino, D.; et al. Early Neolithic genomes from the eastern Fertile Crescent. *Science* **2016**, *353*, 499–503. [CrossRef]
33. Revesz, P.Z. Data mining autosomal archaeogenetic data to determine Minoan origins. In *Proceedings of the 25th International Database Engineering and Applications Symposium*; Desai, B.C., Ed.; ACM: New York, NY, USA, 2021; pp. 46–55.
34. Revesz, P.Z. Spatio-temporal data mining of major European river and mountain names reveals their Near Eastern and African origins. In *Proceedings of the 22nd European Conference on Advances in Databases and Information Systems*; Benczúr, A., Thalheim, B., Horváth, T., Eds.; Springer: New York, NY, USA, 2018; Lecture Notes in Computer Science series; Volume 11019, pp. 20–32.
35. Ancient DNA Database. Available online: https://haplotree.info/maps/ancient_dna (accessed on 15 March 2024).
36. Simoni, L.; Calafell, F.; Pettener, D.; Bertranpetit, J.; Barbujani, G. Geographic patterns of mtDNA diversity in Europe. *Am. J. Hum. Genet.* **2000**, *66*, 262–728. [CrossRef]
37. Skourtanioti, E.; Ringbauer, H.; Gnecchi Ruscone, G.A.; Bianco, R.A.; Burri, M.; Freund, C.; Furtwängler, A.; Gomes Martins, N.F.; Knolle, F.; Neumann, G.U.; et al. Ancient DNA reveals admixture history and endogamy in the prehistoric Aegean. *Nat. Ecol. Evol.* **2023**, *7*, 290–303. [CrossRef]
38. Bramanti, B.; Thomas, M.G.; Haak, W.; Unterländer, M.; Jores, P.; Tambets, K.; Antanaitis-Jacobs, I.; Haidle, M.N.; Jankauskas, R.; Kind, C.J.; et al. Genetic discontinuity between local hunter-gatherers and central Europe's first farmers. *Science* **2009**, *326*, 137–140. [CrossRef]
39. Malyarchuk, B.; Derenko, M.; Grzybowski, T.; Perkova, M.; Rogalla, U.; Vanecek, T.; Tsybovsky, I. The peopling of Europe from the mitochondrial haplogroup U5 perspective. *PLoS ONE* **2010**, *5*, e10285. [CrossRef]
40. Hughey, J.R.; Paschou, P.; Drineas, P.; Mastropaolo, D.; Lotakis, D.M.; Navas, P.A.; Michalodimitrakis, M.; Stamatoyannopoulos, J.A.; Stamatoyannopoulos, G. A European population in minoan Bronze age crete. *Nat. Commun.* **2013**, *4*, 1861. [CrossRef] [PubMed]
41. Revesz, P.Z. Minoan archaeogenetic data mining reveals Danube Basin and western Black Sea littoral origin. *Int. J. Biol. Biomed. Eng.* **2019**, *13*, 108–120.
42. PhyloTree, mtDNA Tree Build 17. Available online: https://www.phylotree.org (accessed on 15 March 2024).
43. Yfull Database. Available online: https://www.yfull.com (accessed on 15 March 2024).
44. Penske, S.; Rohrlach, A.B.; Childebayeva, A.; Gnecchi-Ruscone, G.; Schmid, C.; Spyrou, M.A.; Neumann, G.U.; Atanassova, N.; Beutler, K.; Boyadzhiev, K.; et al. Early contact between late farming and pastoralist societies in southeastern Europe. *Nature* **2023**, *620*, 358–365. [CrossRef] [PubMed]
45. Lazaridis, I.; Alpaslan-Roodenberg, S.; Acar, A.; Açıkkol, A.; Agelarakis, A.; Aghikyan, L.; Akyüz, U.; Andreeva, D.; Andrijašević, G.; Antonović, A.; et al. The genetic history of the Southern Arc: A bridge between West Asia and Europe. *Science* **2022**, *377*, eabm4247. [CrossRef]
46. Revesz, P.Z. Data Science Applied to Discover Ancient Minoan-Indus Valley Trade Routes Implied by Common Weight measures. In *Proceedings of the 26th International Database Engineered Applications Symposium*; Desai, B.C., Revesz, P.Z., Eds.; ACM: New York, NY, USA, 2022; pp. 150–155.
47. Saag, L.; Vasilyev, S.V.; Varul, L.; Kosorukova, N.V.; Gerasimov, D.V.; Oshibkina, S.V.; Griffith, S.J.; Solnik, A.; Saag, L.; D'Atanasio, E.; et al. Genetic ancestry changes in Stone to Bronze Age transition in the East European plain. *Sci. Adv.* **2021**, *7*, eabd6535. [CrossRef]
48. Frangipane, M. Different types of multiethnic societies and different patterns of development and change in the prehistoric Near East. *Proc. Natl. Acad. Sci. USA* **2015**, *112*, 9182–9189. [CrossRef]
49. Hanel, A.; Carlberg, C. Skin colour and vitamin D: An update. *Exp. Dermatol.* **2020**, *29*, 864–875. [CrossRef]
50. Brami, M.; Winkelbach, L.; Schulz, I.; Schreiber, M.; Blöcher, J.; Diekmann, Y.; Burger, J. Was the fishing village of Lepenski Vir built by Europe's first farmers? *J. World Prehistory* **2022**, *35*, 109–133. [CrossRef]
51. de Menocal, P.B. Cultural responses to climate change during the late Holocene. *Science* **2001**, *292*, 667–673.
52. Livarda, A.; Kotzamani, G. The archaeobotany of Neolithic and Bronze Age Crete: Synthesis and prospects. *Annu. Br. Sch. Athens* **2013**, *108*, 1–29. [CrossRef]
53. Ukkonen, P.; Aaris-Sørensen, K.; Arppe, L.; Clark, P.U.; Daugnora, L.; Lister, A.M.; Lougas, L.; Seppä, H.; Sommer, R.S.; Stuart, A.J.; et al. Woolly mammoth (*Mammuthus primigenius* Blum.) and its environment in northern Europe during the last glaciation. *Quat. Sci. Rev.* **2011**, *30*, 693–712. [CrossRef]
54. Klyosov, A.A.; Rozhanskii, I.L. Haplogroup R1a as the Proto Indo-Europeans and the legendary Aryans as witnessed by the DNA of their current descendants. *Adv. Anthropol.* **2012**, *2*, 1–13. [CrossRef]
55. Gimbutas, M. *The Prehistory of Eastern Europe*; Peabody Museum: Cambridge, MA, USA, 1956.
56. Gimbutas, M. *The Civilization of the Goddess: The World of Old Europe*; HarperCollins: San Francisco, CA, USA, 1991.
57. Haarmann, H. *The Mystery of the Danube Civilisation: The Discovery of Europe's Oldest Civilisation*; Marix: Wiesbaden, Germany, 2020.

Disclaimer/Publisher's Note: The statements, opinions and data contained in all publications are solely those of the individual author(s) and contributor(s) and not of MDPI and/or the editor(s). MDPI and/or the editor(s) disclaim responsibility for any injury to people or property resulting from any ideas, methods, instructions or products referred to in the content.

Article

Quickening Data-Aware Conformance Checking through Temporal Algebras †

Giacomo Bergami *, Samuel Appleby and Graham Morgan

School of Computing, Faculty of Science, Agriculture and Engineering, Newcastle University, Newcastle Upon Tyne NE4 5TG, UK
* Correspondence: giacomo.bergami@newcastle.ac.uk
† This paper is an extended version of our paper: Appleby, S.; Bergami, G.; Morgan, G. Running Temporal Logical Queries on the Relational Model. In Proceedings of the IDEAS'22, 26th International Database Engineered Applications Symposium, Budapest, Hungary, 22–24 August 2022.

Abstract: A temporal model describes processes as a sequence of observable events characterised by distinguishable actions in time. Conformance checking allows these models to determine whether any sequence of temporally ordered and fully-observable events complies with their prescriptions. The latter aspect leads to Explainable and Trustworthy AI, as we can immediately assess the flaws in the recorded behaviours while suggesting any possible way to amend the wrongdoings. Recent findings on conformance checking and temporal learning lead to an interest in temporal models beyond the usual business process management community, thus including other domain areas such as Cyber Security, Industry 4.0, and e-Health. As current technologies for accessing this are purely formal and not ready for the real world returning large data volumes, the need to improve existing conformance checking and temporal model mining algorithms to make Explainable and Trustworthy AI more efficient and competitive is increasingly pressing. To effectively meet such demands, this paper offers KnoBAB, a novel business process management system for efficient Conformance Checking computations performed on top of a customised relational model. This architecture was implemented from scratch after following common practices in the design of relational database management systems. After defining our proposed temporal algebra for temporal queries (xtLTL$_f$), we show that this can express existing temporal languages over finite and non-empty traces such as LTL$_f$. This paper also proposes a parallelisation strategy for such queries, thus reducing conformance checking into an embarrassingly parallel problem leading to super-linear speed up. This paper also presents how a single xtLTL$_f$ operator (or even entire sub-expressions) might be efficiently implemented via different algorithms, thus paving the way to future algorithmic improvements. Finally, our benchmarks highlight that our proposed implementation of xtLTL$_f$ (KnoBAB) outperforms state-of-the-art conformance checking software running on LTL$_f$ logic.

Keywords: logical artificial intelligence; knowledge bases; query plan; temporal logic; conformance checking; temporal data mining; intraquery parallelism

Citation: Bergami, G.; Appleby, S.; Morgan, G. Quickening Data-Aware Conformance Checking through Temporal Algebras. *Information* 2023, 14, 173. https://doi.org/10.3390/info14030173

Academic Editor: Peter Revesz

Received: 14 November 2022
Revised: 3 March 2023
Accepted: 5 March 2023
Published: 8 March 2023

Copyright: © 2023 by the authors. Licensee MDPI, Basel, Switzerland. This article is an open access article distributed under the terms and conditions of the Creative Commons Attribution (CC BY) license (https://creativecommons.org/licenses/by/4.0/).

1. Introduction

(Temporal) conformance checking is increasingly at the heart of ARTIFICIAL INTELLIGENCE activities: due to its logical foundation, assessing whether a sequence of distinguishable events (i.e., a *trace*) does not conform to the expected process behaviour (*process model*) reduces to the identification of the specific unfulfilled temporal patterns, represented as logical *clauses*. This leads to Explainable AI, as the process model's violation becomes apparent. Clauses are the instantiation of a specific behavioural pattern (i.e., *template*) that expresses temporal correlation between actions being carried out (*activations*) and their expected results (*targets*). These, therefore, differ from traditional association rules [1], as they can also describe complex temporal requirements: e.g., whether the target should

immediately follow (ChainResponse) or precede (ChainPrecedence) the activation, if the former might happen any time in the future (Response), or if the target should have never happened in the past (Precedence). These temporal constraints can be fully expressed in a LINEAR TEMPORAL LOGIC OVER FINITE TRACES (LTL$_f$) and its extensions; this logic is referred to as linear as it assumes that, in a given sequence of events of interest, only one possible future event exists immediately following a given one. This differs from stochastic process modelling, where each event is associated with a probabilistic distribution of possibly following events [2,3]. Such a formal language can be extended to express correlations between activations and targets through binary predicates correlating data payloads. Events are also associated with either an action or a piece of state information represented as an *activity label*. Collections of traces are usually referred to as *log*.

Despite its theoretical foundations, state-of-the-art conformance checking techniques for entire logs expose sub-optimal run-time behaviour [4]. The reasons are the following: while performing conformance checking over relational databases requires computing costly aggregation conditions [5], tailored solutions do not exploit efficient query planning and data access minimisation, thus requiring scanning the traces multiple times [6]. Efficiency becomes of the uttermost importance after observing that conformance checking's run-time enhancement has a strong impact on a whole wide range of practical use case scenarios (Section 1.1). To make conformance checking computations efficient, we synthesise temporal data derived from a system (be it digital or real) to a simplified representation in the Relational Database model. In this instance, we use xtLTL$_f$, our proposed extension of LTL$_f$, to represent process models. While the original LTL$_f$ merely asserts whether a trace is *conformant* to the model, our proposed algebra returns all of the traces satisfying the temporal behaviour, as well as activated and targeted events. As a temporal representation in the declarative model provides a point-of-relativity in the context of correctness (i.e., time itself may dictate if traces maintain correctness throughout the logical declarations expressed by the model), the considerations of such temporal issues significantly increase the checking requirement. This is due to a need to visit logical declarations for correctness in the context of each temporal instance.

This paper extends our previous work [4], where we clearly showed the disruptiveness of the relational model for efficiently running temporal queries outperforming state-of-the-art model checking systems. While our original work [4] provided just a brief rationale behind the success of KnoBAB (The acronym stands for KNOWLEDGE BASE FOR ALIGNMENTS AND BUSINESS PROCESS MODELLING). The Business Process Mining literature often uses the term Knowledge Base differently from customary database literature: while in the former, the intended meaning is a customary relational representation for trace data, in the latter, we often require that such representation provides a machine-readable representation of data in order to infer novel facts or to detect inconsistencies., this paper wants to dive deep into each possible contribution leading to our implementation success.

1. As an extension from our previous work, we fully formalise the logical data model (Section 3.1) and characterise the physical one (Section 4) in order to faithfully represent our log. This will prelude the full formalisation of the xtLTL$_f$ algebra;
2. Contextually, we also show for the first time that the xtLTL$_f$ algebra (Section 3.2) can not only express declarative languages such as Declare [7] as in our previous work but can express the semantics of LTL$_f$ formula by returning any non-empty finite trace satisfying the latter if loaded in our relational representation (see Appendix A.2). We also show for the first time a formalisation for data correlation conditions associated with binary temporal operators;
3. Differently from our previous work, where we just hinted at the implementation of each operator with some high level, we now propose different possible algorithms for some xtLTL$_f$ operators (Section 6), and we then discuss both theoretically (Supplement II.2) and empirically (Section 7.1) which might be preferred under different trace length ϵ or log size $|\mathcal{L}|$ conditions. This leads to the definition of hybrid algorithms [8];

4. Our benchmarks demonstrate that our implementation outperforms conformance checking techniques running on both relational databases (Section 7.2) and on tailored solutions (Section 7.5) when customary algorithms are chosen for implementing xtLTL$_f$ operators;
5. Finally, this paper considerably extends the experimental section from our previous work. First, we show (Section 7.3) how the query plan's execution might be parallelised, thus further improving with super-linear speed-up our previous running time results. Then, we also discuss (Section 7.4) how different data accessing strategies achievable through query rewriting might affect the query's running time.

Figure 1 provides a graphical depiction of this paper's table of contents, with the mutual dependencies between its sections. Appendices and Supplements start from p. 50.

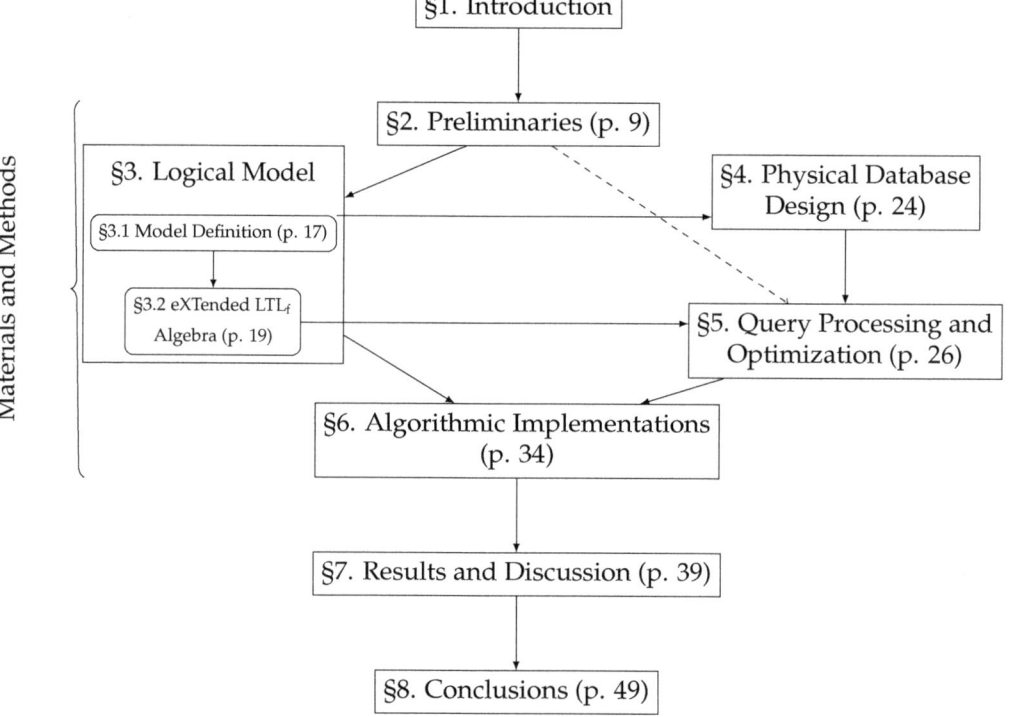

Figure 1. Table of Contents.

Figure 2 provides a bird-eye view of the overall KnoBAB architecture: in the upper half, we show how a log is loaded in our business process management system as a series of distinct tables providing some activity statistics (CountingTable) and full payload information (AttributeTable) in addition to reconstructing the unravelling of the events as described by their traces (ActivityTable). On the other hand, the lower half shows the main steps of the query engine transforming a declarative model into a DAG query plan accessing the previously-loaded relational tables. The most recent version of our system is on GitHub (https://github.com/datagram-db/knobab as accessed the 5 March 2023). When not explicitly stated, all the links were last accessed the 5 March 2023.

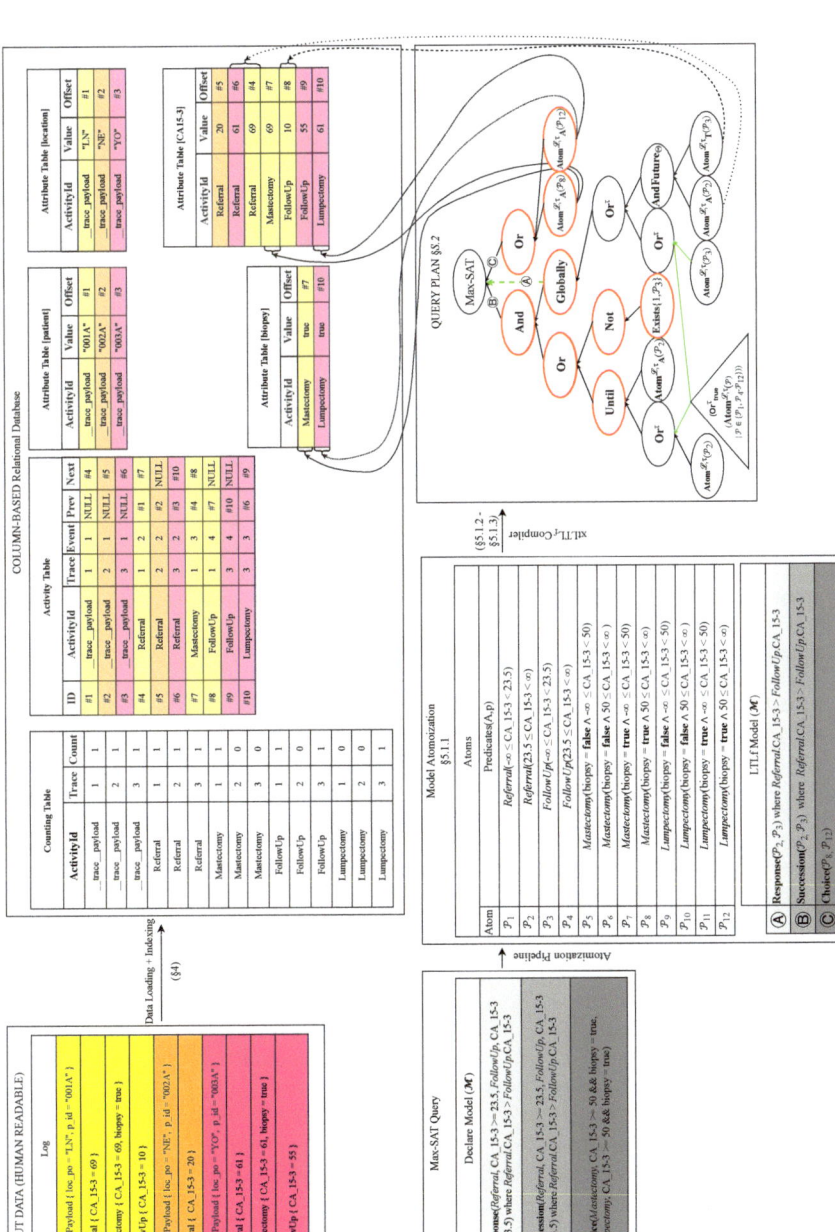

Figure 2. KnoBAB Architecture for Breast Cancer patients. Each trace ①–③ represents one single patient's clinical history, represented with unique colouring, while each Declare clause Ⓐ–Ⓒ prescribes a temporal condition that such traces shall satisfy. Please observe that the atomisation process does not consider data distribution but rather partitions the data space as described by the data activation and target conditions. In the query plan, green arrows indicate access to shared sub-queries as in [9], and thick red ellipses indicate which operators are untimed.

1.1. Case Studies

The present section shows a broad-ranging set of real case studies requiring efficient conformance checking computations in LTL$_f$. This, therefore, motivates the need for our proposed approach in a practical sense.

1.1.1. Cyber Security

Intrusion detection for cyber security aims at auditing an environment for identifying potential flaws that can be remedied and fixed later. While *anomaly-based* approaches raise an alarm if the observed behaviour differs significantly from the expected one, *signature-based* approaches check whether attack patterns might be recognised from the environment. The latter are often used to mitigate the high false-alarm rates of the former [10]. Expected behaviour might be encoded as process models expressed in LTL$_f$, which, when violated, lead to the detection of an attack: such a language can be directly exploited to represent several different kinds of attacks, such as Denial Of Service, Buffer Overflows, and Password Guessing [10]. In his dissertation [11], Ray shows how malware can be detected by determining LTL$_f$ formulae discriminating between system–calls patterns generated by malicious software from expected run-time behaviour. Recent developments [12,13] showed that it is possible to perform prediction (and therefore reasoning) on potentially infinite sequences by analysing a finite subsequence of the overall behaviour within a sliding window; Buschjäger et al. [12] predict future events not covered by the sliding window by correlating them to the patterns observed in such a window. By associating a positive label to each finite subsequence preceding or containing an attack, and a negative one otherwise, we can also extract temporal models detecting subsequences containing attacks [14]. This entails that real-time verification boils down, to some extent, to offline monitoring, as we guarantee that it is sufficient to analyse currently-observed behaviours to predict and detect an attack. The learned model, once validated, can be exploited in the aforementioned real-time verification systems [10].

Example 1. *The Cyber Kill Chain® framework (https://www.lockheedmartin.com/en-us/capabilities/cyber/cyber-kill-chain.html as accessed the 5 March 2023) allows the identification and prevention of intrusion activities on computer systems. This framework is based on a military tactic simply known as a kill chain (https://en.wikipedia.org/wiki/Kill_chain, 5 March 2023), which breaks down the attack into the following phases: target identification, marshalling and organizing forces towards the target, starting an attack, and target neutralisation. Lockheed Martin reformulated these steps to be transferred to the IT world and redirected the attack against a virtual target. These phases were reformulated as follows:*

Reconnaissance (*rec*): *An attacker observes the situation from the outside in order to identify targets and tactics. As the attacker mainly collects information regarding the system's vulnerabilities, this is the hardest part to detect.*

Weaponisation (*weap*): *After gathering the information, the cybercriminal implements his strategy through a software artefact. This detection will have greater chances of success in the future after post-mortem analysis, when either a temporal model is mined over the collected attack data or the strategy is directly inferred from available artefacts (e.g., malaware).*

Payload or Delivery (*del*): *The cybercriminal devises a way to infiltrate the host system that hides the previously produced artefact (e.g., a Trojan). This must sound as harmless as possible to fool the system.*

Exploitation (*expl*): *The cybercriminal exploits the system's vulnerabilities and infiltrates it through the previous "cover". At this stage, the defensive system should raise the alarm if any kind of unusual behaviour is detected while increasing the security level.*

Installation (*inst*): *The weapon escapes the payload and gets installed into the host computer system. At this point, any kind of suspected behaviour might be detected by malicious system calls.*

Command & Control (*comm*): *The weapon establishes a communication with the cybercriminal for receiving orders from the attacker. The system should detect any kind of suspicious network communication and should attempt to break the communication channel.*

Action (*act*): *The intruder starts the attack on the system. At this stage, the attack should be more evident, and the **Industrial IoT Shields** (iiot_sh), such as network devices protection, should be activated.*

Figure 3a describes the actions (and therefore activity labels) of interest. Having defined the actions that should be monitored, records of activities can be stored as traces within a log. This is represented in Figure 3b, where we define three distinct attacks as distinct traces ($\sigma_1, \sigma_2, \sigma_3$). Each trace contains the event information leading up to the completion of an attack attempt (which may be (un)successful). Data payload information is also considered, and here this is provided as the unique timestamp (**ts**) associated with each event. Trace payload information is not simulated here but is described and applied in Example 2.

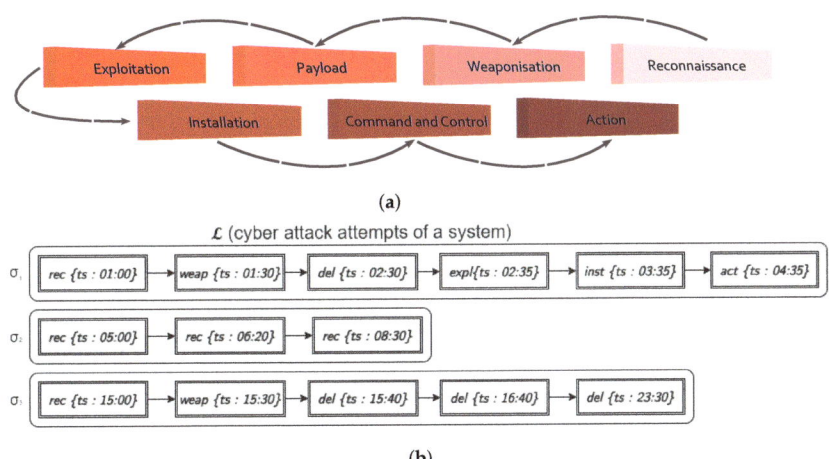

Figure 3. We can express a cyber-security scenario by considering (**a**) possible situations in a Cyber Kill Chain, than are then (**b**) represented in the activity labels' names associated to the events.

A temporal model might describe a completely successful attack. The occurrence of the aforementioned phases can be described through a temporal declarative language Declare [7], where each constraint is an instantiated Declare clause (see Table 1). Our declarative language should be able to state the following requirements: Ⓐ all reconnaissance events should be followed by a weaponisation, Ⓑ there should be no IoT shields in place, and Ⓒ either command and control or action should occur.

On blockchains, each trace event represents a proper blockchain event, thus including function or event invocations issued by one or more smart contracts. In particular, smart contracts are sets of conditions specified in self-executing programs [15], which include protocols within which the parties will fulfil some promises [16]. Given that smart contracts can also be seen as postconditions activated upon the occurrence of specified pre-conditions [17], they are also exploited as security measures reducing malicious and accidental exceptions [15]. As per previous considerations, we can directly encode the smart contract premises in LTL_f, as well as represent the whole smart contract as a whole LTL_f formula under the assumption that the blockchain guarantees its execution [17]. Therefore, we can perform post-mortem analysis checking whether a given run-time abides by the rules imposed by the system.

Table 1. Declare templates illustrated as exemplifying clauses. $A \wedge p$ ($B \wedge q$) represents the *activation* (*target*) condition, A (B) denotes the activity label, and p (q) is the data payload condition.

Type	Exemplifying Clause (c_l)	Natural Language Specification for Traces	LTL$_f$ Semantics ($[\![c_l]\!]$)
Simple	Init(A, p)	The trace should start with an activation	$A \wedge p$
	Exists(A, p, n)	Activations should occur at least n times	$\begin{cases} \Diamond(A \wedge p \wedge \bigcirc([\![\text{Exists}(A, p, n-1)]\!])) & n > 1 \\ \Diamond(A \wedge p) & n = 1 \end{cases}$
	Absence($A, p, n+1$)	Activations should occur at most n times	$\neg[\![\text{Exists}(A, p, n+1)]\!]$
	Precedence(A, p, B, q)	Events preceding the activations should not satisfy the target	$\neg(B \wedge p) \, \mathcal{W} \, (A \wedge p)$
(Mutual) Correlation	ChainPrecedence(A, p, B, q)	The activation is immediately preceded by the target.	$\Box(\bigcirc(A \wedge p) \Rightarrow (B \wedge q))$
	Choice(A, p, A', p')	At least one of the two activation conditions must appear.	$\Diamond(A \wedge p) \vee \Diamond(A' \wedge p')$
	Response(A, p, B, q)	The activation is either followed by or simultaneous to the target.	$\Box((A \wedge p) \Rightarrow \Diamond(B \wedge q))$
	ChainResponse(A, p, B, q)	The activation is immediately followed by the target.	$\Box((A \wedge p) \Rightarrow \bigcirc(B \wedge q))$
	RespExistence(A, p, B, q)	The activation requires the existence of the target.	$\Diamond(A \wedge p) \Rightarrow \Diamond(B \wedge q)$
	ExclChoice(A, p, A', p')	Only one activation condition must happen.	$[\![\text{Choice}(A, p, A', p')]\!] \wedge [\![\text{NotCoExistence}(A, p, A', p')]\!]$
	CoExistence(A, p, B, q)	RespExistence, and vice versa.	$[\![\text{RespExistence}(A, p, B, q)]\!] \wedge [\![\text{RespExistence}(B, q, A, p)]\!]$
	Succession(A, p, B, q)	The target should only follow the activation.	$[\![\text{Precedence}(A, p, B, q)]\!] \wedge [\![\text{Response}(A, p, B, q)]\!]$
	ChainSuccession(A, p, B, q)	Activation immediately follows the target, and the target immediately preceeds the activation.	$\Box((A \wedge p) \Leftrightarrow \bigcirc(B \wedge q))$
	AltResponse(A, p, B, q)	If an activation occurs, no other activations must happen until the target occurs.	$\Box((A \wedge p) \Rightarrow (\neg(A \wedge p) \, \mathcal{U} \, (B \wedge q)))$
	AltPrecedence(A, p, B, q)	Every activation must be preceded by an target, without any other activation in between	$[\![\text{Precedence}(A, p, B, q)]\!] \wedge \Box((A \wedge p) \Rightarrow \bigcirc(\neg(A \wedge p) \, \mathcal{W} \, (B \wedge q))$
Not.	NotCoExistence(A, p, B, q)	The activation nand the target happen.	$\neg(\Diamond(A \wedge p) \wedge \Diamond(B \wedge q))$
	NotSuccession(A, p, B, q)	The activation requires that no target condition should follow.	$\Box((A \wedge p) \Rightarrow \neg\Diamond(B \wedge q))$

1.1.2. Industry 4.0

Smart factories enable the collection and analysis of data through advanced sensors and embedded software for better decision-making. These enable monitoring each phase of the entire production process in both real-time and domain-specific applications where the safety of both autonomous cyber-physical systems as well as human workers is at stake [18]. This is of the uttermost importance, as both humans and machines cooperate in the same environment where a minimal violation of safety requirements might damage the overall production process, thus reflecting in maintenance costs. This calls for logical-based formal methods providing correctness guarantees [19]. Run-time verification [19] and prediction [13] have started gaining momentum against customary static analysis tools: in fact, *real* complex systems such as factories are often hard to predict and analyse before execution. As run-time verification can be deployed as a permanent testing condition on the environment, Mao et al. [19] show that this approach is complete, thus reducing the complicated model-checking problem into a simpler conformance checking one. PROGRAMMABLE LOGIC CONTROLLERS (PLC) are at the heart of this mechanism, where controllers can make decisions over previously-observed events. PLC work is similar to smart contracts in the previous scenario: at each "scan cycle", the controllers perceive through sensors the status change of the environment (e.g., variations of temperature and pressure). This information is then fed to the internal logic, which, on the other hand, might decide to intervene directly in the environment by sending signals to some actuators (e.g., controlling the pressure and temperature on the system). Due to the similarity of PLC to smart contracts, these might also exploit LTL$_f$ for determining security requirements: when a safety condition is violated, the PLC might activate an alarm while ensuring that the system works within safe operation ranges [19]. Please observe that ptLTL, also defined in [19], is a version of LTL allowing reasoning on past events so as to avoid semi-decidable computations for traces of infinite length, might be still represented through an equivalent LTL$_f$ formula evaluated over a finite sliding window [13] bounded by the first and the latest event. Please observe that the difference between LTL and LTL$_f$ is that only the latter considers traces of finite length.

In some other industrial scenarios, we might be interested in detecting unexpected variations in time series reflecting the fluctuation of some perceived variables (e.g., variations in temperature and pressure). The latest developments [13] showed that (industrial) time series could also be represented as traces: we might assign to each event an activity label ϖ if the current event has a data payload whose values upper bound the ones from immediately preceding event's payload, and $\neg\omega$ otherwise. Consequently, we can encode disparate data variation patterns in LTL_f reflecting different types of data volatility or steep increases/decreases [13]. This shows how LTL_f can also represent *anomaly-based* problems by reducing them to the identification of anomaly patterns [20].

1.1.3. Healthcare

A medical process describes clinical-related procedures as well as organisational management ones (e.g., registration, admission, and discharge) [21]. The renowned openEHR (https://www.openehr.org/ accessed the 5 March 2023) standard distinguishes the former in four main archetypes: an observation, recording patients' clinical symptoms (e.g., body temperature, blood pressure); an evaluation, providing preliminary diagnosis and assessing the patient's health based on the former results; and an instruction, the execution of the treatment plan proposed by a physician (e.g., prescribing, examining, and testing). An action describes the way to intervene or treat medical patients according to the treatment plan (e.g., drug administration, blood matching). Once encoded as such, each process representing an instantiation of a medical process, i.e., a patient's clinical course, can be then collected and represented in a log. As such, each action is going to be represented as a distinct activity label of a given event [22] that might contain relevant payload information recording the outcome of the clinical procedures, as well as demographical information related to the patient [21] for future socio-clinical analyses [23].

Declarative temporal languages such as Declare can then be exploited to provide a descriptive approach specifying temporal constraints among activities without strictly enforcing their order of completion, thus restricting the order of application of a specific set of activities [21]. As these models come with temporal semantics expressed in LTL_f, these are, for all intents and purposes, process models. As such, these might be applied to detect discrepancies between clinical guidelines, expressed by the aforementioned model, and the actual process executions collected in a log. This is of the utmost concern as often deviations represent errors compromising the patient recovery [22], which, if efficiently and identified in advance, lead to an increased patient satisfaction as well a reduction of healthcare costs (e.g., due to mismanagement) [21].

Example 2. *To minimise costs and unrequired procedures, only ill patients should receive treatment. Thus, sufferers not receiving treatment (false negatives) and non-sufferers receiving treatment (false positives) need to be minimised. Figure 2 proposes a simplified scenario where we consider two event payload keys: CA 15-3 (cancer antigen concentration in a patient's blood) and biopsy (biopsies should be taken before any procedure is acted upon). Our model targets only breast cancer patients with successful therapies that describe a medical protocol and the desired patients' health condition at each step. Ⓒ states that two possible surgical operations for breast tumours are mastectomy or lumpectomy if the biopsy is positive and the CA-13.5 is way above (≥ 50) the guard level, being 23.5 units per mL, and Ⓐ–Ⓑ any successful treatment should decrease the CA-13.5 levels, which should be below the guard level; such correlation data condition is expressed via a Θ condition (introduced by a where). A twinned negative model (not in Figure) might better discriminate healthy patients from patients where the therapy was unsuccessful. Novel situations can be represented as a log. For example, in Figure 2, we have three patients: ① a cancer patient with a successful mastectomy, ② a healthy patient, and ③ an unsuccessful lumpectomy, thus suggesting that the patient might still have some cancerous cells. Given the aforementioned model, patient ① will satisfy the model as the surgical operation was successful, ② will not satisfy the model because neither a mastectomy nor a lumpectomy was required ($\mathcal{M}$ is only fulfilled for successful procedures), and ③ will not satisfy the target condition, even though the correlation condition was met. Our model of interest should only return ① as an outcome of the conformance checking process.*

2. Preliminaries

eXtensible Event Stream (XES). This paper relies on temporal data represented as a temporally ordered sequence of events (*trace* or *streams*), where events are associated with at most one action described by a single *activity label* [24]. In this paper, we formally characterize payloads as part of both events and traces while, in our previous work, we only considered payloads from events [25].

Given an arbitrarily ordered set of keys K and a set of values V, a **tuple** [26] is a finite function $p \colon K \to V$ (also $p \in V^K$), where each key is either associated with a value in V or is undefined. After denoting $\bot$ as a null element missing from the set of values ($\bot \notin V$), we can express that κ is not associated with a value in p as $p(\kappa) = \bot$, thus $\kappa \notin \operatorname{dom}(p)$. An empty tuple ε has an empty domain.

(Data) *payloads* are tuples, where values can represent either categorical data or numerical data. An *event* σ^i_j is a pair $\langle \mathsf{a}, p \rangle \in \Sigma \times V^K$, where Σ is a finite set of activity labels, and p is a finite function describing the data payload. A *trace* σ^i is an ordered sequence of distinct events $\sigma^i_1, \ldots, \sigma^i_n$, which is distinguished from the other traces by a case id i; n represents the trace's length ($n = |\sigma^i|$). If a payload is also associated with the whole trace, this can be easily mimicked by adding an extra initial event containing such a payload with an associated label of `__trace_payload`. A *log* $\mathcal{L}$ is a finite set of traces $\{\sigma^1, \ldots, \sigma^m\}$. In this paper, we further restrict our interest to the traces containing at least one event, as empty traces are meaningless as they are not describing any temporal behaviour of interest. Finally, we denote as $\beta \colon \Sigma \leftrightarrow \{1, \ldots, |\Sigma|\}$ the bijection mapping each activity label occurring in the log to an unique id.

Example 3. *The log $\mathcal{L}$ in Figure 2 comprises three distinct traces $\mathcal{L} = \{\sigma^1, \sigma^2, \sigma^3\}$. In particular, the second trace comprises two events $\sigma^2 = \sigma^2_1 \sigma^2_2$, where the first event represents the trace payload, and therefore $\sigma^2_1 = \langle$`__trace_payload`, $p\rangle$ having $p(loc\_po) = NE$ and $p(p\_id) = 002A$. The other event is $\sigma^2_2 = \langle$`Referral`, $\tilde{p}\rangle$, where payload $\tilde{p}$ is only associated with the CA-13.5 levels as $\tilde{p}(CA\text{-}13.5) = 20$. Similar considerations can be carried out for the other log traces.*

Linear Temporal Logic over finite traces (LTL$_f$). LTL$_f$ is a well-established extension of *modal logic* considering the possible worlds as finite traces, where each event represents a single relevant instant of time. The time is thereby linear, discrete, and future-oriented. This entails that that the events represented in each trace are totally ordered and, as LTL$_f$ quantifies only on events reported in the trace, all the events of interest are fully observable. The *syntax* of an well-formed LTL$_f$ formula φ is defined as follows:

$$\varphi ::= \mathsf{a} \mid \neg \varphi \mid \varphi \vee \varphi' \mid \varphi \wedge \varphi' \mid \bigcirc \varphi \mid \Box \varphi \mid \Diamond \varphi \mid \varphi \, \mathcal{U} \, \varphi' \tag{1}$$

where $\mathsf{a} \in \Sigma$. Its *semantics* is usually defined in terms of First Order Logic [27] for a given trace σ^i at a current time j (e.g., for event σ^i_j) as follows:

- An event satisfies the **activity label** a iff. its activity labels is a: $\sigma^i_j \vDash \mathsf{a} \Leftrightarrow \sigma^i_j = \langle \mathsf{a}, p \rangle$;
- An event satisfies the **negated** formula iff. the same event does not satisfy the non-negated formula: $\sigma^i_j \vDash \neg \varphi \Leftrightarrow \sigma^i_j \nvDash \varphi$;
- An event satisfies the **disjunction** of LTL$_f$ sub-formulæ iff. the event satisfies one of the two sub-formulæ: $\sigma^i_j \vDash \varphi \vee \varphi' \Leftrightarrow \sigma^i_j \vDash \varphi \vee \sigma^i_j \vDash \varphi'$;
- An event satisfies the **conjunction** of LTL$_f$ formulæ iff. the event satisfies all of the sub-formulæ: $\sigma^i_j \vDash \varphi \wedge \varphi' \Leftrightarrow \sigma^i_j \vDash \varphi \wedge \sigma^i_j \vDash \varphi'$;
- An event satisfies a formula at the **next** step iff. the formula is satisfied in the incoming event if present: $\sigma^i_j \vDash \bigcirc \varphi \Leftrightarrow i < |\sigma_j| \wedge \sigma^i_{j+1} \vDash \varphi$;
- An event **globally** satisfies a formula iff. the formula is satisfied in all the following events, including the current one: $\sigma^i_j \vDash \Box \varphi \Leftrightarrow \forall j \leq x \leq |\sigma^i|. \sigma^i_x \vDash \varphi$;

- An event **eventually** satisfies a formula iff. the formula is satisfied in either the present or in any future event: $\sigma_j^i \models \Diamond \varphi \Leftrightarrow \exists j \leq x \leq |\sigma^i|.\sigma_x^i \models \varphi$;
- An event satisfies φ **until** φ' holds iff. φ holds at least until φ' becomes true, which must hold at the current or a future position: $\sigma_j^i \models \varphi \, \mathcal{U} \, \varphi' \Leftrightarrow \exists j \leq y \leq |\sigma^i|.\sigma_y^i \models \varphi' \wedge (\forall x \leq z < y.\sigma_z^i \models \varphi)$.

Other operators can be seen as syntactic sugar: Weak-Until is denoted as $\varphi \, \mathcal{W} \, \varphi' := \varphi \, \mathcal{U} \, \varphi' \vee \Box \varphi$, while the implication can be rewritten as $\varphi \Rightarrow \varphi' := (\neg \varphi) \vee (\varphi \wedge \varphi')$. Generally, binary operators bridge activation and target conditions appearing in two distinct sub-formulæ. The semantics associated with activity labels, consistently with the literature on business process execution traces [25], assumes that, in each point of the sequence, one and only one element from Σ holds. We state that a trace σ^i is *conformant* to an LTL$_f$ formula iff. it satisfies it starting from the first event: $\sigma^i \models \varphi \Leftrightarrow \sigma_1^i \models \varphi$, and is *deviant* otherwise [25]. The Declare language described in the next paragraph provides an application for such logic. As relational algebra describes the semantics for SQL [28,29], LTL$_f$ is extensively applied [30] as a semantics for formally expressing temporal and human-readable declarative constraints such as Declare.

At the time of the writing, different authors proposed several extensions for representing data conditions in LTL$_f$. The simplest extensions are *compound conditions* a $\wedge$ q, which are the conjunction of data predicate $q \in \mathbf{Prop}$ to the activity label a [25]. Nevertheless, this straightforward solution is not able to express correlation conditions in the data which might be relevant in business scenarios [31], as representing correlations as single atoms requires decomposing the former into disjunctions of formulae [32]. Despite prior attempts to define a temporal logic expressing correlation conditions, no explicit formal semantics on how this can be evaluated was provided [6]. This poses a problem to the current practitioner, as this hinders the process of both checking formally the equivalence among two languages expressing correlation conditions, as well as providing a correct implementation of such an operator. We, on the other hand, propose a relational representation of xtLTL$_f$, where the semantics of all of the operators, thus including the ones requiring correlation conditions, is clearly laid out and implemented.

Declare. Temporal declarative languages pinpoint highly variable scenarios, where state machines provide complicated graph models that can be hardly understandable by the common business stake-holder [33]. Among all possible temporal declarative languages, we constrain our interest to Declare, originally proposed in [7]. Every single temporal pattern is expressed through *templates* (i.e., an abstract parameterised property: Table 1 column 2), which are parametrised over activation, target, or correlation conditions. Template names induce the semantic representation in LTL$_f$ $[\![c_l]\!]$ of each model clause c_l. Therefore, a trace σ^i is conformant to a Declare clause iff. it satisfies its associated semantic representation in LTL$_f$ ($\sigma^i \models c_l \Leftrightarrow \sigma^i \models [\![c_l]\!]$). At this stage, *activation* (and *target*) *conditions* are predicates A $\wedge$ p (and B $\wedge$ q) in such a clause asserting required properties for the events' activity label (A and B) and payload (p and q). An event in a given trace *activates* (or *targets*) a given clause if they satisfy the activation (or target) condition. Please observe that neither activation nor target conditions postulate the temporal (co)occurrence between activating or targeting events, as this is duty is transferred to the specific LTL$_f$ semantics of the clause. A trace *vacuously satisfies* a clause if the trace satisfies the clause despite no event in the trace satisfied the activation condition. After this, we state that a trace *non-vacuously satisfies* the declarative clause if the trace satisfies the clause and one of the following conditions is satisfied:

- The clause provides no target condition and it exists at least one activating event;
- The clause provides a target condition but no binary (payload) predicate Θ, and the declarative clause establishes a temporal correlation between (at least one) activating event and (at least one) targeting one;
- The clause provides both a target condition and a binary predicate Θ, while the activating and targeting events satisfying the temporal correlation as in the previous

case also satisfy a binary Θ predicate over their payloads; in this situation, we state that the activating and targeting event *match* as they jointly satisfy the *correlation condition* Θ.

Finally, the presence of activating events is a necessary condition for non-vacuous satisfiability.

We can then categorize each Declare template from [30] through these conditions and the ability to express correlations between two temporally distant events happening in one trace: simple templates (Table 1, rows 1–3) only involving activation conditions; (mutual) correlation templates (rows from 4 to 15), which describe a dependency between activation and target conditions, thus including correlations between the two; and negative relation templates (last 2 rows), which describe a negative dependency between two events in correlation. Despite these templates possibly appearing quite similar, they generate completely different finite state machines, thus suggesting that these conditions are not interchangeable (http://ltlf2dfa.diag.uniroma1.it/, 5 March 2023). Figure 4 exemplifies the behavioural difference between two clauses differing only on the template of choice.

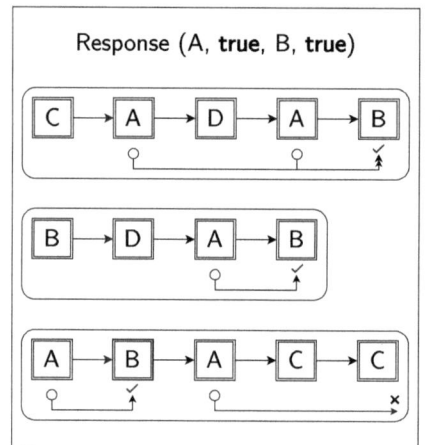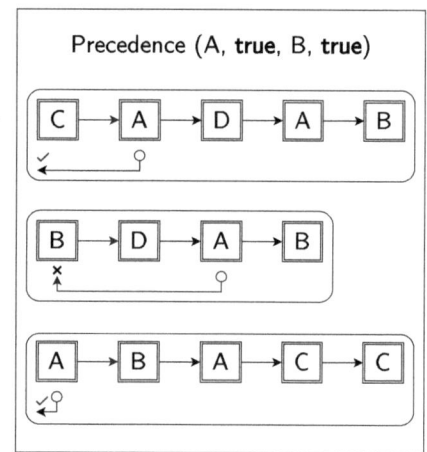

Figure 4. Two exemplifying clauses distinguishing Response and Precedence behaviours. Traces are represented as temporally ordered events associated with activity labels (boxed). Activation (or target) conditions are circled here (or ticked/crossed). Ticks (or crosses) indicate a (un)successful match of a target condition. For all activations, there must be an un-failing target condition; for precedence, we shall consider at most one activation. These conditions require the usage of multiple join tests per trace.

A Declare Model is composed of a set of clauses $\mathcal{M} = \{c_l\}_{l \leq n, n \in \mathbb{N}}$ which have to be contemporarily satisfied in order to be true. A trace σ^i is conformant to a model $\mathcal{M}$ iff. such a trace satisfies each LTL$_f$ formula $[\![c_l]\!]$ associated with the model clause $c_l \in \mathcal{M}$. Consequently, a Declare model can be represented as a finitary conjunction of the LTL$_f$ representation of each of its clauses, $[\![\mathcal{M}]\!] := \bigwedge_{c_l \in \mathcal{M}} [\![c_l]\!]$: for this, the MAXIMUM-SATISFIABILITY PROBLEM (Max-SAT) for each trace counts the ratio between the satisfied clauses over the whole model size. This consideration can be extended later on to also data predicates through predicate atomisation [25], as discussed in the next paragraph.

Relational Models and Algebras. The *relational model* was firstly introduced by Codd [34] to compactly operate over tuples grouped into tables. Such tables are represented as mathematical n-ary relations $\Re$ that can be handled through a relational algebra. Upon the effective implementation of the first RELATIONAL DATABASE MANAGEMENT SYSTEMS (RDBMS), such algebra expressed the semantics of the well-known declarative query language, SQL. The rewriting of SQL in algebraic terms allowed the efficient execution of the declarative queries through abstract syntax tree manipulations [28]. Our proposed xtLTL$_f$ (Section 3.2) takes inspiration from this historical precedent, in order to run conformance checking and temporal model mining queries over an relational representation of the log via relational tables (Section 3.1).

More recently, *column-oriented DBMS* such as MonetDB [35] proposed a new way to store data tables: instead of representing these per row, these were stored by column. There are several advantages to this approach, including better access to data when querying only a subset of columns (by eliminating the need to read columns that are not relevant) as well as discarding null-valued cells. This is achieved by representing each relation $\Re(\mathtt{id}, A_1, \ldots, A_n)$ in the database schema as distinct binary relations $\Re_{A_i}(\mathtt{id}, A_i)$ for each attribute A_i in $\Re$. As this decomposition guarantees that the full-outer natural join $\bowtie_{1 \leq i \leq n} \Re_{A_i}$ over the decomposed tables is equivalent to the initial relation $\Re$, we can avoid representing NULL values in each single binary relation, thus limiting our space allocation to the values effectively present in the data. We therefore took inspiration from this intuition for representing the payload information, thus storing one single table per payload attribute. To further optimise the query engine, it is also possible to boost the query performance by guaranteeing that the results always have a fixed schema, mainly listing the record ids satisfying the query conditions [36]. As we will see while introducing our temporal operators (Section 3.2), we will also guarantee that each operator returns the output in the same schema, thus guaranteeing time and memory optimality.

Finally, the *nested relational model* [37] extends the relational model by relaxing its first normal form (1NF), thus allowing table cells to contain tables and relations as values. Relaxing this 1NF allows for storing data in a hierarchical way in order to access an entire sub-tree with a single read operation. We will leverage this representation for our intermediate result representation, in order to associate multiple activation, target, or correlation conditions to one single event, thus including any relevant future event occurring after it.

Common Subquery Problem. Query caching mechanisms [38] are customary solutions for improving query runtime by holding partially-computed results in temporary tables referred to as materialised views, under the assumption that the queries sharing common data are pipelined [39]. Recently, Kechar et al. [9] proposed a novel approach that can also be run when queries are run contemporarily: it is sufficient to find the shared subqueries before actually running them so that, when they are run, their result is stored into materialised views thus guaranteeing that these are computed at most once.

Example 4. *Figure 2 shows how this idea might be transferred to our use case scenario requiring running multiple declarative clauses:* RESPONSE *is both a subquery of* SUCCESSION *as well as a distinct declarative clause of interest. Green arrows indicate operators' output shared among operators expressed in our proposed xt LTL$_f$ extension of xt LTL$_f$. Please also observe that operators with the same name and arguments but marked either with activation, target, or no specification are considered different as they provide different results, and therefore are not merged together. This includes distinctions between timed and untimed operators, which will be discussed in greater detail in Section 3.2.*

To further minimize tables' access times, it is possible to take this reasoning to its extreme by minimising the data access per data predicate in order to avoid accessing the same table multiple times. In order to do so, we need to partition the data space according to the queries at our disposal as in our previous work [25]. This process can be eased if we assume that each payload condition p and p' for the declarative clauses within a model $\mathcal{M}$ is represented in Disjunctive Normal Form (DNF) [40]: in this scenario, data predicates q are in DNF if they are a disjunction of one or more conjunctions of one or more data intervals referring to just one payload key.

Example 5. *The model illustrated in Figure 3a and discussed in former Example 1 comes with data conditions associated with neither activation nor target conditions. Therefore, no atomisation process is performed. Thus, each event in a log might just be distinguished by its activity label [25].*

Given an LTL$_f$ expression φ containing compound conditions, we denote $\mathcal{D}_\varphi$ as the set of distinct compound conditions in φ. We refer to the items in $\mathcal{D}_\varphi$ as *atoms* iff. for each pair of distinct compound conditions in it, they never jointly satisfy any possible payload p (More formally, $\forall p. \forall a \in \Sigma. \forall a \wedge q, a \wedge q' \in \mathcal{D}_\varphi.(q \neq q') \Rightarrow (q(p) \Rightarrow \neg q'(p)))$. Ref. [25] shows a procedure showing how any formula φ can be rewritten into an equivalent one φ' by ensuring that $\mathcal{D}_{\varphi'}$ contains atoms. This can be achieved by constructing $\mathcal{D}_{\varphi'}$ first from φ (Algorithm 1), and then converting each compound conditions in φ as disjunctions of atoms in $\mathcal{D}_{\varphi'}$, thus obtaining φ'.

Algorithm 1 Atomisation: $\mathcal{D}_\varphi$-encoding pipeline.

1: **global** $\mu \leftarrow \{\}; ad \leftarrow \{\}; ak \leftarrow \{\}$

2: **procedure** COLLECTINTERVALS(a, DNF)   ▷ DNF:= $\bigvee_{1 \leq i \leq n} \bigwedge_{1 \leq k \leq m(i)} low_{i,k} \leq k_{i,k} \leq up_{i,k}$
3:    **for all** $conj \in$ DNF **and** $low \leq k \leq up \in conj$ **do**
4:      $\mu(a,k)$.put$([low, up])$
5:    **end for**

6: **procedure** COLLECTINTERVALS($\mathcal{M}$)   ▷ $\mathcal{M} := \bigwedge_{1 \leq i \leq |\mathcal{M}|}$ clause$_i(A, p, B, p')$
7:    **for all** clause$_i(A, p, B, p') \in \mathcal{M}$ **do**
8:      **if** $p \neq$ **True then** COLLECTINTERVALS(A, p)
9:      **if** $p' \neq$ **True then** COLLECTINTERVALS(B, p')
10:    **end for**

11: **procedure** $\mathcal{D}_\varphi$-ENCODING( )
12:    **for all** $a \in \Sigma$ **do**
13:      **for all** $k \in K$ **do**
14:        $\mu(a,k) \leftarrow$ SEGMENTTREE($\mu(a,k)$)
15:      **end for**
16:      **for all** $partition \in \bigtimes_{k \in K} \mu(a,k).$elementaryIntervals() **do**   ▷ $partition := (low_k \leq k \leq up_k)_{k \in K}$
17:        $p_i \leftarrow$ **new** atom()
18:        $p_i := a \wedge partition$
19:        $ak(a)$.put(p_i)
20:        **for all** $low_k \leq k \leq up_k \in partition$ **do**
21:          $ad(a, low_k \leq k \leq up_k)$.put$(p_i)$
22:        **end for**
23:      **end for**
24:      **if** $ak(a) = \emptyset$ **then**
25:        $ak(a) \leftarrow \{a\}$
26:      **end if**
27:    **end for**

We collect all the conjunctions referring to the same payload key into a map $\mu(a, \kappa)$ (Line 4). After doing so, we can construct a Segment Tree [41] from the intervals in $\mu(a, \kappa)$, thus identifying the *elementary intervals* partitioning the collected intervals (Line 14). These elementary intervals also partition the payload data space associated with events for each activity label a. This can be achieved by combining each elementary interval in each dimension κ for a (Line 16) and then associating it with a new atom representing such a partition (Line 18) that is then guaranteed to be an atom by construction. This entails that each interval $low_\kappa \leq \kappa \leq up_\kappa$ will be characterised by the disjunction of all of the atoms p_i comprising such interval (Line 21). Given this, we can then associate to each activation condition A that is associated with an activation payload condition p the disjunction of atoms that are collected by the following formula:

$$\text{Atom}_{\mu,ad}(\mathsf{A}, p) := \bigcup_{conj \in p} \bigcap_{(low \leq \kappa \leq up) \in conj} \bigcup_{I \in \mu(\mathsf{A},\kappa).\texttt{findElementaryIntervals}(low,up)} ad(\mathsf{A}, I) \quad (2)$$

If we assume that the dimension of $\mu(\mathsf{a}, \kappa)$ for each $\mathsf{a} \in \Sigma$ and $\kappa \in K$ is at most m, our implementation available at https://github.com/datagram-db/knobab/blob/main/incl ude/yaucl/structures/query_interval_set/structures/segment_partition_tree.h (5 March 2023) builds such trees in $\sum_{1 \leq i \leq m} \log(i) + m \in O(m \cdot \log(m))$ time, as we first insert the intervals into the data structure and then we guarantee to minimise the tree representation, requiring a linear visit cost to the whole tree data structure. The time complexity of $\mathcal{D}_\varphi$-ENCODING() is $m|K|(1 + \log m + |\Sigma|) \in O(m|K||\Sigma|)$.

Example 6. *Each distinct payload conditions associated with either activation or target conditions in Figure 2 can be expressed as one single atom, as there are no overlapping data conditions associated with the same activity label, and each data condition can be mapped into one single elementary interval associated with an activity label. The next example will provide another use case example and a different model on the same dataset leading to a decomposition of payload conditions into a disjunction of several atoms. Table 2 shows the partitioning of the data payloads associated with each activity label in the log by the elementary interval of interest.*

Table 2. Definition of the atoms from Figure 2 in terms of partitioning over the elementary intervals.

Referral	CA-15.3 < 23.5	CA-15.3 ≥ 23.5
	p_1	p_2
Mastectomy	CA-15.3 < 50	CA-15.3 ≥ 50
biopsy = false	p_5	p_6
biopsy = true	p_7	p_8
FollowUp	CA-15.3 < 23.5	CA-15.3 ≥ 23.5
	p_3	p_4
Lumpectomy	CA-15.3 < 50	CA-15.3 ≥ 50
biopsy = false	p_9	p_{10}
biopsy = true	p_{11}	p_{12}

Example 7. *Let us suppose to return all the false negative and false positive* Mastectomy *cases that are not caused by data imputation errors. For this, we want to obtain all of the negative biopsies having CA15.3 levels greater than the guard level of 50 and positive biopsies having CA15.3 below the same threshold. Under the assumption that biopsy values were imputed through numerical numbers thus leading to more imputation errors, we are ignoring cases where both CA15.3 and biopsy values are out of scale, that is, we want to ignore the data where CA15.3 levels are negative or above 1000, and where the biopsy values are neither true (1.0) nor false (0.0). For this, we can outline the following model:*

$$\mathcal{M}' = \{\textit{Choice}(\textsf{Mastectomy}, \textit{biopsy} = 0.0 \wedge \textit{CA15.3} \geq 50, \textsf{Mastectomy}, \textit{biopsy} = 1.0 \wedge \textit{CA15.3} < 50),$$
$$\textit{Absence}(\textsf{Mastectomy}, \textit{CA15.3} > 1000 \vee \textit{CA15.3} < 0),$$
$$\textit{Absence}(\textsf{Mastectomy}, \textit{biopsy} \neq 1.0 \vee \textit{biopsy} \neq 0.0)\}$$

(3)

This implies that we are interested in decomposing the intervals pertaining to both CA-15.3 and biopsy into elementary intervals: Table 3a shows that only CA-15.3 < 50 and CA-15.3 ≥ 50 are decomposed into two elementary intervals, as the former also includes the range CA-15.3 < 0, while the latter also includes CA-15.3 > 1000. Elementary intervals not occurring in the initially collected ones are not reported in this graphical representation. Table 3b shows the partitioning of the Mastectomy *data payload induced by the elementary intervals of interest; the former data conditions can be now rewritten after Equation* (2) *in the Supplement as follows:*

1. $\bigvee Atom_{\mu,ad}(\text{Mastectomy}, biopsy=0.0 \land CA15.3 \geq 50) = p_{12} \lor p_{17}$
2. $\bigvee Atom_{\mu,ad}(\text{Mastectomy}, biopsy=1.0 \land CA15.3 < 50) = p_4 \lor p_9$
3. $\bigvee Atom_{\mu,ad}(\text{Mastectomy}, CA15.3 > 1000 \lor CA15.3 < 0) = p_1 \lor \cdots \lor p_5 \lor p_{16} \lor \cdots \lor p_{20}$
4. $\bigvee Atom_{\mu,ad}(\text{Mastectomy}, biopsy \neq 0.0 \lor biopsy \neq 1.0) = p_1 \lor p_3 \lor p_5 \lor p_6 \lor p_8 \lor p_{10} \lor p_{11} \lor p_{13} \lor p_{15} \lor p_{16} \lor p_{18} \lor p_{20}$

where each atom is defined as a conjunction of compound conditions defined upon the previously collected elementary intervals. Some examples are then the following:

- $p_1 := biopsy < 0 \land CA\text{-}15.3 < 0$
- $p_2 := biopsy = 0 \land CA\text{-}15.3 < 0$

This decomposition will enable us to reduce the data access time while scanning the tables efficiently.

Table 3. Intermediate steps to generate distinct atoms for the Referral data predicates from Example 7.

(a) Interval decomposition in *basic intervals* $\mu(\text{Mastectomy}, \cdot)$.			
$\mu(\text{Mastectomy}, CA\text{-}15.3)$			
$CA\text{-}15.3 < 0$	$CA15.3 < 0$		
$CA\text{-}15.3 < 50$	$CA15.3 < 0,$	$0 \leq CA\text{-}15.3 < 50$	
$CA\text{-}15.3 \geq 50$	$50 \leq CA15.3 \leq 1000,$	$CA\text{-}15.3 > 1000$	
$CA\text{-}15.3 > 1000$	$CA15.3 > 1000$		
$\mu(\text{Mastectomy}, biopsy)$			
$biopsy = 0$	$biopsy = 0$		
$biopsy = 1$	$biopsy = 1$		
$biopsy \neq 0$	$biopsy < 0,$	$0 < biopsy < 1,$	$biopsy = 1,$	$biopsy > 1$
$biopsy \neq 0$	$biopsy < 0,$	$biopsy = 0,$	$0 < biopsy < 1,$	$biopsy > 1$

(b) Atom generation by partitioning the data space $\bigtimes_{\kappa \in K} \mu(\text{Mastectomy}, \kappa).\texttt{elementaryIntervals}()$ with $K = \{ biopsy, CA\text{-}15.3 \}$.

	$biopsy < 0$	$biopsy = 0$	$0 < biopsy =< 1$	$biopsy = 1$	$biopsy > 1$
$CA15.3 < 0$	p_1	p_2	p_3	p_4	p_5
$0 \leq CA15.3 < 50$	p_6	p_7	p_8	p_9	p_{10}
$50 \leq CA15.3 \leq 1000$	p_{11}	p_{12}	p_{13}	p_{14}	p_{15}
$CA15.3 > 1000$	p_{16}	p_{17}	p_{18}	p_{19}	p_{20}

Further Notation. We represent relational tables as a sequence of records indexed by id as per the physical relational model: given a relational table T, $T[i]$ represents the i-th record in T counting from 1. We denote $f = [x \mapsto y, z \mapsto t]$ as a finite function such that $f(x) = y$ and $f(z) = t$. Table 4 collects the notation used throughout the paper.

Table 4. Table of Notation for symbols $\chi \in \mathcal{T}$ defined as $(\chi := \mathcal{E})$ or characterised by $(\mathcal{E}(\chi))$ $\mathcal{E}$.

Symbol (χ)	Definition ($\mathcal{E}$)	Type ($\mathcal{T}$)	Comments
		Set Theory	
$\emptyset$		Set	An empty set contains no items.
$(\preceq, S)$			A partially ordered set (*poset*) is a relational structure for which $\preceq$ is a partial ordering over S [40]. $\preceq$ over S might be represented as a lattice, referred to as the Hasse diagram.
$\top_S$	$\forall a \in S. a \preceq \top_S$	S	Given a poset $(\preceq, S)$, $\top_S$ is the unique greatest element of S.
$\complement C$	$\mathcal{U} \setminus C$	Set	*Complement* set: given an universe $\mathcal{U}$, the complement returns all of the elements that do not belong to C.
$\bigtimes_{\kappa \in K} f(\kappa)$	$f(\kappa_1) \times \cdots \times f(\kappa_n)$	$\text{dom}(f)^{\vert K \vert}$	*Generalised cross product* for ordered sets K where $\kappa_1 \prec \cdots \prec \kappa_n$
$\vert C \vert$	$\sum_{c \in C} 1$	$\mathbb{N}$	The *cardinality* of a finite set indicates the number of contained items.
$\wp(C)$	$\{T \mid T \subseteq C\}$	Set	The *powerset* of C is the set whose elements are all of the subsets of C.
		XES Model & LTL$_f$	
Σ		Set	Finite set of activity labels
K		Set	Finite set of ordered (payload) keys, κ
V		Set	Finite set of (payload) values
p	$[\kappa_1 \mapsto v_1, \ldots]$	V^K	Tuple (or finite function) mapping keys $\kappa_1 \in K$ to values in $v_1 \in V$
$\bot$	$\bot \notin V$		NULL value
σ_j^i	$\langle p, a \rangle$	$\Sigma \times V^K$	Event
σ^i	$\sigma_1^i, \ldots, \sigma_n^i$	Sequence	Trace, sequence of temporarily ordered events.
$\mathcal{L}$	$\{\sigma^1, \ldots, \sigma^m\}$	Set	Log, set of traces.
β		$\Sigma \leftrightarrow \{1, \ldots, \vert \Sigma \vert\}$	Bijection mapping each activity label to its unique identifier.
φ	Equation (1)	Expression	An LTL$_f$ expression.
$\vDash$			$\Gamma \vDash \varphi$ denotes that φ is *satisfied* for the world/environment Γ.
		xtLTL$_f$	
ψ	Section 3.2	Expression	eXTended LTL$_f$ Algebra expression.
$A(k)/T(k)/M(h,k)$		ω	Marks associated with activation/target/matching conditions.
ρ	$\{\langle i, j, L \rangle, \ldots\}$	$\Omega = \{\wp(\mathbb{N} \times \mathbb{N} \times S) \mid S \in \wp(\omega)\}$	*Intermediate representation* returned by each xtLTL$_f$ operator
$T[i]$		$T[i] \in T$	Accessing the i-th record of a sequence T.
$\Theta(x, y)$		Binary Predicate	*Correlation condition* between activated and targeted events.
$\Theta^{-1}(y, x)$	$\Theta(x, y)$	Binary Predicate	*Inverted/Flipped* correlation condition.
True		Binary Predicate	Always-true binary predicate.
$E_\Theta^i(M_1, M_2)$	Equation (S1)	Algorithm 7	*Existential matching condition* for which there exists at least one event in M_1, M_2 providing a match.
$A_\Theta^i(M_1, M_2)$	Equation (S2)	Algorithm 9	*Universal matching condition* returning a non-empty set if each event expressed in the maps M_1, M_2 provides a match.
$\mathcal{T}_\Theta^{F,i}(M_1, M_2)$	Equation (S3)	$\mathcal{T}_\Theta^{F,i}(M_1, M_2) \in \wp(\omega) \cup \{\textbf{False}\}$	*Testing functor* returning **False** iff., despite the maps containing activated and targeted events, the matching condition $F_\Theta^i(M_1, M_2)$ is empty. It returns $F_\Theta^i(M_1, M_2)$ otherwise.
		Pseudocode	
$\uparrow$			Null pointer or terminated iterator.
Iterator(ρ)		POINTER	On ρ non-empty, it returns the iterator pointing to the first record in ρ
current(it)		DEREFERENCE	Element pointer by the pointer/iterator it.
LOWERBOUND(d, b, e, ν)		BINARY SEARCH	Given a beginning b and end e iterator range within a sequential and sorted data structure by increasing order, LOWERBOUND returns either the first location in this range pointing at a value greater or equal to ν or e otherwise.
UPPERBOUND(d, b, e, ν)		BINARY SEARCH	Given a beginning b and end e iterator range within a sequential and sorted data structure by increasing order, UPPERBOUND returns either the first location in this range pointing to a value strictly less to ν or e otherwise.
		Time Complexity	
ϵ		$\mathbb{N}$	Maximum trace length.
ℓ		$\mathbb{N}$	Maximum length of the third component of the intermediate representation.

https://en.cppreference.com/w/cpp/algorithm/lower_bound as accessed the 5 March 2023. https://en.cppreference.com/w/cpp/algorithm/upper_bound as accessed the 5 March 2023.

3. Logical Model

Differently from our previous work [4], we provide a full definition of the (logical) model, thus describing the relational schema and how such tables are instantiated in order to fully represent the original log $\mathcal{L}$ (Section 3.1). This is a required preliminary step, as this will provide the required background to understand the definitions for the xt LTL$_f$ operators (Section 3.2). These operators, differently from the LTL$_f$ ones, are defined over the aforementioned model and assess the satisfiability of multiple traces loaded in such a model.

The discussion on how such tables are loaded and indexed is postponed when discussing the physical model (Section 4), as well as the different algorithms associated with the different operators (Section 6).

3.1. Model Definition

KnoBAB provides a tabular (i.e., relational) representation of the log $\mathcal{L}$, in order to efficiently query it through tailored relational operators (xtLTL$_f$). If the log does not contain data payloads, the entire log can be represented in two relational tables, CountingTable$_\mathcal{L}$(Activity, Trace,Count) and ActivityTable$_\mathcal{L}$(Activity,Trace,Event,Prev,Next). While the former can efficiently assess how many events in the same given trace share the same activity label, the latter allows a faithful reconstruction of the activity label associated with the traces. In particular, we use the former to assess whether a trace contains a given activity label at all. Such tables are then defined as follows:

Definition 1 (CountingTable). *Given a log $\mathcal{L}$, the CountingTable$_\mathcal{L}$(Activity, Trace, Count) counts for each trace in $\mathcal{L}$ how many times each activity label occurs. More formally:*

$$\text{CountingTable}_\mathcal{L} = \left[\langle \beta(a), i, |\{\sigma^i_j \in \sigma^i | \sigma^i_j = \langle a, p \rangle\}| \rangle \ \Big| \ a \in \Sigma, \sigma^i \in \mathcal{L} \right]$$

A record $\langle \beta(a), i, n \rangle$ states that the i-th trace from the log $\sigma^i \in \mathcal{L}$ contains n occurrences of a-labelled events with id $\beta(a)$.

Definition 2 (ActivityTable). *Given a log $\mathcal{L}$, the ActivityTable (Activity, Trace, Event, Prev, Next) lists all of the possible events occurring in each log trace, where Prev (π) and Next (ϕ) are offsets pointing to the row representing the immediately preceding or following event in the trace if any. More formally:*

$$\text{ActivityTable}_\mathcal{L} = \left[\langle \beta(a), i, j, \pi, \phi \rangle \ \Big| \ a \in \Sigma, \sigma^i \in \mathcal{L}, \sigma^i_j \in \sigma^i, \sigma^i_j = \langle a, p \rangle \right]$$

A record $\langle \beta(a), i, j, \pi, \phi \rangle$ states that the j-th event of the i-th log trace ($\sigma^i_j \in \sigma^i$, $\sigma^i \in \mathcal{L}$) has an activity label a and that its preceding and following events (if any) are respectively located on the π-th and ϕ-th record of the same table. Each record of this table should also satisfy the following integrity constraints:

- $(j = 1 \wedge \pi = \bot) \vee (\exists h, \pi', \phi'. \langle h, i, j-1, \pi', \phi' \rangle \in \text{ActivityTable}_\mathcal{L}[\pi])$;
- $(j = |\sigma^i| \wedge \phi = \bot) \vee (\exists h, \pi', \phi'. \langle h, i, j+1, \pi', \phi' \rangle \in \text{ActivityTable}_\mathcal{L}[\phi])$

Please observe that Prev and Next are computed after bulk inserting while loading and indexing the data (see LOADINGANDINDEXING from Algorithm 2). If a log is associated with either trace or event payloads, we must store for each payload the values associated with keys k in an AttributeTable$_\mathcal{L}^k$(Activity,Value,Offset), where Offset points to the event described in the ActivityTable$_\mathcal{L}$.

Definition 3 (AttributeTable). *Given a log $\mathcal{L}$, for each attribute $\kappa \in K$ associated with at least one value in a payload, we define a table AttributeTable$_\mathcal{L}^\kappa$(Activity, Value, Offset) associating each value to the pertaining event's payload as follows:*

$$\text{AttributeTable}_\mathcal{L}^\kappa = \left[\langle \beta(a), p(\kappa), \pi \rangle \ \Big| \ \sigma^i \in \mathcal{L}, \sigma^i_j \in \sigma^i, \sigma^i_j = \langle a, p \rangle, p(\kappa) \neq \bot \right]$$

A record $\langle \beta(a), v, \pi \rangle$ states that the event $\sigma^i_j = \langle a, p \rangle$ stored in the ActivityTable associated with the π-th offset contains a payload p associating κ to a value v ($p(\kappa) = v$).

Please observe that, similarly to the former table, the offset π is also computed while loading and indexing the data: this is discussed in greater detail in Section 4.2.2.

Algorithm 2 Populating the Knowledge Base (Section 4.2)

```
 1: procedure BULKINSERTION(L)
 2:     Σ, K ← ∅
 3:     for all σⁱ ∈ L do
 4:         Σ ← Σ ∪ {a}
 5:         for all σⱼⁱ = ⟨a, p⟩ ∈ σⁱ do
 6:             CountBulkMap[β(a)][i] = CountBulkMap[β(a)][i] + 1
 7:             ActToEventBulkVector[β(a)].put(⟨i,j⟩)
 8:             TraceToEventBulkVector[i][j] = j
 9:             for all κ ∈ dom(p) do
10:                 K ← K ∪ {κ}
11:                 AttBulkMapₖ[β(a)][p(κ)].put(⟨i,j⟩)
12:             end for
13:         end for
14:     end for

15: procedure LOADINGANDINDEXING(L)
16:     actTableOffset ← 1
17:     for all β(a) ∈ {1,...,|Σ|} do
18:         ActivityTable_L.primary_index[β(a)] ← actTableOffset
19:         for all σⁱ ∈ L do
20:             CountingTable_L.load(⟨β(a), i, CountBulkMap[β(a)][i]⟩)
21:         end for
22:         for all ⟨i,j⟩ ∈ ActToEventBulkVector[β(a)] do
23:             ActivityTable_L.load[⟨β(a), i, j, ↑, ↑⟩]
24:             TraceToEventBulkVector[i][j] = actTableOffset
25:             actTableOffset ← actTableOffset + 1
26:         end for
27:     end for
28:     for all κ ∈ K and β(a) ∈ {1,...,|Σ|} do
29:         begin ← |AttributeTable_L^κ|,  map ← {}
30:         for all ⟨v, lst⟩ ∈ AttBulkMapₖ[β(a)] and ⟨i,j⟩ ∈ lst do         ▷ σⱼⁱ = ⟨a, p⟩ with v = p(κ)
31:             offset ← TraceToEventBulkVector[i][j]
32:             AttributeTable_L^κ.load(⟨β(a), v, offset⟩)
33:             AttributeTable_L^κ.secondary_index[offset] ← |AttributeTable_L^κ|
34:         end for
35:         AttributeTable_L^κ.primary_index[β(a)] ← ⟨begin, |AttributeTable_L^κ|⟩
36:     end for
37:     for all σⁱ ∈ L and σⱼⁱ ∈ σⁱ do
38:         curr ← TraceToEventBulkVector[i][j]
39:         if j = 1 then
40:             ActivityTable_L.secondary_index[i] ← ⟨curr, TraceToEventBulkVector[i][|σⁱ|]⟩
41:         else
42:             ActivityTable_L[curr](Prev) ← TraceToEventBulkVector[i][j − 1]
43:         end if
44:         if j < |σⁱ| then
45:             ActivityTable_L[curr](Next) ← TraceToEventBulkVector[i][j1]
46:         end if
47:     end for

48: function RECONSTRUCTLOG(L)
49:     L' ← ∅
50:     for all ⟨i, ⟨begin, end⟩⟩ ∈ ActivityTable_L.secondary_index do
51:         ςⁱ ← [];  j ← 1
52:         repeat
53:             r ← ActivityTable_L[begin]
54:             a ← β⁻¹(r(Activity))
55:             p ← {}
56:             for all κ ∈ K s.t. ∃o. ⟨begin, o⟩ ∈ AttributeTableₖ.secondary_index do
57:                 p(κ) ← AttributeTableₖ[o](Value)                          ▷ AttributeTableₖ[o](Offset) = begin
58:             end for
59:             ςⱼⁱ ← ⟨a, p⟩;   σⁱ.put(ςⱼⁱ)
60:             begin ← r(Next);  j ← j + 1
61:         until begin ≠ ↑
62:         L'.put(ςⁱ)
63:     end for
64:     return L'
```

Example 8. *Figure 2 provides a graphical depiction of the tables storing our data. The records are also sorted by ascending order induced by the first three cells of each record, as required by our Physical Database Design (Section 4). For representation purposes, the first cell of each row shows the activity label* a *rather than its associated unique id* β(a).

3.2. eXTended LTL$_f$ Algebra (xt LTL$_f$)

We extend the operators provided in our previous work [4] into more minimal ones, thus better describing the data access on the relational model. Furthermore, we provide a full formal characterisation for each of these operators via their access to the aforementioned relational tables. Please observe that, similarly to the relational algebra, each xt LTL$_f$ operator might come with different possible algorithms [42], which are discussed in Section 6.

Our operators, assessing the behaviour of non-empty traces, come in two flavours: timed and untimed. While the former are marked by a τ and return all of the traces' events for which a given condition holds, the latter guarantee that such a condition will hold any time from the beginning of the trace. Furthermore, these operators assess the satisfiability of all the log traces simultaneously and not only one trace at a time as per LTL$_f$.

Each xtLTL$_f$ operator returns a nested relational table ρ with schema IntermediateResult(Trace, Event, MarkList(Mark)) implemented as an ordered set of triplets $\langle i, j, L \rangle$, where each triplet states that an event σ_j^i from trace σ^i satisfies a condition specified by the returning operator. If L (MarkList(Mark)) is not empty, the current event σ_j^i might have observed events σ_k^i and σ_h^i satisfying either an activation ($A(k) \in L, k \geq j$), a target ($T(k) \in L, k \geq j$), or a correlation condition ($M(h,k) \in L, k, h \geq j$). The nested relation L is implemented as a vector ordered by *mark* type and referenced event id. ρ is implemented as a vector and sorted by increasing Trace and Event id, as sorted vectors guarantee efficient intersection and union operations, as well as efficient event counting within the same trace through linear scanning. Binary operators associated with a non-**True** binary predicate Θ return matching/correlation conditions $M(h,k) \in L$ if at least one activation and one target condition were matched, depending on the definition of the operator. As we are going to see next, if the output comes from a base operator, as defined in the next section, L might contain a single activation or target corresponding to the immediately returned event.

3.2.1. Base Operators

First, we discuss the base operators directly accessing the tables. These might have an associated marker specifying whether the event of interest is considered an activation (A) or a target (T) condition; if none is required, the mark can be omitted from the operator. The Activity$^\tau$(a)$_{A/T}^{\mathcal{L}}$ operator lists all of the events associated with an activation label a. As the ActivityTable$_\mathcal{L}$ directly provides this information, this operator is defined as follows:

$$\text{Activity}_{A/T}^{\mathcal{L},\tau}(\mathsf{a}) = \{ \langle i, j, \{A/T(j)\} \rangle \mid \exists \pi, \phi. \langle \beta(\mathsf{a}), i, j, \pi, \phi \rangle \in \text{ActivityTable}_\mathcal{L} \}$$

We can also make similar considerations for single *elementary interval* representable as an LTL$_f$ *compound condition* a $\wedge$ *lower* $\leq \kappa \leq$ *upper*, which can be run as a single range query over an AttributeTable$_\mathcal{L}^\kappa$. As each of its records has an offset π to the ActivityTable$_\mathcal{L}$, this resolves the trace id and event id information required for the intermediate result. This operator can therefore be formalised as follows:

$$\text{Compound}_{A/T}^{\mathcal{L},\tau}(\mathsf{a}, \kappa, [lower, upper]) = \{ \langle i, j, \{A/T(j)\} \rangle \mid \exists \pi, \pi', \phi, v. \, lower \leq v \leq upper, \langle \beta(\mathsf{a}), v, \pi \rangle \in \text{AttributeTable}_\mathcal{L}^\kappa,$$
$$\text{ActivityTable}_\mathcal{L}[\pi] = \langle \beta(\mathsf{a}), i, j, \pi', \phi \rangle \}$$

If we want to list all of the initial (or terminal) events of a trace, we can directly access the ActivityTable and provide a linear scan over the number of the possible traces through its associated secondary index. If we are not interested in whether the trace starts with a specific activity label, then we can define the First$_A^{\mathcal{L},\tau}$ (and Last$_A^{\mathcal{L},\tau}$) operators as follows:

$$\text{First}_A^{\mathcal{L},\tau} = \{ \langle i, 1, \{A(1)\} \rangle \mid \exists \mathsf{a}, \phi. \langle \beta(\mathsf{a}), i, 1, \bot, \phi \rangle \in \text{ActivityTable}_\mathcal{L} \}$$

$$\text{Last}_A^{\mathcal{L},\tau} = \{ \langle i, |\sigma^i|, \{A(|\sigma^i|)\} \rangle \mid \exists \mathsf{a}, \pi. \langle \beta(\mathsf{a}), i, |\sigma^i|, \pi, \bot \rangle \in \text{ActivityTable}_\mathcal{L} \}$$

On the other hand, Init (and Ends) are the specific refinements of the former operators if we are also interested in retrieving events with a specific activity label. These can be defined as follows:

$$\text{Init}_A^{\mathcal{L}}(a) = \{ \langle i, 1, \{A(1)\}\rangle \mid \exists \phi. \langle \beta(a), i, 1, \bot, \phi \rangle \in \text{ActivityTable}_{\mathcal{L}} \}$$

$$\text{Ends}_A^{\mathcal{L}}(a) = \{ \langle i, 1, \{A(|\sigma^i|)\}\rangle \mid \exists \pi. \langle \beta(a), i, |\sigma^i|, \pi, \bot \rangle \in \text{ActivityTable}_{\mathcal{L}} \}$$

Given a natural number n, $\text{Exists}(a, n)_A^{\mathcal{L}}$ lists the traces containing at least n events with an activity label a. As $\text{Absence}(a, n)_A^{\mathcal{L}}$ is the substantial negation of the former, this lists the traces containing at most $n - 1$ events with an activity label a. Please observe that these operators directly provide the formal semantics for the homonym Declare template. As the CountingTable precisely contains the counting information required to solve this query, these operators can be formalised as follows for $n \in \mathbb{N}_{>0}$:

$$\text{Exists}_A^{\mathcal{L}}(a, n) = \{ \langle i, 1, \{A(1)\}\rangle \mid \exists m \geq n. \langle \beta(a), i, m \rangle \in \text{CountingTable}_{\mathcal{L}} \}$$

$$\text{Absence}_A^{\mathcal{L}}(a, n) = \{ \langle i, 1, \{A(1)\}\rangle \mid \exists m < n. \langle \beta(a), i, m \rangle \in \text{CountingTable}_{\mathcal{L}} \}$$

The following paragraph shows how these last two operators can be generalised for counting the salient event information returned by any sub-expression returning an operand ρ.

3.2.2. Unary Operators

The unary xtLTL_f operators come in two flavours: the first ones extend some of the former operators for compound conditions or atoms not necessarily associated with activity labels, while the second ones directly extend the unary operators from LTL_f.

Base Operators' generalisations. We extend the definition of Init/Ends or Exists/Absence for any possible set of events of interest listed in an intermediate result ρ, not necessarily associated with the same activity label. We first define Exists and Absence operator as such: instead of exploiting the counting table, we now actually need to count the events returned in ρ for each trace and return an intermediate result triplet iff. they satisfy the counting condition. These can be then defined as follows for $n \in \mathbb{N}_{>0}$:

$$\text{Exists}_n(\rho) = \{ \langle i, 1, \cup_{\langle i,j,L_j \rangle \in \rho} L_j \rangle \mid n \leq |\{\langle i, j, L'\rangle \in \rho\}| \}$$

$$\text{Absence}_n(\rho) = \{ \langle i, 1, \cup_{\langle i,j,L_j \rangle \in \rho} L_j \rangle \mid n > |\{\langle i, j, L'\rangle \in \rho\}| \}$$

Similarly, while the operators accessing the CountingTable (Exists/Absence) return the result by linearly scanning such a table, their generalised counterparts require scanning their operand ρ as returned from a subexpression of choice, and then creaming them off depending on how many events per trace were in ρ. As we might observe, we might exploit the previously provided operators when we want to evaluate conditions only associated with activity labels, while we might need to exploit the former if we are interested in results associated with compound conditions whose evaluation is returned in ρ.

Finally, we refine Init and Ends for a given operand ρ, to keep only the events at the beginning or end of a given trace:

$$\text{Init}(\rho) = \{ \langle i, j, L \rangle \in \rho \mid j = 1 \}$$

$$\text{Ends}(\rho) = \{ \langle i, 1, L \rangle \mid \langle i, |\sigma^i|, L \rangle \in \rho \}$$

Further details on our intended notion of these operators' generality if compared to the corresponding base operators can be found in Appendix A.1.

LTL_f extensions. The unary xtLTL_f operators work differently from the corresponding ones in LTL_f: while the latter compute the semantics from the first occurring operator appearing in the formula towards the leaves, the former assume to receive intermediate results from the leaves.

This structural difference also imposes an explicit distinction between timed and untimed operators. This is required as each operator is completely agnostic from the semantics associated with the upstream operator, and therefore the downstream operator has to combine the incoming intermediate results appropriately. This motivates why LTL$_f$ operators do not have to provide such an explicit distinction from their syntactical standpoint.

Such a premise motivates the counter-intuitive definition of the timed Next$^\tau$ operator if compared to the homonym in LTL$_f$: as this needs to return the events for which desired temporal constraints happen immediately after them, it needs to assume that the desired forthcoming temporal behaviour is the one received as an input ρ, for which all the events preceding the ones listed in ρ are the ones of interest. As per the previous statement, it also follows that this operator shall never possess an equivalent untimed flavour. From these considerations, Next$^\tau$ is formally defined as follows:

$$\text{Next}^\tau(\rho) = \{\, \langle i, j-1, L\rangle \mid \langle i,j,L\rangle \in \rho, j > 1 \,\}$$

where L fulfils the role of preserving the information of the events satisfying an activation, target, or correlation condition independently from the event stated in the second component of the intermediate representation record. Therefore, $\langle i, j, L\rangle$ shall be interpreted as follows: σ_j^i witnesses the satisfaction of any activation, target, or correlation condition by the events collected in L.

We now discuss the definition of "globally". As per previous considerations, checking that all of the events in a trace satisfy a given condition corresponds to retrieving all of the events satisfying such a condition, for then counting if the length of the returned events corresponds to the trace length. Similarly, the timed version of the same operator shall test the same condition for each possible event and return the points in the trace after which the desired condition always happens in the future. These operators are therefore defined as follows:

$$\text{Globally}^\tau(\rho) = \left\{ \langle i, j, \bigcup_{\substack{j \leq k \leq |\sigma^i| \\ \langle i,k,L_k\rangle \in \rho}} L_k \rangle \;\middle|\; \langle i,j,L_j\rangle \in \rho,\; |\sigma^i| - j + 1 = \left|\{\langle i,k,L_k\rangle \in \rho \mid j \leq k \leq |\sigma^i|\}\right| \right\}$$

$$\text{Globally}(\rho) = \left\{ \langle i, 1, \bigcup_{\langle i,j,L_k\rangle \in \rho} L_k \rangle \;\middle|\; |\sigma^i| = \left|\{\langle i,k,L_k\rangle \in \rho\}\right| \right\}$$

The operators expressing the eventuality that a condition shall happen in the future undergo similar considerations, with the only difference that these do not require to test that all of the trace events from a given point in time will satisfy a given condition, as it suffices that at least one event will satisfy it. The Future operator with its timed counterpart are then formally defined as follows:

$$\text{Future}^\tau(\rho) = \left\{ \langle i, j, \bigcup_{\substack{j \leq k \leq |\sigma^i| \\ \langle i,k,L_k\rangle \in \rho}} L_k \rangle \;\middle|\; \exists h \geq j. L. \langle i,h,L\rangle \in \rho \right\}$$

$$\text{Future}(\rho) = \left\{ \langle i, 1, \bigcup_{\langle i,k,L_k\rangle \in \rho} L_k \rangle \;\middle|\; \exists j, L. \langle i,j,L\rangle \in \rho \right\}$$

Timed and untimed negations are implemented dissimilarly by design. While the timed negation returns all of the events that are in the log but which were not returned in the previous computation ρ, the untimed version returns the traces containing no events associated with the provided input. These operators are therefore defined as follows:

$$\text{Not}^\tau(\rho) = \{\, \langle i,j,\varnothing\rangle \mid (\nexists L. \langle i,j,L\rangle \in \rho) \wedge \exists \alpha, \pi, \phi. \langle \alpha, i, j, \pi, \phi\rangle \in \text{ActivityTable}_{\mathcal{L}} \,\}$$

$$\text{Not}(\rho) = \{\, \langle i,1,\varnothing\rangle \mid (\nexists j, L. \langle i,j,L\rangle \in \rho) \wedge \exists \alpha, j, \pi, \phi. \langle \alpha, i, j, \pi, \phi\rangle \in \text{ActivityTable}_{\mathcal{L}} \,\}$$

3.2.3. Binary Operators

Differently from the LTL$_f$ binary operators, the xtLTL$_f$ binary operators are specifically tailored to express data correlation conditions Θ between activation and target payloads. This requires that one of the two operands, either ρ or ρ', returns activated events while the other provides targeted ones. Supplement I discusses the formal definition of predicates assessing whether an event $\langle i, j, L\rangle \in \rho$ matches with another event $\langle i, j', L'\rangle \in \rho'$ on the basis of their matched and activated events in L and L'. After this, we have the definition of our required binary operators.

The until operators work similarly to the other LTL$_f$-derived unary operators. The timed until returns all of the events within the trace satisfying the until condition, expressed by returning all of the "activated" events σ_j^i listed in the right operand (as they trivially satisfy the until condition) alongside all of the "targeted' events σ_j^i from the left operand with $k < j$ at a distance $j - k + 1$ from the second operand's event while guaranteeing that all the events in $\sigma_k^i, \ldots, \sigma_{j-1}^i$ appear in the first operand while satisfying the matching condition within this temporal window. The untimed version of this operator performs such considerations only from the beginning of the trace. These are defined as follows:

$$\text{Until}_\Theta^\tau(\rho_1, \rho_2) = \rho_2 \cup$$
$$\left\{ \langle i, k, \tau\rangle \;\middle|\; \exists j > k.\, \langle i, j, L\rangle \in \rho_2, (\forall k \leq h < j.\, \langle i, h, L\rangle \in \rho_1), \right.$$
$$\left. \tau := \mathcal{T}_\Theta^{A,i}([k \mapsto L]_{k \leq h < j}, [h \mapsto L_h]_{k \leq h < j, \langle i, h, L_h\rangle \in \rho_1}), \tau \neq \textbf{False} \right\}$$

$$\text{Until}_\Theta(\rho_1, \rho_2) = \{\langle i, j, L\rangle \in \rho_2 \mid j = 1\} \cup$$
$$\left\{ \langle i, 1, \tau\rangle \;\middle|\; \exists j > 1, L.\, \langle i, j, L\rangle \in \rho_2, (\forall 1 \leq k < j.\, \langle i, k, L_k\rangle \in \rho_1), \right.$$
$$\left. \tau := \mathcal{T}_\Theta^{A,i}([k \mapsto L]_{i \leq k < j}, [k \mapsto L_k]_{i \leq k < j, \langle i, k, L_k\rangle \in \rho_1}), \tau \neq \textbf{False} \right\}$$

where $\mathcal{T}_\Theta^{A,i}$ performs (Please see Supplement I for more details.) the correlation tests and returns the set of the matches if any and, if no match was successful, it returns **False**. Differently from Until$_\Theta^\tau$ and Until$_\Theta$, the rest of the binary operators assume to receive "activated" (or "targeted") events from the left (right) operand. The timed conjunction states that a join condition effectively happens in a given event σ_j^i if both operands return such an event and their associated activation and target conditions match. Thus, we only care for activation and target conditions at the same event σ_j^i. For its untimed counterpart, we state that a trace satisfies the conjunction of events if at least one activation condition from the left operand matching with a target from the right operand, if any, exists; this corresponds to coalescing the activations and target conditions on the first event while requiring that at least one of them occurs. These two operators can then be defined as follows:

$$\text{And}_\Theta^\tau(\rho_1, \rho_2) = \left\{ \langle i, j, \tau\rangle \;\middle|\; \exists L_1, L_2.\, \langle i, j, L_1\rangle \in \rho_1, \langle i, j, L_2\rangle \in \rho_2, \tau := \mathcal{T}_\Theta^{E,i}([j \mapsto L_1], [j \mapsto L_2]), \tau \neq \textbf{False} \right\}$$

$$\text{And}_\Theta(\rho_1, \rho_2) = \left\{ \langle i, 1, \tau\rangle \;\middle|\; \exists j, j', L, L'. (\langle i, j, L\rangle \in \rho_1 \wedge \langle i, j', L'\rangle \in \rho_2), \right.$$
$$\tau := \mathcal{T}_\Theta^{E,i}([1 \mapsto \cup\{L_j \mid \langle i, j, L_j\rangle \in \rho_1\}], [1 \mapsto \cup\{L_j \mid \langle i, j, L_j\rangle \in \rho_2\}]),$$
$$\left. \tau \neq \textbf{False} \right\}$$

The disjunctive version of the timed conjunctive operator returns either the result of the conjunctive operator or the events that did not temporally match from each respective operator. The only difference with its untimed version is that the latter merges all potential activation or target conditions from either of the two operands:

$$\text{Or}^\tau_\Theta(\rho_1,\rho_2) = \text{And}^\tau_\Theta(\rho_1,\rho_2) \cup \{ \langle i,j,L\rangle \in \rho_1 \mid \nexists L'.\, \langle i,j,L'\rangle \in \rho_2 \}$$
$$\cup \{ \langle i,j,L\rangle \in \rho_2 \mid \nexists L'.\, \langle i,j,L'\rangle \in \rho_1 \}$$

$$\text{Or}_\Theta(\rho_1,\rho_2) = \text{And}_\Theta(\rho_1,\rho_2) \cup \{ \langle i,1,\cup\{L|\exists j.\, \langle i,j,L\rangle \in \rho_1\}\rangle \mid \nexists j,L'.\, \langle i,j,L'\rangle \in \rho_2 \}$$
$$\cup \{ \langle i,1,\cup\{L|\exists j.\, \langle i,j,L\rangle \in \rho_2\}\rangle \mid \nexists j,L'.\, \langle i,j,L'\rangle \in \rho_1 \}$$

As we will see, the choice of characterizing Or with an E^i_Θ match while coalescing the activation and target conditions on the first trace event allows us to express the Choice template from Declare with one single operator while preserving its expected LTL$_f$ semantics.

3.2.4. Derived Operators

Similarly to relational algebra, we can now compose some frequently occurring operators together for enhancing the overall time complexity associated with the execution of frequently appearing subqueries in Declare.

Appendix A.3 will show that computing these operators is equivalent to computing their semantically equivalent xtLTL$_f$ expression containing multiple operators.

$$\text{AndFuture}^\tau_\Theta(\rho_1,\rho_2) = \Big\{ \langle i,j,\tau\rangle \ \Big|\ \exists L.\, \langle i,j,L\rangle \in \rho_1, (\exists L', k \geq j.\, \langle i,k,L'\rangle \in \rho_2),$$
$$\tau := \mathcal{T}^{E,i}_\Theta([j \mapsto L], [j \mapsto \cup_{h \geq j, \langle i,h,L_h\rangle \in \rho_2} L_h]), \tau \neq \textbf{False} \Big\}$$

$$\text{AndGlobally}^\tau_\Theta(\rho_1,\rho_2) = \Big\{ \langle i,j,\tau\rangle \ \Big|\ \exists L.\, \langle i,j,L\rangle \in \rho_1, (\forall |\sigma^i| \geq k \geq j.\exists L'.\, \langle i,k,L'\rangle \in \rho_2),$$
$$\tau := \mathcal{T}^{A,i}_\Theta([j \mapsto L], [j \mapsto \cup_{h \geq j, \langle i,h,L_h\rangle \in \rho_2} L_h]), \tau \neq \textbf{False} \Big\}$$

For easing the pseudocode readability, we can also define an $\text{Atom}^{\mathcal{L},\tau}_{A/T}(p_i)$ operator computing the conjunction of all of the compound conditions characterizing each atom:

$$\text{Atom}^{\mathcal{L},\tau}_{A/T}(p_i) = \text{And}^\tau_{\textbf{True}} \underset{\kappa \in K}{} \text{Compound}^{\mathcal{L},\tau}_{A/T}(a,\kappa,[low_\kappa, up_\kappa]) \text{ s.t. } p_i := a \wedge \bigwedge_{\kappa \in K} low_\kappa \leq \kappa \leq up_\kappa$$

Properties of the xtLTL$_f$ Algebra. We furnish the previous definitions with some formal proofs, which, so as not to burden the reader, are postponed to the Appendix A. We show that xtLTL$_f$ is as expressive as traditional LTL$_f$, as we can show that each LTL$_f$ expression evaluated over a finite and non-empty trace σ corresponds to an xtLTL$_f$ expression evaluated over the representation of such a trace within the proposed logical model; as the proofs of Lemmas A5 and A6 in Appendix A.2 are constructive, they show the translation process from LTL$_f$ formulæ to equivalent xtLTL$_f$ expressions.

Next, we also show that the timed and untimed operators correspond to the intended semantics: that is, for each timed operator having a corresponding untimed operator if the former states that the timed formula is satisfied by the i-th trace starting from time j, it follows that the sub-trace of i starting from time j will satisfy the corresponding untimed formula. This shows the correctness of the untimed operators concerning their timed definitions (Lemma A7).

In Appendix A.3, we show that the Declare template Choice can be fully implemented by exploiting an untimed Or operator (Corollary A1) while the latter still abides to the rules of LTL$_f$ semantics. We also motivate the need of the derived operators in terms of equivalence to the intended xtLTL$_f$ expressions (Lemmas A9 and A10) as well as in terms of improved computational complexity (Section 6.4) and run time (Section 7.1). The latter is discussed after describing the physical model in more detail alongside the algorithms associated with each operator, which is introduced in the following section.

4. Physical Database Design

This section shows how the defined model (Section 3.1) is represented in primary memory in terms of indices and data structures (Section 4.1). We also illustrate the algorithm loading a log in such representation of choice (Section 4.2).

4.1. Primary Memory Data Structures

At the time of the writing, KnoBAB is primarily an in-memory database. This is a common assumption in the conformance checking domain where most of the log datasets are quite compact and nicely fit in primary memory.

In order to be both memory and time efficient in our operations, the sub-record referring to the first three columns of both the CountingTable$_\mathcal{L}$ and the ActivityTable$_\mathcal{L}$ are fully stored in primary memory as an unsigned 64-bit unsigned integer, while the Prev and Next are more efficiently stored as pointers to the table records rather than being an offset. After sorting the CountingTable$_\mathcal{L}$, we directly obtain the occurrence of each activity label a within the log by accessing the records in the range $[|\mathcal{L}| \cdot (\beta(a) - 1) + 1, |\mathcal{L}| \cdot \beta(a)]$.

Indexing data structures, on the other hand, eases the access to the ActivityTable$_\mathcal{L}$, as different traces might have different lengths, and activity labels might be differently distributed among the traces. Therefore, we exploit a clustered and sparse primary index for determining which is the first event associated with a given activity label; as the traces in such a table are represented as a doubly linked list, its secondary index maps each trace-id to a block that, in turn, points to the head (first event of the trace) and the tail (last event of the trace) of such a doubly linked list.

The deduplication of trace and event payloads in distinct AttributeTable$_\mathcal{L}^\kappa$ for each key κ follows the prescriptions of the query and memory-efficient representation of columnar-based storages [35]. In our implemnetation, such tables are sorted in ascending order by their three first columns. Each AttributeTable$_\mathcal{L}^\kappa$ is also associated with two indices: the clustered and sparse primary index maps each activity label's id $\beta(a)$ to the records referring to values contained in a-labelled events, and a dense secondary index associates an ActivityTable$_\mathcal{L}$ record offset to an AttributeTable$_\mathcal{L}^\kappa$ record offset if and only if the event described in ActivityTable$_\mathcal{L}$ has a payload containing a value associated with a key κ. While data range queries leverage the former, the latter is used for reconstructing the payload associated with a given event when identified by its offset in the ActivityTable$_\mathcal{L}$. A relevant use case for doing so is the reconstruction of the event payload information while performing the Θ correlation condition, as well as reconstructing the original log leading to the loading of the internal database. RECONSTRUCTLOG function in Algorithm 2 shows the computation of the latter.

Example 9. *With reference to Figure 2, let us consider some events with activity label* Mastectomy *associated with an unique id* $\beta(\text{Mastectomy}) = 3$. *The offsets for accessing the records in the* CountingTable$_\mathcal{L}$ *defining the number of events per trace with such a label is* $[3 \cdot (3-1) + 1, 3 \cdot 3] = [7, 9]$.

The ActivityTable$_\mathcal{L}$'s *primary index allows the access to the first record within the table recording a* Mastectomy *event, i.e.,* ActivityTable$_\mathcal{L}$.*primary_index*$[\beta(\text{Mastectomy})] = 7$; *the index implicitly returns the last event associated with such an activity label by decreasing the offset to the following activity label by one, i.e.,* ActivityTable$_\mathcal{L}$.*primary_index*$[\beta(\text{Mastectomy}) + 1] - 1 = 7$: *please remember that, if the activity label is such that* $\beta(a) = |\Sigma|$, *then the final offset to be considered corresponds to the* ActivityTable$_\mathcal{L}$ *size. This indicates that there exists only one event throughout the whole log associated with such an activity label. We will exploit this mechanism for returning the events associated with* Activity$_{A/T}^{\mathcal{L},\tau}$(Mastectomy). *As the seventh record of such a table refers to the third event of the first trace,* Activity$_A^{\mathcal{L},\tau}$(Mastectomy) *will then return* $\{ \langle 1, 3, \{A(3)\} \rangle \}$.

Finally, we discuss how we can leverage AttributeTable$_{\mathcal{L}}^{\kappa}$'s primary indices for returning results associated with an Atom$_{A/T}^{\mathcal{L},\tau}$ operator. Let us consider atom p_{12}: we can see that this is associated with the elementary intervals Lumpectomy $\wedge$ biopsy = **true** and Lumpectomy $\wedge$ 50 $\leq$ CA_15.3 < +∞. By definition of the operator of interest, we then have that:

$$\text{Atom}_A^{\mathcal{L},\tau}(p_{12}) = \text{And}_{\text{True}}^{\tau}(\text{Compound}_A^{\mathcal{L},\tau}(\text{Lumpectomy} \wedge \text{biopsy} = \textbf{true}),$$
$$\text{Compound}_A^{\mathcal{L},\tau}(\text{Lumpectomy} \wedge \text{CA\_15.3} \geq 50))$$

The first *Compound* operator will access the primary index from AttributeTable$_{\mathcal{L}}^{biopsy}$ while the second one will access the one from AttributeTable$_{\mathcal{L}}^{CA\_15.3}$. Then, the primary index of each table maps each activity to the offsets of the first and last record: AttributeTable$_{\mathcal{L}}^{biopsy}$.`primary_index` $[\beta(\text{Lumpectomy})] = \langle 2,2 \rangle$ and AttributeTable$_{\mathcal{L}}^{CA\_15.3}$.`primary_index`$[\beta(\text{Lumpectomy})] = \langle 7,7 \rangle$. Then, within these returned record offsets, we perform range queries respectively looking for records satisfying biopsy = **true** and $50 \leq \text{CA\_15.3} < +\infty$. All of the ActivityTable$_{\mathcal{L}}\kappa$'s records satisfying these conditions point to the tenth record of the ActivityTable$_{\mathcal{L}}$ referring to the third event of the third trace. Therefore, Atom$_A^{\mathcal{L},\tau}(p_{12})$ returns { $\langle 3,3, \{A(3)\} \rangle$ }.

4.2. Populating the Database

We discuss two subsequent steps for loading a log in our proposed relational model: we preliminarily sort the data by activity label id, event id, and values (Section 4.2.1) for then loading the sorted record in the tables while generating their primary and secondary indices (Section 4.2.2). These are computed in quasi-linear time with respect to the full log size.

4.2.1. Bulk Insertion

KnoBAB uses BULKINSERTION to pre-load the tables' data into an intermediate representation by pre-sorting it according to the ascending order induced by the first column of the tables of interest. Algorithm 2 shows the loading of the following three maps referring to the aforementioned tables. *(i)* `CountBulkMap` counts the occurrence of each activity label per track, implying that the absence of a trace identifier for a given $\beta(a)$ value presupposes the absence of a given activity label a within a trace; as the name suggests, we use this to later on populate the CountingTable$_{\mathcal{L}}$. *(ii)* The `ActToEventBulkVector` prepares the insertion of sorted data in ActivityTable$_{\mathcal{L}}$ by associating an activity label to each event and its associated trace containing it. *(iii)* Similarly to the `ActToEventBulkMap`, the `AttBulkMap`$_k$ associates to each key κ the values $p(\kappa)$ for each event σ_j^i with payload p and activity label a, in order to prepare the insertion of sorted records in AttributeTable$_{\mathcal{L}}^{\kappa}$. Please observe that, by construction, the set of pairs associated with each activity id $\beta(a)$ is already sorted by increasing trace and event id.

We also pre-allocate a `TraceToEventBulkVector` map (represented as a vector of vectors) which will later associate each event trace to an offset on the ActivityTable$_{\mathcal{L}}$ where such event is stored. KnoBAB will later use this to calculate Prev and Next in the ActivityTable. After this, KnoBAB knows the number of the traces within the log $|\mathcal{L}|$, the length $|\sigma^j|$ for each trace σ^j, and the number of distinct activity labels $|\Sigma|$ is known, as well as their associated unique id $\beta(a)$ for each a $\in \Sigma$. We can show that this procedure might be computed in quasi-linear time with respect to the full log size (Lemma S1).

4.2.2. Loading and Indexing

We continue our discussion with LOADINGANDINDEXING. *First*, we can iterate over the activity labels in ascending order of appearance (Line 17). All the tables including the CountingTable$_{\mathcal{L}}$ have activity ids $\beta(a)$ as their first cell: by further iterating by increasing trace id, we can immediately orderly store the records in CountingTable$_{\mathcal{L}}$ (Line 20).

Second, we start populating the ActivityTable$_{\mathcal{L}}$ (Line 23) where each record is associated with an increasing offset of the table (Line 25). We can populate its primary index in order to point at the record representing the first event of the first trace with the currently considered activity label. We store such information in the pre-allocated traceToEventBulkVector (Line 24), in order to later set the currently null pointer (↑) Next and Prev fields.

Third, we start populating each AttributeTable$_\mathcal{L}^\kappa$ for each key $\kappa \in K$ associated with at least one value in a payload: as per the previous discussion, each record associates the offset of event $\sigma_j^i = \langle a, p \rangle$ in the ActivityTable$_\mathcal{L}$ with a value $\nu = p(\kappa)$ and the activity label id $\beta(a)$ (Line 32). We also populate its secondary index by associating each event offset in the ActivityTable$_\mathcal{L}$ to the current position in the AttributeTable$_\mathcal{L}^\kappa$ (Line 33). The last iteration finally populates each ActivityTable$_\mathcal{L}$'s secondary index (Line 50) and sets the Next (Line 45) and Prev (Line 42) fields through the offset via TraceToEventBulkVector. After this, the relational database is fully loaded in primary memory. The overall time complexity grows linearly to the whole log representation (Lemma S2).

5. Query Processing and Optimisation

This section shows how a declarative model $\mathcal{M}$ is compiled to a query plan consisting of xt LTL$_f$ operators (Section 5.1) so it can be run (Section 5.2) on top of the primary memory data described in the previous section.

5.1. Query Compiler

The conversion of a declarative model $\mathcal{M}$ into its corresponding xt LTL$_f$ query plan is structured into three main phases. First, the atomisation pipeline calls the preliminary $\mathcal{D}_\varphi$-encoding from [25] for rewriting the data predicates appearing in each declarative clause as a disjunction of mutually exclusive atoms (Section 5.1.1). Second, we (ii) rewrite each Declare constraint as a xt LTL$_f$ formula from which we obtain a preliminary query plan represented as a DIRECT ACYCLIC GRAPH (DAG) (Section 5.1.2). Third, we compute the scheduling order for the operators' execution over the DAG, thus preparing the execution to a potential parallel evaluation of the query (Section 5.1.3).

5.1.1. Atomisation Pipeline

The atomisation pipeline (Algorithm 3) represents each activation and target condition as a set of disjunct atoms or activity labels. KnoBAB can always be configured in two ways: to either fully represent each possible activation (or target) condition with activity label a as a disjunction of atoms (or activity labels) if there exists at least one declarative clause where a is also associated with a non-trivial payload condition (strategy=AtomizeEverything), or to restrict atomisation to data conditions appearing in a clause (strategy=AtomizeOnlyOnDataPredicates). Both can be set through the Atomisation-Pipeline procedure in Algorithm 3. The $\mathcal{D}_\varphi$-encoding step guarantees that each activation or target condition will be associated with at least one atom or activity label. While the former approach will maximise the access to the AttributeTable$_\mathcal{L}$, the latter will maximise the access to the ActivityTable$_\mathcal{L}$. Correlation conditions do not undergo this rewriting step. We discuss the effects of each different strategy on the query runtime via empirical benchmarks in Section 7.4. We can show that this step has a polynomial complexity with respect to the model, key set, and element intervals' maximum size (Lemma S3).

Example 10. *With reference to Figure 3a, we might observe that, as no activation or target is ever associated with payload conditions, the atomisation pipeline will never express each activation or target condition as a disjunction of atoms, as no elementary interval is collected. Therefore, these will be only associated with activity labels.*

Example 11. *With reference to Example 6, Figure 2 shows the atomised version of the declarative model, where each activation and target condition is associated, in this case, with just one atom.*

Algorithm 3 Atomisation Pipeline (Section 5.1.1)

```
 1: procedure ATOMISATIONPIPELINE(M, strategy)
 2:     COLLECTINTERVALS(M)                                              ▷ See Algorithm 1
 3:     D_φ-ENCODING( )                                                  ▷ See Algorithm 1
 4:     for all clause_l(A, p, B, p') where Θ ∈ M do
 5:         if p=True and (strategy=AtomizeOnlyOnDataPredicates or ak(A) = {A}) then   ▷ Defining S_A for clause_l
 6:             clause_l.left ← {A}
 7:         else
 8:             clause_l.left ← ak(A) ∩ Atom_{μ,ad}(A, p)                 ▷ Equation (2)
 9:         end if
10:         if p'=True and (strategy=AtomizeOnlyOnDataPredicates or ak(B) = {B}) then  ▷ Defining S_T for clause_l
11:             clause_l.right ← {B}
12:         else
13:             clause_l.right ← ak(B) ∩ Atom_{μ,ad}(B, p')                ▷ Equation (2)
14:         end if
15:     end for
```

Example 12. *Continuing with Example 7, where we discussed the outcome of the D_φ-encoding phase for a model $\mathcal{M}'$ in Equation (3), we obtain the following atomisation:*

$$\{\text{Choice}_1(\texttt{left} = \{p_{12}, p_{17}\}, \texttt{right} = \{p_4, p_9\}),$$
$$\text{Absence}_2(\texttt{left} = \{p_1, \ldots, p_5, p_{16}, \ldots, p_{20}\}, n = 1),$$
$$\text{Absence}_3(\texttt{left} = \{p_1, p_3, p_5, p_6, p_8, p_{10}, p_{11}, p_{13}, p_{15}, p_{16}, p_{18}, p_{20}\}, n = 1)\}$$

5.1.2. Query Optimiser

The query optimiser consists of three steps: (i) loading the $xt\text{LTL}_f$ formulæ associated with each declarative clause at warm-up, (ii) exploiting the outcome of the Atomisation Pipeline to instantiate the $xt\text{LTL}_f$ formulæ, (iii) and coalescing the single $xt\text{LTL}_f$ into one compact abstract syntax DAG. Our query plan will not be represented as a tree as we merge as many nodes computing the same result as possible, thus computing the same sub-expression at most once.

First, we load the translation map *xtTemplates* (Table 5) at warm-up through an external script providing the temporal semantics associated with the clauses of interest via partially-instantiated $xt\text{LTL}_f$ expressions. Such representation also supports negated activation or target conditions, thus avoiding the need to compute a Not operator stripping the information of either activation or target conditions. These are marked in the previous table via set complementation, $\complement$. At the time of the writing, the scripts provide the $xt\text{LTL}_f$ semantics for Declare templates. Future investigations will express other temporal declarative languages such as [43] in $xt\text{LTL}_f$, as well as other LTL_f extensions including "past" operators [17,19].

Second, we exploit the aforementioned map to convert each declarative clause into its $xt\text{LTL}_f$ semantics ψ. If the clause is met for the first time, we proceed with its instantiation by recursively visiting ψ until the leaves are reached: at this level, we potentially replace the activation and target placeholders with the associated set of atoms. Disjunctions of atoms and activity labels associated with leaf nodes as returned by the previous pipeline are minimised by ensuring that each shared Or_True computation across all of the atoms and activity labels is computed at most once. If an atom is met, we decompose it into its defining compound conditions (Line 14), thus guaranteeing that each compound condition is evaluated via Compound$_{A/T}^{\mathcal{L},\tau}$ at most once across all of the atoms occurring in the $xt\text{LTL}_f$ formula when running the query (Section 5.2.1).

Third, we complete the process by coalescing shared disjunct sub-expressions via a map (*queryCache*) guaranteeing that all of the equivalent sub-expressions are all replaced by just one instance of these. Finally, we associate each sub-expression referring to each clause to the final query operator representing the expression's root (*queryRoot*), either presenting an aggregation or a conjunctive query.

Table 5. Declare templates illustrated as their associated xtLTL_f semantics. S_A (and S_T) denote the disjunction of collected atoms and activity labels (represented as sets) associated with the activation (and target) condition. The Atomisation Pipeline will return these sets. For declarative clauses that can be directly represented as xtLTL_f operators, we might have two different possible operators depending on the atomisation result.

Exemplifying Clause (c_l)	xtLTL_f Semantics	
	$S_A = \{A\}, S_T = \{B\}$ $A, B \in \Sigma$	Otherwise (e.g., Atomisation)
$\text{Init}(S_A)$	$\text{Init}_A^{\mathcal{L}}(A)$	$\text{Init}(S_A)$
$\text{Exists}(S_A, n)$	$\text{Exists}_A^{\mathcal{L}}(A, n)$	$\text{Exists}_n(S_A)$
$\text{Absence}(S_A, n+1)$	$\text{Absence}_A^{\mathcal{L}}(A, n)$	$\text{Absence}_{n+1}(S_A)$
$\text{Precedence}(S_A, S')$		$\text{Or}_{\text{True}}(\text{Until}(\complement S', S_A), \text{Absence}(S', 1))$
$\text{ChainPrecedence}(S_A, S_T)$ **where** Θ		$\text{Globally}(\text{Or}_{\text{True}}^{\tau}(\text{Or}_{\text{True}}^{\tau}(\text{Last}^{\mathcal{L},\tau}, \text{Next}^{\tau}(\complement S_T)), \text{And}_{\Theta}^{\tau}(\text{Next}^{\tau}(S_A), S_T)))$
$\text{Choice}(S_A, S_{A'})$		$\text{Or}_{\text{True}}(S_A, S_{A'})$
$\text{Response}(S_A, S_T)$ **where** Θ		$\text{Globally}(\text{Or}_{\text{True}}^{\tau}(\complement S_A, \text{AndFuture}_{\Theta}^{\tau}(S_A, S_T)))$
$\text{ChainResponse}(S_A, S_T)$ **where** Θ		$\text{Globally}(\text{Or}_{\text{True}}^{\tau}(\complement S_A, \text{And}_{\Theta}^{\tau}(S_A, \text{Next}^{\tau}(S_T))))$
$\text{RespExistence}(S_A, S_T)$ **where** Θ		$\text{Or}_{\text{True}}(\text{Absence}(S_A, 1), \text{And}_{\Theta}(S_A, S_T))$
$\text{ExclChoice}(S_A, S_{A'})$		$\text{And}_{\text{True}}(\text{Or}_{\text{True}}(\text{Exists}(S_A, 1), \text{Exists}(S_{A'}, 1)), \text{Or}_{\text{True}}(\text{Absence}(S_A, 1), \text{Absence}(S_{A'}, 1)))$
$\text{CoExistence}(S_A, S_T)$ **where** Θ		$\text{And}_{\text{True}}(\text{RespExistence}(S_A, S_T)$ **where** Θ, $\text{RespExistence}(S_{A'}, S_{T'})$ **where** Θ^{-1}) s.t. $S_{A'} = S_T$ and $S_{T'} = S_A$
$\text{Succession}(S_A, S_T)$ **where** Θ		$\text{And}_{\text{True}}(\text{Precedence}(S_A, S'), \text{Response}(S_A, S_T)$ **where** Θ) s.t. $S' = S_T$
$\text{ChainSuccession}(S_A, S_T)$ **where** Θ		$\text{Globally}(\text{And}_{\text{True}}^{\tau}(\text{Or}_{\text{True}}^{\tau}(\text{Or}_{\text{True}}^{\tau}(\text{Last}^{\mathcal{L},\tau}, \text{Next}^{\tau}(\complement S_{T'})), \text{And}_{\Theta-1}^{\tau}(\text{Next}^{\tau}(S_{A'}), S_{T'})),$ $\text{Or}_{\text{True}}^{\tau}(\complement S_A, \text{And}_{\Theta}^{\tau}(S_A, \text{Next}^{\tau}(S_T)))))$ s.t. $S_{A'} = S_T$ and $S_{T'} = S_A$
$\text{AltResponse}(S_A, S_T)$ **where** Θ		$\text{Globally}(\text{Or}_{\text{True}}^{\tau}(\complement S_A, \text{And}_{\Theta}^{\tau}(S_A, \text{Next}^{\tau}(\text{Until}_{\text{True}}^{\tau}(\complement S_A, S_T)))))$
$\text{AltPrecedence}(S_A, S_T)$ **where** Θ		$\text{And}_{\text{True}}(\text{Precedence}(S_A, S_T), \text{Globally}(\text{Or}_{\text{True}}^{\tau}(\complement S_A, \text{And}_{\Theta}^{\tau}(S_A, \text{Next}^{\tau}(\text{Or}_{\text{True}}^{\tau}(\text{Until}^{\tau}(\complement S_A, S_T), \text{Globally}^{\tau}(\complement S_A)))))))$
$\text{NotCoExistence}(S_A, S_T)$ **where** Θ		$\text{Not}(\text{And}_{\Theta}(S_A, S_T))$
$\text{NotSuccession}(S_A, S')$		$\text{Globally}(\text{Or}_{\text{True}}^{\tau}(\complement S_A, \text{AndGlobally}_{\text{True}}^{\tau}(S_A, \complement S_T)))$

Example 13. *The model in Figure 3a, when compiled and associated with a conjunctive query, might produce the following xLTLf expression:*

$$\text{And}_{\text{True}}\Big(\text{Globally}\Big(\text{Or}^{\tau}(\text{Not}^{\tau}(\text{Activity}^{\mathcal{L},\tau}(\text{rec})), \text{AndFuture}_{\text{True}}^{\tau}(\text{Activity}_A^{\mathcal{L},\tau}(\text{rec}), \text{Activity}_T^{\mathcal{L},\tau}(\text{weap}))\Big)\Big),$$
$$\text{And}_{\text{True}}\Big(\text{Absence}(\text{iiot\_sh}, 1),$$
$$\text{Or}_{\text{True}}(\text{Activity}^{\mathcal{L}}(\text{comm}), \text{Activity}^{\mathcal{L}}(\text{act}))\Big)\Big)$$

We might observe that this expression cannot be further minimised, as there are neither shared atoms nor sub-expression in common. This can neither be achieved by rewriting $\text{Not}^{\tau}(\text{Activity}^{\mathcal{L},\tau}(\text{rec}))$ as $\text{Or}_{\text{True}}^{\tau}{}_{\substack{a \in \Sigma \\ a \neq \text{rec}}} \text{Activity}^{\mathcal{L},\tau}(a)$, as the comm and act atoms associated with the choice clause are untimed, while the former rewriting only included timed Activity operators. As these two different flavours of operators do not necessarily return the same result, these nodes are not merged.

Example 14. *With reference to Example 11 and Table 1, as the Response clause was associated with the same activation and target condition to Succession, the former is indeed a subquery of the latter. For this reason, these queries are fused together, thus guaranteeing that the result for Response is computed at most once. As the query root requires the computation of Max-SAT, this one is always going to be linked to the sub-expression being the representation of an original declarative clause. Green arrows in Figure 2 indicate operators' output shared among operators.*

Example 15. *This last example shows the effect of the reduction of the number of shared timed union operators at the leaf level. By recalling the atomised model discussed in Example 12, we need to represent each set of atoms as a timed disjunction of Atom operators. While doing so, we observe that Choice and the first Absence condition share atoms p_4 and p_{17}, while the two Absence clauses share all the atoms in $\{p_1, p_3, p_5, p_{16}, p_{18}, p_{20}\}$. Not ensuring that the timed unions associated with these last elements are computed only once will result in both multiple data access to our relational tables, as well as a considerable increase in run time as union operations are run twice. The detection and minimisation of such kind of shared sub-queries cannot be merely computed through a simple* caching mechanism, thus requiring a more sophisticated algorithm for determining the maximal common subset shared among all of the possible sets of atoms (and potentially activity labels).

Algorithm 4 provides additional details on the implementation of such an approach. Line 11 refers to the *first* phase and shows the point in the code where we associate each negated leaf with the complementary set of atoms appearing after the decomposition process. With respect to the *second* phase, Line 65 shows the rewriting of the Declare clause into an intermediate xtLTL$_f$ by recursively visiting it in each of its operands until the leaves are reached (Line 5). If during this visit we meet a binary operator marked as being the "tester" for the correlation condition, we associate to it the Θ coming from the declarative clause (Line 4); otherwise, the operator keeps the default **True**. Concerning the leaves, for unary clauses, we consider the sole activation condition, while for binary clauses, we might also consider target conditions. If the leaf node is associated with an S_A (or S_T) containing more than one activity label or atom, we need to keep track of all of these while representing such a leaf as a disjunction of such atoms

(Lines 18–25). Next, we optimize each disjunction of atoms and activity labels in order to minimize the number of shared union computations (Line 48); such optimisation is performed after fully visiting the xtLTL$_f$ expression, thus ensuring that each appearing disjunction is actually collected (Line 69).

Line 14 shows where we collect atoms representing compound conditions while guaranteeing that its associated Compound$_{A/T}^{\mathcal{L},\tau}$ operator is computed only once, as well as decomposing it in its constituent compound conditions.

Finally, the method PUTINCACHE extends the *queryCache* map by guaranteeing that each distinct disjunction of atoms is also represented at most once within the query plan.

Example 16. *Figure 5 showcases the result of the application of such an algorithm while generating unique* xt LTL$_f$ *expressions. Such an algorithm also guarantees the non-repetition of single-leaf operators appearing in different clauses. Its upper box shows a query plan where common union operations are shared across sub-trees by representing each sub-tree at most once. These are actually represented in the query plan as opposed to the evaluation associated with the atoms, which is discussed in Supplement III.1.*

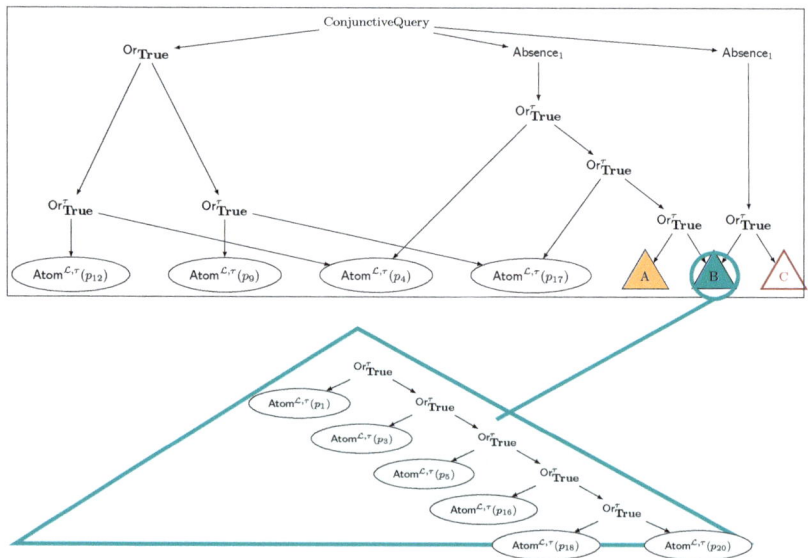

Figure 5. In-depth representation of the query plan associated with the model described in Example 15.

Algorithm 4 Query Optimiser

```
 1: global declare2xtLTL_f ← {}; queryCache ← {}; collectUnions ← {}; Q ← {}; atomQ ← ∅
 2: global keyToLabelToSortedIntervals ← {}; S_Σ ← {}; Results ← {}

 3: function INSTANTIATE(ψ, Θ, S_A, S_T)
 4:     if ψ.hasTheta then ψ.theta ← Θ
 5:     if ψ.arg= ∅ then                                                       ▷ ψ is a leaf
 6:         if ψ.isActivation or ψ.isNeither then
 7:             ψ.atom ← S_A
 8:         else if ψ.isTarget then
 9:             ψ.atom ← S_T
10:         end if
11:         if ψ.negated then ψ.atom ← ∁ ψ.atom         ▷ Complementing the atoms from the universe set upon negation
12:         for all atom ∈ ψ.atom do
13:             if atom ∈ ⋃_{a∈Σ} ak(a) then                                   ▷ The atom is generated from D_φ-encoding
14:                 RETRIEVEINTERVALS(atom)
15:             else atomQ.put(atom)
16:             end if
17:         end for
18:         if |ψ.atom| > 1 then
19:             disj ← ∅
20:             for all atom ∈ ψ.atom do
21:                 ψ' ← new xtLTL_f ()
22:                 ψ'.atom = {atom}
23:                 disj.put(atom)
24:             end for
25:             collectUnions[disj].put(ψ)
26:         else
27:         end if
28:     else
29:         for all arg ∈ ψ do
30:             arg ← INSTANTIATE(arg, Θ, S_A, S_T)
31:         end for
32:     end if

33: procedure COLLECTUNIONS( )                                                ▷ DAG over the leaves undergoing union operations.
34:     for all ⟨atomSet, ψ'⟩ ∈ FINITARYSETOPERATIONS(collectUnions, Or_True) do   ▷ Algorithm S4
35:         for all ψ ∈ collectUnions[atomSet] do
36:             queryCache[ψ] ← ψ'
37:         end for
38:     end for

39: procedure PUTINCACHE(ψ)
40:     if ∃ψ'. ⟨ψ, ψ'⟩ ∈ queryCache then
41:         return ψ'
42:     else
43:         for all arg ∈ ψ.args do
44:             arg ← PUTINCACHE(arg)
45:         end for
46:         ψ' ← new xtLTL_f ()
47:         ψ' ← ψ
48:         queryCache[ψ] ← ψ'
49:         return ψ'
50:     end if

51: procedure RETRIEVEINTERVALS(p_i)                                           ▷ p_i := a ∧ partition
52:     for all low_κ ≤ κ ≤ up_κ ∈ partition do                                ▷ p_i = ⋀_{κ∈K} low_κ ≤ κ ≤ up_κ
53:         if ∃h. ⟨low_κ ≤ κ ≤ up_κ, h⟩ ∈ keyToLabelToSortedIntervals[κ][a] then
54:             S_Σ[p_i].put(h)
55:         else
56:             Results.put(∅)
57:             S_Σ[p_i].put(|Results|)
58:             keyToLabelToSortedIntervals[κ][a].put(⟨low_κ ≤ κ ≤ up_κ, |Results|⟩)
59:         end if
60:     end for

61: function QUERYOPTIMISER(M, queryRoot)
62:     for all clause_l(A, p, B, q) where Θ ∈ M do
63:         if ∃ψ : xtLTL_f. ⟨clause_l(A, p, B, q) where Θ, ψ⟩ ∈ declare2xtLTL_f then Q.push(ψ)
64:         else
65:             ψ ← INSTANTIATE(xtTemplates[clause_l], Θ, clause_l.left, clause_l.right)
66:             Q.push(ψ)
67:         end if
68:     end for
69:     COLLECTUNIONS( )
70:     queryRoot.args ← { PUTINCACHE(ψ) | ψ ∈ Q }
71:     return queryRoot
```

5.1.3. Enabling Intraquery Parallelism

The query scheduler (Algorithm 5) takes as an input the query compiled in the previous phase and returns the scheduling order for achieving *intraquery parallelism* [42]. The previously generated expression might not be considered as an abstract syntax tree, rather than an abstract syntax DIRECT ACYCLIC GRAPH (DAG) rooted in the entry-point operator *queryRoot*, as we guarantee that sub-expressions appearing multiple times are replaced by unique instances of them.

Therefore, we can freely represent the query plan as a DAG $\mathcal{G}$ in our pseudocode notation, where each root operator in ψ is a single node while edges connect parent operators to the siblings' (ψ.args) root operator. Graph edges induce the execution order, where any ancestor node needs to be run after all of its immediate siblings. A reversed topological sort (Line 3) induces the order in which the operations should be run. To know which of these operators can be run contemporarily (i.e., scheduled together [44]) as they share no interdependencies, we compute for each node its maximum distance from *queryRoot* (Line 6). This generates a *layer*ing [45] guaranteeing that all of the nodes at the same levels share no mutual dependencies (Line 10). This enables the level-wise parallelisation of the tasks' execution (also referred to as Intraquery Parallelism [42]), thus showing how such a problem can be reduced into an embarrassingly parallel problem by parallelising the computation of each operator in the same given layer. This procedure runs in linear time with respect to the number of operators appearing in the $xtLTL_f$ query plan (Lemma S4). We benchmark query plan parallelisation with different task scheduling policies in Section 7.3.

Algorithm 5 Query Scheduler (Section 5.1.3)

1: **function** QUERYSCHEDULER($\mathcal{G}$)
2:    $layer \leftarrow \{\}$
3:    $V \leftarrow$ REVERT(TOPOLOGICALSORT($\mathcal{G}$))
4:    **for all** $\psi \in V$ **do**
5:      **for all** $\psi' \in \psi$.args **do**
6:        ψ'.distance $\leftarrow \max(\psi'$.distance, ψ.distance $+ 1)$
7:      **end for**
8:    **end for**
9:    **for all** $\psi \in V$ **do**
10:      $layer[\psi$.distance$]$.put(ψ)
11:    **end for**
12:    **return** *layer*

Example 17. *The DAG in Figure 2 depicts a query plan, where operators' dependencies are suggested as arrows starting from the ancestors. The graph is also already represented as a layered graph, as all of the nodes having the same maximum distance from the query root are aligned horizontally. We might observe that none of the nodes within each layer shares dependencies.*

5.2. Execution Engine

The execution engine (Algorithm 6) runs the previously compiled query (Section 5.1) on top of the relational model populated from the XES log (Section 4.2). The computation will start from the DAG query leaves directly accessing the relational database (Section 5.2.1) for then propagating the results until the root of the DAG is reached (Section 5.2.2). At this point, we can perform the final conjunctive or aggregation queries (Section 5.2.3).

At each stage, we exploit a functor $\mathcal{A}$ associating to each $xt\,LTL_f$ operator an algorithm which will take the result from the ψ's operands as an input while returning the expected output by formal definition in an intermediate result ρ. This abstraction enables the separation between $xt\,LTL_f$ syntax and multiple possible algorithmic implementations. Some algorithmic implementations for such operators are discussed in Section 6.

Algorithm 6 Execution Engine (Section 5.2)

```
 1: function EXECUTIONENGINE(layer, L, A)
 2:     for all ψ ∈ atomQ (parallel) do ψ.result ← A(ψ)
 3:     RUN D_φ-ENCODINGATOMS(L)                                    ▷ Algorithm S5
 4:     for all ⟨distance, Ψ⟩ ∈ layer do
 5:         for all ψ ∈ Ψ (parallel) do
 6:             if ψ.atom = {p_i} and p_i ∈ ⋃_{a∈Σ} ak(a) then
 7:                 ψ.result ← A(ATOM^{L,τ})(ψ)                     ▷ Algorithm S5
 8:             else if ψ.atom = {a} ∧ a ∈ Σ then
 9:                 continue                                        ▷ Already run in Line 2
10:             else
11:                 ψ.result ← (A(ψ))({ψ'.result|ψ' ∈ ψ.args})
12:             end if
13:         end for
14:     end for
15:     queryRoot ← layer[0]
16:     return queryRoot.result
```

For this step, we will not discuss the computational complexity of evaluating the query plan as this is heavily dominated by the computation of every single operator, the model of choice, and the log size. For this reason, we only conducted empirical analysis by benchmarking the run time of the whole execution engine, where models either only contain $\text{Activity}_{A/T}^{L,\tau}$ (Section 7.2) operators or mainly $\text{Atom}_{A/T}^{L,\tau}$ ones (Section 7.5).

5.2.1. Basic Operators' Execution

Among all of the possible DAG node leaves, we first (Line 2) execute the leaves either *(i)* directly associated with an activity label, or *(ii)* First and Last. For the former *(i)*, each activity label a is run through its correspondent $\text{Activity}_{A/T}^{L,\tau}(a)$ operator, whether either A or T or none are going to be set depending on the fact that such atom refers to an activation (ψ.isActivation) or target (ψ.isTarget) condition, or whether the associated result should be ignored as a whole (ψ.isNeither). For the latter *(ii)*, we directly access the data tables and retrieve the data from them. As the tables are already sorted by trace and event id, no further post-processing besides the insertion of activation or target label in the nested component L of the intermediate representation is required.

Next, we evaluate the intermediate result associated with each atom generated by the $D_φ$-encoding (Line 3). Intuitively (Please refer to Supplement III.1 for a more in-depth discussion with pseudocode.), this requires three subsequent phases. *First*, we obtain the compound conditions grouped by key and activity label as collected at query compile time, and we exploit them to pipeline multiple range queries over each $\text{AttributeTable}_L^k$. The associated results are cached. *Second*, we compute the results for each atom by intersecting the previously cached results before actually computing the actual $\text{Atom}_{A/T}^{L,\tau}$. This also guarantees that shared intersections are run at most once across all of the previously cached results. *Third*, we exploit the former result to compute the $\text{Atom}_{A/T}^{L,\tau}$ operator at the leaf level on our DAG, while associating either an activation or a target mark in L depending on the prior definition of our leaf-level operator.

5.2.2. Results Propagation

After running the basic operators and their derived counterparts (e.g., $\text{Atom}_{A/T}^{L,\tau}$), the only xtLTL_f operators that KnoBAB runs are the ones not accessing the relational tables. KnoBAB implements three different A-s which are only sharing the implementation for the aforementioned operators: one set is either strictly abiding by the formal definition and completely ignoring the fact that the intermediate results are provided as an ordered set of tuples or providing slower algorithms overall, one will leverage appropriate data representation, thus outperforming the former operations, while the other will implement hybrid algorithms for selecting the best performant implementation depending on the data conditions through hybrid algorithms. An in-depth discussion of how different operators might have different algorithmic implementations is postponed to a specific section (Section 6).

5.2.3. Conjunctive and Aggregation Queries

The first version of KnoBAB supports the CONJUNCTIVE QUERY of the model as well as three aggregation queries: MAX-SAT, CONFIDENCE, and SUPPORT. While the former requires a further untimed And$_{\text{True}}$ among all the intermediate results associated with the computation to each clause, the aggregation requires just an iteration over the provided results. The conjunctive query is formulated as follows:

$$\text{CONJUNCTIVEQUERY}(\rho_1,\ldots,\rho_n) = \text{And}_{\text{True}}(\rho_1,\ldots\text{And}_{\text{True}}(\rho_{n-1},\rho_n))$$

The Max-SAT will calculate the ratio of the intermediate results ρ_l associated with each clause c_l, over the total number of model clauses $|\mathcal{M}|$. ActLeaves(ρ_l) is the untimed union of the intermediate results yielded by activation conditions for the Declare clause $c_l \in \mathcal{M}$. For c_l, the CONFIDENCE represents the ratio between the number of traces returned by ρ_l and the total number of traces that contain activation conditions. When the same numerator is on the other hand divided by the total log traces, we have SUPPORT. Following the computation of each ρ_l per clause c_l, the aggregation functions can be expressed as follows:

$$\text{Max-SAT}(\rho_1,\ldots,\rho_n) = \left(\frac{|\{\, l \mid \exists j, L.\ \langle i,j,L \rangle \in \rho_l \,\}|}{|\mathcal{M}|} \right)_{\sigma^i \in \mathcal{L}}$$

$$\text{CONFIDENCE}(\rho_1,\ldots,\rho_n) = \left(\frac{|\{\, i \mid \exists j, L.\ \langle i,j,L \rangle \in \rho_l \,\}|}{|\text{ActLeaves}(\rho_l)|} \right)_{c_l \in \mathcal{M}}$$

$$\text{SUPPORT}(\rho_1,\ldots,\rho_n) = \left(\frac{|\{\, i \mid \exists j, L.\ \langle i,j,L \rangle \in \rho_l \,\}|}{|\mathcal{L}|} \right)_{c_l \in \mathcal{M}}$$

The execution of such queries is performed in a non-parallel way, as each aggregation query will appear at the top of the query plan, and this will be associated with the latest execution run of the scheduler (Line 15). We then return and prompt the result associated with the root node of our query plan (Line 16).

Example 18. *As per previous discussions, the satisfaction of a model requires the satisfaction of all constituent clauses. The model described as the bottom table in Figure 6 is the result of further elaborating on the requirements from Example 1. This is only one example of a myriad of possible solutions, which can either be manually defined (as here), or generated through mining/learning techniques. Such model can be now used to compute the degree to which the model is satisfied, or per trace, each requiring different metrics. An example of a trace-wise metric is Max-SAT while Support and Confidence values can be computed per clause. By providing the trace metrics, we are able to analyse the scenarios with respect to the model, and therefore help provide insight into the exhibits of any backdoors in the software. On the contrary, providing model metrics allows us to establish the suitability of a model and its constituent clauses; for example, clauses with low Support but high Confidence may indicate a correlation between events. Finally, a conjunctive query will return all the traces satisfying all the model clauses. From Figure 6, it is evident that the only trace where a successful attack occurred is σ_1, as returned by the Conjunctive Query, providing the grounds that we have a suitable model. By exploiting the previous formulæ, we can compute the metrics as Table 6. These metrics may provide some insight of correlations between events. For example, clause Ⓑ had Support(Confidence) values as 1.0, while clause Ⓒ had 1/3 (1.0). This therefore indicates that the activation of the latter occurred much less than that of the former; however, every time the activation occurred, the clause was always fulfilled. Conclusions such as these can help to identify* *any weaknesses/strengths within the model and the system itself (here, the metrics obtained from ⓒ may suggest that comm/act contain a correlation that needs investigating).*

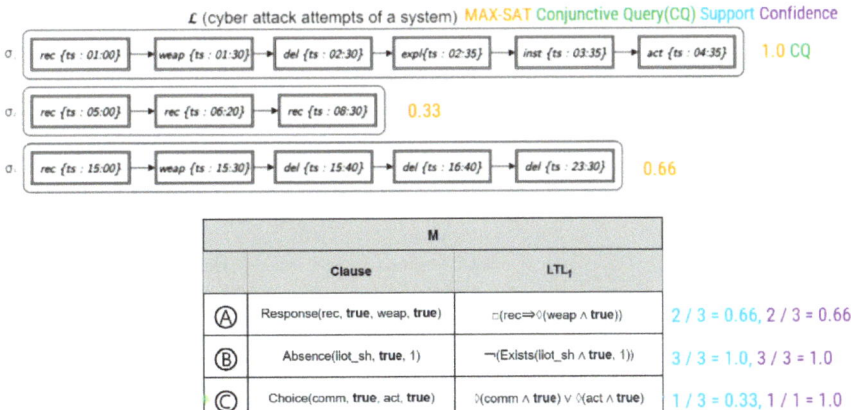

Figure 6. Assessing a high-level use case of an intrusion attack on a software system through a declarative model.

Table 6. Conjunctive and Aggregation queries for Figure 6.

	(a) Metric calculations per trace.	
Trace	MAX-SAT	in Conjunctive Query
σ_1	$\frac{|\{c_1,c_2,c_3\}|}{|M|} = 1.0$	true
σ_2	$\frac{|\{c_2\}|}{|M|} = 1/3$	false
σ_3	$\frac{|\{c_1,c_2\}|}{|M|} = 2/3$	false
	(b) Metric calculations per clause.	
Clause	Support	Confidence
Ⓐ	$\frac{|\{\sigma_1,\sigma_3\}|}{|\mathcal{L}|} = 2/3$	$\frac{|\{\sigma_1,\sigma_3\}|}{|\{\sigma_1,\sigma_2,\sigma_3\}|} = 2/3$
Ⓑ	$\frac{|\{\sigma_1,\sigma_2,\sigma_3\}|}{|\mathcal{L}|} = 1.0$	$\frac{|\{\sigma_1,\sigma_2,\sigma_3\}|}{|\{\sigma_1,\sigma_2,\sigma_3\}|} = 1.0$
Ⓒ	$\frac{|\{\sigma_1\}|}{|\mathcal{L}|} = 1/3$	$\frac{|\{\sigma_1\}|}{|\{\sigma_1\}|} = 1.0$

6. Algorithmic Implementations

In this section, we show how the relational model and the proposed intermediate result representation enable the definition of different operators boosting the query performance compared to an equivalent xt LTL$_f$ expression obtained through the straightforward translation procedure entailed by the lemmas in Appendix A.2 (LTL$_f$-rewriting). Each subsection is going to discuss different possible algorithms for implementing some operators, as well as discussing its associated pseudocode and computational complexity.

6.1. Timed and Untimed Or/And

Algorithm 7 shows the implementation of the timed version of the And$_\Theta^\tau$ (Line 27) and Or$_\Theta^\tau$ (Line 28) operators, for then generalising this concept for the implementation of the untimed And$_\Theta$. We omit the discussion related to the implementation of the untimed Or$_\Theta$ operator for the sake of conciseness.

As we see from their formal definition, any binary xtLTL$_f$ operator supports Θ conditions. And (and Or) resembles a sorted set intersection (or union, Line 11), where we use both trace (i) and event (j) id information from the intermediate result triplet as preliminary equality condition for the match. We also use a Θ binary predicate to be tested over the activated and targeted events in the third component (L). The event shared among the operands is returned if either Θ is always true (Line 7) or, from this point in time, if there exists one activated future activated event (in a L coming from the left operand) as well as a targeted one (in a L coming from the right operand) satisfying the correlation (Line 4). The match is then represented as a marked correlation condition $M(h, k)$, which is then collected in the L associated with the returned event (Line 5).

For the untimed And$_\Theta$ operator, we require to return one single trace i as $\langle i, 1, L \rangle$ if either Θ is true and each operator has an event from σ^i, or if there exists at least one event per operand from the same trace performing the match. This can be implemented in two different ways: we can either group the records by trace id (Lines 31 and 32) and then scan the intermediate results' records (Line 38) associated with the same trace id (Line 36, SLOWUNTIMEDAND) or straightforwardly scan them by trace id without exploiting the preliminary aggregation (FASTUNTIMEDAND). This latter implementation is possible as the intermediate results records are already sorted, thus allowing the results' aggregation while scanning the intermediate results without the need for any preliminary aggregation. We show that the faster version is always faster than computing it with its slower counterpart in Corollary S1.

Similar considerations can be also applied for the untimed Or operation, for which we implemented equivalent SLOWUNTIMEDOR and FASTUNTIMEDOR, as we only need to pay an additional linear scan for the unmatched traces.

6.2. Choice and Untimed Or

We prelude our analysis of derived operators by firstly discussing the difference in computational complexity between providing the straightforward translation from LTL$_f$ to xtLTL$_f$ and to exploiting equivalent expression rewriting in xtLTL$_f$. We remind the reader that the definition of Choice (see Table 1) states that either one condition or another should occur anytime in the trace.

This requirement can be interpreted in two distinct ways: by either returning all the traces satisfying the first condition or the second separately and then merging them, or instead collecting all of the events satisfying either the former or the latter condition while jointly scanning both operands, and then returning the traces where any one of them is met. After observing (Please also refer to the experiments in Section 7.1 for the empirical evidence of such theoretical claims.) that the SLOWUNTIMEDOR is actually slower than FASTUNTIMEDOR and that the latter actually implements the Choice declarative clause (Corollary A1), the time complexity of computing the LTL$_f$ rewriting of Choice in its LTL$_f$-rewriting is almost equivalent to the time complexity of FASTUNTIMEDOR, as we can have an asymptotic constant speed-up in the best case scenario (Corollary S3). As the untimed Or$_\Theta$ behaves by computing a Future operator (Algorithm 8) on each of its operands, the computation of an additional Future operator for each of its operands becomes an omittable overhand.

6.3. Untimed Until(s)

We show how different data access policies for scanning the intermediate results affect the overall computational complexity as well as their associated run time. Algorithm 9 provides two possible variants for the untimed until:

Algorithm 7 xtLTL$_f$ pseudocode implementation for And$_\Theta$ and Or$_\Theta$ operators

1: **function** $\mathcal{T}_\Theta^{E,i}(L, L')$
2:   $L'' \leftarrow \varnothing$; *hasMatch*$\leftarrow \Theta =$ **True**   ▷ (Explicitly) computing $\mathcal{T}_\Theta^{E,i}$
3:   **if** $\Theta \neq$ **True and** $L \neq \varnothing$ **and** $L' \neq \varnothing$ **then**
4:     **for all** $A(m) \in L$ **and** $T(n) \in L'$ s.i. $\Theta(m, n)$ **do**
5:       $L'' \leftarrow L'' \cup \{ M(m, n) \}$; *hasMatch*$\leftarrow$ **true**
6:     **end for**
7:   **else**
8:     $L'' \leftarrow L'' \cup L' \cup L$
9:   **end if**
10: **if** *hasMatch* **then return** L'' **else return False**

11: **function** TIMEDINTERSECTION$_\Theta(\rho, \rho',$ *isUnion*$)$
12:   $it \leftarrow$**Iterator**(ρ), $it' \leftarrow$**Iterator**(ρ')
13:   **while** $it \neq \uparrow$ **and** $it' \neq \uparrow$ **do**
14:     $\langle i, j, L \rangle \leftarrow$ current(it), $\langle i', j', L' \rangle \leftarrow$ current(it')
15:     **if** $i = i'$ **and** $j = j'$ **then**
16:       $tmp \leftarrow \mathcal{T}_\Theta^{E,i}(L, L')$
17:       **if** $tmp \neq$ **False then yield** $\langle i, j, tmp \rangle$
18:       next(it); next(it');
19:     **else if** $i < i'$ **or** $(i = i'$ **and** $j < j')$ **then**
20:       **if** *isUnion* **then yield** $\langle i, j, L \rangle$ **end if**
21:       next(it)
22:     **else**
23:       **if** *isUnion* **then yield** $\langle i', j', L' \rangle$ **end if**
24:       next(it')
25:     **end if**
26:   **end while**

27: **function** AND$_\Theta^\tau(\rho, \rho')$ TIMEDINTERSECTION$_\Theta(\rho, \rho',$ **false**$)$

28: **function** OR$_\Theta^\tau(\rho, \rho')$ TIMEDINTERSECTION$_\Theta(\rho, \rho',$ **true**$)$

29: **function** SLOWUNTIMEDAND$_\Theta(\rho, \rho')$
30:   *leftOperand* $\leftarrow \{\}$; *rightOperand* $\leftarrow \{\}$
31:   **for all** $\langle i, j, L \rangle \in \rho$ **do** *rightOperand*$[i]$.put$(\langle i, j, L \rangle)$
32:   **for all** $\langle i, j, L \rangle \in \rho'$ **do** *rightOperand*$[i]$.put$(\langle i, j, L \rangle)$
33:   $it \leftarrow$**Iterator**$(leftOperand)$, $it' \leftarrow$**Iterator**$(rightOperand)$
34:   **while** $it \neq \uparrow$ **and** $it' \neq \uparrow$ **do**
35:     $\langle i, R \rangle \leftarrow$ current(it); $\langle i', R' \rangle \leftarrow$ current(it')
36:     **if** $i = i'$ **then**
37:       $L'' \leftarrow \varnothing$; *hasMatch*$\leftarrow \Theta =$ **True**
38:       **for all** $\langle i, j, L \rangle \in R$ **and** $\langle i, j', L' \rangle \in R'$ **do**
39:         $tmp \leftarrow \mathcal{T}_\Theta^{E,i}(L, L')$
40:         **if** $tmp \neq$ **False then**
41:           *hasMatch*$\leftarrow$ **true**; $L'' \leftarrow L'' \cup tmp$
42:         **end if**
43:       **end for**
44:       **if** *hasMatch* **then yield** $\langle i, 1, L'' \rangle$;
45:     **else if** $i < i'$ **then** next(it)
46:     **else** next(it')
47:     **end if**
48:   **end while**

49: **function** FASTUNTIMEDAND$_\Theta(\rho, \rho')$
50:   $it \leftarrow$**Iterator**(ρ), $it' \leftarrow$**Iterator**(ρ')
51:   **while** $it \neq \uparrow$ **and** $it' \neq \uparrow$ **do**
52:     $\langle i, \iota, \lambda \rangle \leftarrow$ current(it); $\langle i', \iota', \lambda' \rangle \leftarrow$ current(it')
53:     **if** $i = i'$ **then**
54:       $L'' \leftarrow \varnothing$; *canOptimize*$\leftarrow$ **false**
55:       $it_* \leftarrow it$
56:       **while** $it_* \neq \uparrow$ **do**
57:         $\langle i, j, L \rangle \leftarrow$ current(it_*); $it'_* \leftarrow it'$
58:         **if not** *canOptimize* **then**
59:           **while** $it'_* \neq \uparrow$ **do**
60:             $\langle i', j', L' \rangle \leftarrow$ current(it'_*)
61:             $tmp \leftarrow \mathcal{T}_\Theta^{E,i}(L, L')$
62:             **if** $tmp \neq$ **False then**
63:               *hasMatch*$\leftarrow$ **true**; $L'' \leftarrow L'' \cup tmp$
64:             **end if**
65:             next(it'_*)
66:           **end while**
67:           **if** $\Theta =$ **True then** *canOptimize*$\leftarrow$ **true**
68:         **else** $L'' \leftarrow L'' \cup L$
69:         **end if**
70:         next(it_*)
71:       **end while**
72:       **if** *hasMatch* **then yield** $\langle i, 1, L'' \rangle$;
73:       $it \leftarrow it_*$; $it' \leftarrow it'_*$;
74:     **else if** $i < i'$ **then** next(it)
75:     **else** next(it')
76:     **end if**
77:   **end while**

Algorithm 8 xtLTL_f pseudocode implementation for Future and Globally

```
1: function FUTURE(ρ)                                                          ▷ O(|L|ε²)
2:     for all ⟨i, j, L⟩ ∈ ρ do yield ⟨i, j, ⋃{ L' | ⟨i, j', L'⟩ ∈ ρ and j' ≥ j }⟩
3:     end for

4: function GLOBALLY(ρ)
5:     for all ⟨i, j, L⟩ ∈ ρ do
6:         E ← { j' | ⟨i, j', L'⟩ ∈ ρ and j' ≥ j }
7:         if |E| = ℓ_t − j then yield ⟨i, j, ⋃{ L' | ⟨i, j', L'⟩ ∈ ρ and j' ∈ E }⟩ end if
8:     end for
```

All optimisations happen when the activation condition coming from the second operand does not occur at the beginning of a trace (Lines 34 and 61). In the first variant, we calculate, for all of the events in the first operand starting from the beginning of the trace (Line 29, and Line 51 for the second variant), the position of the last activated event preceding the current target condition with a logarithmic scan with respect to the length of the first operand (Line 34). On the other hand, the second variant directly discards the traces not starting with a target condition (Line 59) and, otherwise, it moves the scan of the first operand—from that initial position—by an offset equal to the distance from the event preceding activation (Line 61): if that position does not correspond to an activation condition preceding the current activation condition, then we completely discard the trace (Line 65). The matching conditions between activations and target are implemented similarly (Lines 37–40 and 67–69). Lemma S7 shows that the second variant is better asymptotically only for bigger datasets.

6.4. Derived Operators

Our previous observation for the untimed Or_Θ led us to the definition of additional derived operators with the hope of easing the overall computational complexity. We walked in the same footsteps of relational algebra, where it was customary to merge multiple operators into one single new operator if the latter might be implemented through a more performant algorithm than computing an equivalent expression being the straightforward translation of LTL_f formulae into LTL_f (LTL_f *rewriting*).

For example, we can implement TIMEDANDFUTURE by extending the fast implementation of the timed AND operator, and considering all of the trace events from the second operand succeeding the events from the first operand within the same trace. Similar considerations can be carried out with TIMEDANDGLOBALLY, where in the former we need to count whether all of the events from the current time until the end of the trace are present in the rightmost operand, while in the latter we also need to skip the matched event from the rightmost operand and start scanning from the following ones.

For simplicity's sake, we postpone the discussion of these operands' pseudocode as well as the discussion of their computational complexity in Supplement II.2, where we show that these two operators might come with two different algorithms, for which there always exists one of them having a lower running time with respect to the equivalent xtLTL_f expression containing no derived operators. We can show formally that, while the first implementation (*variant*) works better for smaller datasets, the second works better for reasonably long traces when the number of the traces is upper bounded by an exponential number of events (Corollary S2).

Algorithm 9 Two implementations for the untimed xtLTL$_f$ Until$_\Theta$.

```
 1: function A^i_Θ(⟨it', bEnd⟩, ⟨it, aEnd⟩)
 2:     ⟨i', j', L'⟩ ← current(it'); L'' ← ∅;
 3:     if Θ ≠ True and L' ≠ ∅ then
 4:         for all A(k), M(k, k') ∈ L' do
 5:             aBeg ← it
 6:             while aBeg ≠ aEnd do
 7:                 ⟨i, j, L⟩ ← current(aBeg)
 8:                 if L = ∅ then L'' ← L'' ∪ L
 9:                 else
10:                     anyMatch ← false
11:                     for all T(h) ∈ L s.t. Θ(σ^i_k, σ^i_h) do anyMatch ← true; L'' ← L'' ∪ {M(k, h)}
12:                     end for
13:                     if not anyMatch then return False
14:                 end if
15:             end while
16:         end for
17:     else
18:         while aBeg ≠ aEnd do
19:             ⟨i, j, L⟩ ← current(aBeg++); L'' ← L'' ∪ L
20:         end while
21:         L' ← L'' ∪ L'
22:     end if
23:     return L''

24: function UntimedUntil^1_Θ(ρ, ρ')
25:     it ← Iterator(ρ), it' ← Iterator(ρ')
26:     while it' ≠ ↑ do
27:         ⟨i', j', L'⟩ ← current(it'); bend ← UpperBound(ρ', it', ↑, ⟨i', |σ^{i'}| + 1, ∅⟩)
28:         it ← LowerBound(ρ, it, ↑, ⟨i', 1, ∅⟩)
29:         atLeastOneResult ← false; L'' ← ∅
30:         while it' < bend do
31:             if j' = 1 then
32:                 atLeastOneResult ← true; L'' ← L'' ∪ L; it'++
33:             else
34:                 aEnd ← UpperBound(ρ, it, ↑, ⟨i', j' − 1, ⊤_Ω⟩)
35:                 if it = aEnd or Distance(aEnd − 1, it) + 1 ≠ j' − 1 then break
36:                 else                                                                             ▷ i = i'. Computing partial T^{A,i}_Θ
37:                     tmp ← A^i_Θ(⟨it', bend⟩, ⟨it, aEnd⟩)
38:                     atLeastOneResult ← atLeastOneResult or tmp ≠ False
39:                     if tmp ≠ False then L'' ← L'' ∪ tmp;
40:                     it'++
41:                 end if
42:             end if
43:         end while
44:         if atLeastOneResult then yield ⟨i, 1, L''⟩
45:         it' ← bend
46:     end while

47: function UntimedUntil^2_Θ(ρ, ρ')
48:     it ← Iterator(ρ), it' ← Iterator(ρ')
49:     while it' ≠ ↑ do
50:         ⟨i', j', L'⟩ ← current(it'); bend ← UpperBound(ρ', it', ↑, ⟨i', |σ^{i'}| + 1, ∅⟩)
51:         it ← LowerBound(ρ, it, ↑, ⟨i', 1, ∅⟩)
52:         atLeastOneResult ← false; L'' ← ∅
53:         while it' < bend do
54:             if j' = 1 then
55:                 atLeastOneResult ← true; L'' ← L'' ∪ L; it'++
56:             else if it = ↑ then break
57:             else
58:                 ⟨i, j, L⟩ ← current(it);
59:                 if j > 1 then break
60:                 else
61:                     aEnd ← MoveForward(it, j' − 1);                                              ▷ (it) + j' − 1
62:                     if aEnd = ↑ then break
63:                     else
64:                         ⟨i_e, j_e, L_e⟩ ← current(aEnd)
65:                         if i_e > i' or j_e ≠ j' − 1 then break
66:                         else                                                                     ▷ i = i' = i_e. Computing partial T^{A,i}_Θ
67:                             tmp ← A^i_Θ(⟨it', bend⟩, ⟨it, aEnd⟩)
68:                             atLeastOneResult ← atLeastOneResult or tmp ≠ False
69:                             if tmp ≠ False then L'' ← L'' ∪ tmp;
70:                             it'++
71:                         end if
72:                     end if
73:                 end if
74:             end if
75:         end while
76:         if atLeastOneResult then yield ⟨i, 1, L''⟩
77:         it' ← bend
78:     end while
```

Table 7 shows the range of datasets used for benchmarking.

Table 7. Range of datasets used for benchmarking.

Competitor	Dataset	Traces $|\mathcal{L}|$	Events	Distinct Activities $|\Sigma|$
SQL Miner	BPIC 2011 (original)	1143	150,291	624
	BPIC 2011 (10)	10	2613	158
	BPIC 2011 (100)	100	12,195	276
	BPIC 2011 (1000)	1000	133,935	607
Declare Analyzer	BPIC 2012 (original)	13,087	262,200	24

7. Results and Discussion

Our benchmarks exploited a Razer Blade Pro on Ubuntu 20.04: Intel Core i7-10875H CPU @ 2.30 GHz–5.10 GHz, 16GB DDR4 2933 MHz RAM, 450 GB free disk space. All of our datasets used for benchmarking (synthetic data generation (Section 7.1), BPIC_2011 (Sections 7.2 and 7.3), BPIC_2012 (Sections 7.4 and 7.5) and our proposed cancer example (Section 1.1) are publicly available (https://dx.doi.org/10.17605/OSF.IO/2CXR7). Table 7 summarises these datasets' features.

7.1. Comparing Different Operators' Algorithms

We advocate that the choice of representing the intermediate representation as an ordered record set allows the exploitation of efficient algorithms through which we might avoid costly counting and aggregation operations [46]. From these comparisons, the operators fully assuming that the data are sorted greatly outperform naïve operators. Walking in the footsteps of relational algebra, we show that the computational complexity of so-called derived operators outperforms the computation of an equivalent expression evaluated through either naïve or fast algorithms. The experiments are discussed in order of presentation of the algorithms in the previous section.

To create a suitable testing environment, we synthetically generate data-less logs, where the trace and log lengths are increased 10-fold at a time from 10^1–10^4, with the resulting sets $|\mathcal{L}| \in \{10, 100, 1000, 10,000\}$ $\epsilon \in \{10, 100, 1000, 10,000\}$, with the most extreme log consisting of 10^8 events. In some cases, we exceeded 16 GB of primary memory on the testing machine; in the following results (Figure 7–10), M+ denotes an out of memory exception. We chose to generate our data in place of using existing real-world logs (https://dx.doi.org/10.17605/OSF.IO/2CXR7), as the controlled scenario allows for identifying the location and extent of any possible speed-ups. These data were up-sampled, guaranteeing that a given log configuration was always a subset of the larger. The data generation randomly assigned events from the universal alphabet ($\Sigma = \{A, B, C, D, E\}$), up to the maximum length for the set in consideration, and we stored the resulting logs as tab-separated files.

Our operators consider correlations between timed events A and B, where the computed speed-up is *per* operator. Given this, we denote $\rho_1 = \text{Activity}_A^{\mathcal{L},\tau}(A)$, $\rho_2 = \text{Activity}_T^{\mathcal{L},\tau}(B)$, prior to benchmarking, and we ignore the time required for accessing the data on the knowledge base, as the focus of the present benchmarks is solely on the operators. Details of how the custom clauses/derived operators are run are demonstrated in Table 8, while singular operators are run sequentially.

Table 8. Proposed operator semantics vs. traditional.

Operator	LTL$_f$ Rewriting	Optimised
Choice	$\text{Or}_\Theta(\text{Future}(\rho_1), \text{Future}(\rho_2))$	$\text{Or}_\Theta(\rho_1, \rho_2)$
TIMEDANDFUTURE	$\text{And}_\Theta(\rho_1, \text{Future}^\tau(\rho_2))$	$\text{AndFuture}_\Theta^\tau(\rho_1, \rho_2)$
TIMEDANDGLOBALLY	$\text{And}_\Theta(\rho_1, \text{Globally}^\tau(\rho_2))$	$\text{AndGlobally}_\Theta^\tau(\rho_1, \rho_2)$

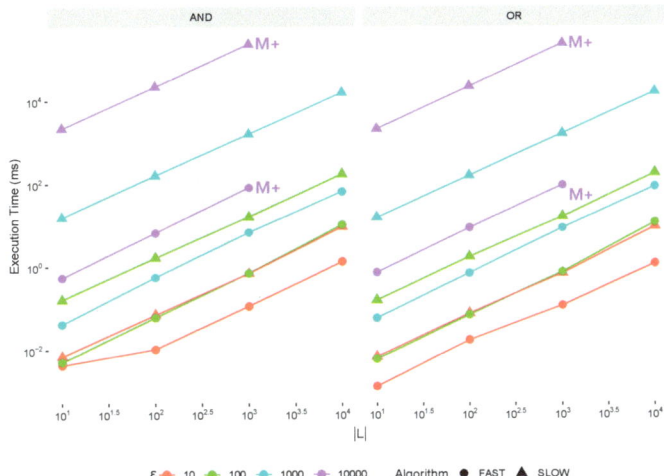

Figure 7. Results for the fast set operations Section 6.1 against the traditional logical implementation.

Untimed Or/And. The first group of experiments aim to challenge different possible algorithms for the same xtLTL$_f$ operators, And$_{True}$ and Or$_{True}$, as discussed in Section 6.1. The outcome of such experiments is given in Figure 7: our experiments reveal that, in every case, the FAST- operators are *always* more performant than their logical counterparts. Our benchmark confirms the cost of overhead encumbered by the SLOW- implementation, which conforms **linearly** to increased log size, almost **polynomially** with trace length. This aggregation is upper bounded with a **quadratic** with respect to trace length ϵ (Lines 31 and 32); in the most extreme case ($\epsilon = 10^4$), the cost is over one order of magnitude versus the algorithm without aggregation. From now on, we always exploit our FAST- operators in place of the SLOW- equivalent for representing non-derived xtLTL$_f$ operators, which usually suffer the cost caused by the preliminary aggregation as per previous experiments.

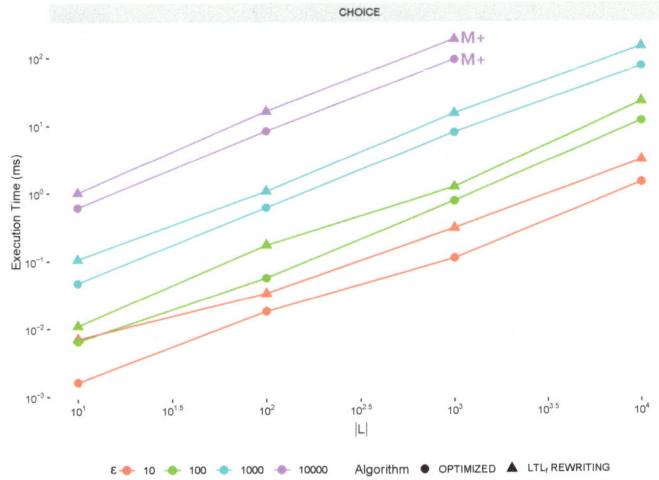

Figure 8. Results for the custom declarative clause implementations Section 6.2 against the traditional logical implementation.

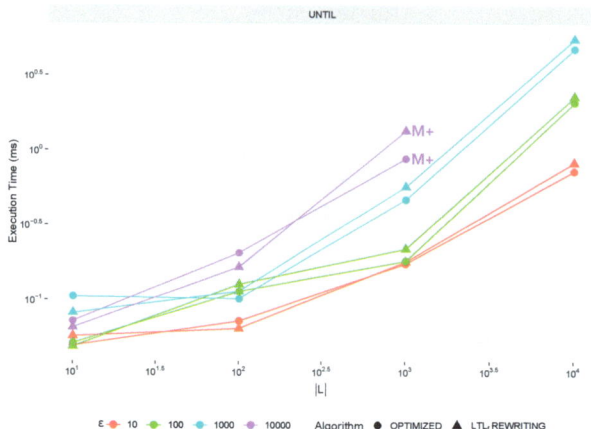

Figure 9. Results for the UNTIL operator (Section 6.3).

Choice and Untimed Or. The next set of experiments is to evaluate the customary declarative clause implementation, where we hypothesise reformulating the semantics associated with Choice to provide performance gains from the absence of preliminary aggregations via the UntimedFuture operator. In fact, the proposed optimisation derives from the omittance of the Future operators for ρ_1, ρ_2, which formally comply with the logical definition. For the untimed Future Section 3.2.2 operator, bounded scans can be exploited, as the data are sorted with respect to trace id, and all the events that satisfy ρ for the current trace id are included in the result. Therefore, we expect an overhead that grows linearly with log size. Figure 8 shows that, in the best case ($\epsilon = 10$), we gain 0.5 orders of magnitude in performance. The findings affirm that log size has a greater influence on computational overhead than trace length. For $\epsilon \geq 10^3$, the overhead resulting from the Future operators steadily increases while both the trace length ϵ and the log size $|\mathcal{L}|$ grows, albeit this is negligible in the logarithmic scale.

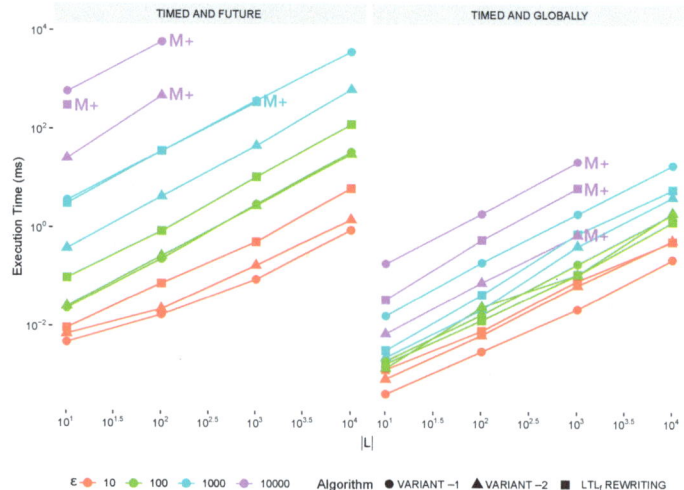

Figure 10. Results for the derived operators TIMEDANDFUTURE and TIMEDANDGLOBALLY Section 6.4. We include both variants of the fast implementations to analyse the environments where each thrive.

Untimed Until(s). Benchmarks from Figure 9 show that the first variant is almost always more performant than the second one for considerably short traces, while the latter becomes more efficient when ϵ increases. With significant increases to log size, the latter becomes more performant; when $|\mathcal{L}| = 10^4$, all cases show improved running times, regardless of ϵ. The plots also show that the operator's running time is polynomial with respect to the number of traces in the log, as a consequence of the increased scans within every single trace.

Derived Operators. The final set of experiments is to test whether the newly proposed derived operators achieve more optimised results than those from their LTL$_f$ rewriting counterpart (Table 8). For example, TIMEDANDGLOBALLY can be optimised with the customary algorithms replacing one single operator with the execution of multiple pipelined operators. Computations from LTL$_f$ rewriting demonstrate worse performance than the derived counterparts across all operators; in the most extreme case TIMEDANDGLOBALLY, there is over $10^{1.5}$ speed-up for $\epsilon = 10^4$. We were able to conclude that different impersonations to the internal data storage of the optimised algorithm may provide better results depending on the log size. As for UntimedUntil, we provide two implementations for TIMEDANDGLOBALLY and TIMEDANDFUTURE, VARIANT-1 (Algorihtm S1) and VARIANT-2 (Algorihtm S2), with the latter exploiting bounded reversed scans on the data.

TIMEDANDGLOBALLY: by merging the AND join operation with Globally, we only consider elements within the same trace **after** the first operand. The logical implementation performs these operations separately, and so cannot reap the benefits of a merged join [47]. Figure 10 shows that, in most cases, there is a linear performance gain with log size. VARIANT-2 aims to exploit potential gains from a reversed scan of a trace while VARIANT-1 provides a forwards scan for every activation. By performing a reverse scan, the latter is able to prune further events from any activations happening in the past, as the condition did not hold for the current time. For smaller trace lengths ($\epsilon \leq 10^1$), the VARIANT-1 demonstrates better performance than VARIANT-2. With increased trace length, the latter operators outperform the former, sometimes by over an order of magnitude ($\epsilon = 10^4$). In some cases, the VARIANT-1 performs slower than their LTL$_f$-rewriting counterparts ($\epsilon \geq 10^3$).

TIMEDANDFUTURE: the principal optimisation gains from this operator follow the same reasoning as TIMEDANDGLOBALLY; however, the implementations of the variants follow a unique approach. By exploiting the allocation of intermediate data structures in reverse, VARIANT-2 also provides improved performance for larger $|\mathcal{L}|$. As with TIMEDANDGLOBALLY, VARIANT-1 outperforms the former for smaller trace lengths.

We conclude that VARIANT-1 (VARIANT-2) of TIMEDANDFUTURE and TIMEDANDGLOBALLY outperform each other for small (large) trace lengths. In addition, the first variant of Until proves to be more performant than our second variant for smaller log lengths. We design a mechanism for always running the fastest algorithm under the previously-observed circumstances. We then need to calculate the average trace length and the log size at data loading time (this only needs to happen once per log). Then, at query time, the most optimal operator is chosen based on these values. We define a HYBRID TRACE QUERY THRESHOLD γ of $10^2/2$ (Lines 5 and 9) and a HYBRID LOG QUERY THRESHOLD η of $10^3/2$ (Line 1); values exceeding these thresholds will execute the operators more tailored towards large trace (log) sizes. The pseudocode provided as Algorithm 10 demonstrates how two different variants can be engulfed in one single parametric algorithm.

Algorithm 10 Hybrid Algorithms

1: **function** HYBRIDUNTIMEDUNTIL$_\Theta^\eta(\rho, \rho')$
2:    **if** $|\mathcal{L}| \geq \eta$ **then return** UNTIMEDUNTIL$_\Theta^2(\rho, \rho')$ ▷ Algorithm 9
3:    **else return** UNTIMEDUNTIL$_\Theta^1(\rho, \rho')$ ▷ Algorithm 9
4:    **end if**

5: **function** HYBRIDANDFUTURE$_\Theta^\gamma(\rho, \rho')$
6:    **if** $\epsilon > \gamma$ **then return** ANDFUTURE$_\Theta^2(\rho, \rho')$ ▷ Algorithm S1
7:    **else return** ANDFUTURE$_\Theta^1(\rho, \rho')$ ▷ Algorithm S2
8:    **end if**

9: **function** HYBRIDANDGLOBALLY$_\Theta^\gamma(\rho, \rho')$
10:    **if** $\epsilon \geq \gamma$ **then return** ANDGLOBALLY$_\Theta^2(\rho, \rho')$ ▷ Algorithm S1
11:    **else return** ANDGLOBALLY$_\Theta^1(\rho, \rho')$ ▷ Algorithm S2
12:    **end if**

7.2. Relational Temporal Mining

We now move from synthetic data, required to tune hybrid algorithms and thoroughly test our operators, towards real data benchmarks with no data payload conditions. We contextualise our experiments for data-intensive model mining operations that can also be run on a relational model. While doing so, we compare our runtimes both with hybrid operators with the one from the previous paper [4], as well as run times from the relational model with traditional SQL queries.

SQLMiner, provided by Schonig et al. [5], utilises database architectures for declarative process mining. We chose to test our hypothesis of engineering a custom database architecture against state-of-the-art traditional relational databases (**PostgreSQL 14.2**). For this set of experiments, we exploited the BPIC 2011 (Dutch academic hospital log) dataset (https://dx.doi.org/10.17605/OSF.IO/2CXR7), as used in [5]. This log contained data payload information, though the queries executed as [5] were comprised of data-less events. The original dataset was sampled into sub-logs containing 10, 100, and 1000 traces, and the sampling approach adopted the same behaviour as the synthetic dataset from the previous set-up, where each sub-log is guaranteed to be a subset of the greater ones. Increased sizes of datasets exhibited exponential increases in primary memory requirements and thus justifies our sampling approach. Schönig [48] provides the templated implementations for mining eight declarative clauses. As these are only templates, the models were instantiated from the resulting combinations of the five most occurring events. Therefore, we generated eight models, each consisting of 25 clauses. SQLMiner simulated this by creating a secondary Actions table, with each row containing the instantiated Declare template. SQLMiner provides the Support values associated with each clause. We extend this to also provide trace information, where each clause also contains the traces satisfying it. We also want to test our hypothesis that our proposed hybrid operator pipeline (Section 7.1) can outperform the pipeline set up from our previous work [4] that does not exploit the potential gains that can be made from picking the best algorithm according to the data conditions, and only uses our defined VARIANT-1 operators. The outcomes of these experiments are shown in Figure 11, where each plot represents the execution times for a given elected template, with the more complex queries located on the first row.

SQLMiner results. In the worst case, our running time is comparable with SQLMiner (Response). Even for this case, SQLMiner returns only the Support information, while KnoBAB also returns (for the same execution time) trace information. In SQL, providing the least possible query alterations to provide the trace information causes $10^{1.5}$ run time increase, thus demonstrating that we are more performant on the same conditions. Conversely, in the best case, we outperform SQLMiner by over five orders of magnitude. By exploiting efficient database design, our custom query plan can minimise data access and our computation avoided explicit computations of aggregations. In addition, guaranteeing that the intermediate results are always sorted allows for **linear** scanning cost for counting operations. Responded Existence is a clear candidate for demonstrating the gains from custom database design: with access to our proposed CountingTable$_\mathcal{L}$, our solution requires only a table look-up, while SQLMiner requires an aggregation requiring an **entire** scan of the Log table. Combining this with the extended xtLTL$_f$ operators allows for much more optimised query times; this is shown in the results, where KnoBAB is consistently at least two orders of magnitude more performant with queries returning trace information. As $|\mathcal{L}|$ increases beyond 10^2, the more complex queries were unable to finish to completion for SQLMiner, exceeding the 16 GB primary memory of the benchmarking machine.

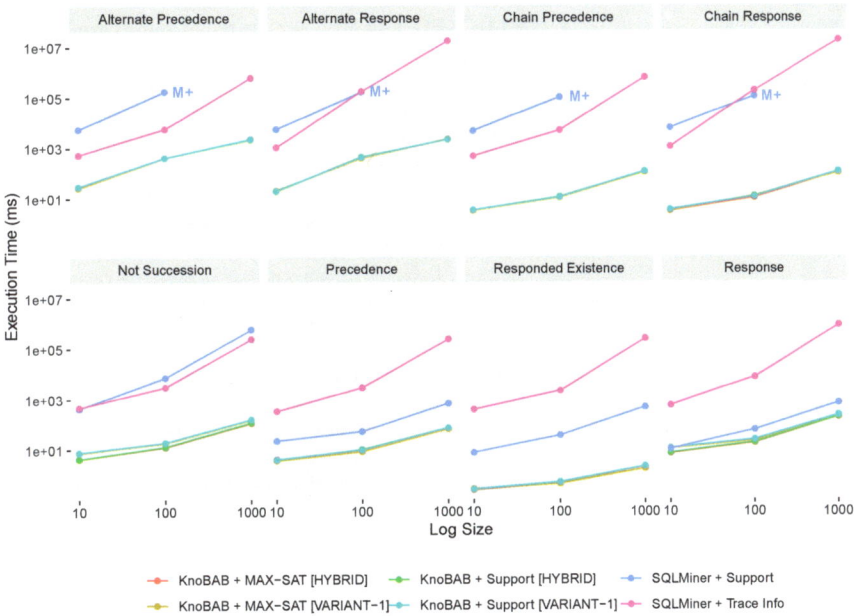

Figure 11. Results for relational temporal mining Section 7.2.

Pipeline results. The execution times for KnoBAB + Support and KnoBAB + Max-SAT are comparable, while there is much greater variation for SQLMiner + Support and SQLMiner + Trace Info. As support requires only an aggregation over intermediate results (Section 5.2.3), we guarantee that we suffer at most a cost proportional to the model size, so we expect a constant overhead based on model size. The large fluctuation in results for SQLMiner is a culprit of the query rewriting provided by the PostgreSQL query engine; in some cases, returning trace information yielded better results. In these experiments, we combined the alternate ensemble methods with our proposed HYBRID operators. The results demonstrate that, for most operators, there is a marginal improvement in time complexity. For NotSuccession and Response, the improvement is more apparent, with the former, for $|\mathcal{L}| = 10$ providing 20% improvement against VARIANT-1. The reader is encouraged to refer back to Figure 10 to explain this. The faster operators thrive with $|\mathcal{L}| > 10^3$, while, for traces within the region of 10^2, the gain is much less apparent. The BPIC_2011 dataset has a corresponding average trace length of ~220: exploiting the VARIANT-2 operators within this region will therefore yield lesser benefit than much larger $|\mathcal{L}|$.

7.3. Query Plan Parallelisation

By keeping the immediately preceding experimental setting while considering the whole log as well as extending the model size, we now benchmark our solution in a multithreaded environment, where we perform intra-query parallelism by running each operator laying in the same layer in parallel as per previous discussions.

The correctness of our proposed parallelisation approach is guaranteed by the fact that each thread in a given layer can operate independently with no interdependencies requiring costly mutual exclusions. In place of directly using the pthread C++ library on multiple tasks, we utilised a thread pool proposed by [49], to minimise the thread creation overhead, while feeding the pool with the tasks denoted by **for ... (parallel) do** statements in our pseudocode Algorithm 6. We extended the library to support both static and dynamic scheduling approaches proposed by the OpenMP specifications [50]; these are:

- **BLOCKED STATIC** : aims to balance the chunk sizes per thread by distributing any leftover iterations;
- **BLOCK-CYCLIC STATIC**. Does not utilise balancing as the former. Instead, work blocks are cyclically allocated over the threads;
- **GUIDED DYNAMIC**: aims to distribute large chunks when there is a lot of work still to be completed; tasks are split into smaller chunks as the work load diminishes;
- **MONOTONIC DYNAMIC**: uses a single centralised counter that is incremented when a thread performs an iteration of work. The schedule issues iterations to threads in an increasing manner.

In addition to these, we also implemented two different scheduling policies splitting the tasks to be run in parallel while estimating the running time that each operator will take depending on the size of its associated operands (if any).

- **TASK SIZE PREDICTION BLOCK STATIC** provides an estimation of work required per chunk. Then, these chunks are sorted in ascending work load, with the last providing the greatest amount of computation. Threads are then assigned chunks through a distribution algorithm, distributing the first and last chunk of the sorted work to the first thread, the second and penultimate to the second, etc.. The algorithm aims to distribute equal amounts of work to each thread, though assumes that the workload is strictly increasing while workload sizes are evenly distributed;
- **TASK SIZE PREDICTION UNBALANCED DYNAMIC**: unlike the former, we assume that the incoming work is not balanced. Instead, a chunk is taken, its work size estimated and assigned to a thread. Then, the next thread will recursively receive chunks until the summed work load is approximate to that of the former. The next thread is then pulled from the pool and the process repeated until all chunks are assigned.

For this set of experiments, we exploited the full BPIC 2011 (Dutch hospital log) dataset. We want to determine how varying the total number of threads affects execution time, and therefore use only the original dataset with no sampling. This also demonstrates the performance against the real-world scenario. Similarly to the previous mining approach in Section 7.2, we generated models from the most occurring events labels. Here, we extended the model size to consider the top 15 events for the same eight Declare templates, thus resulting in 225 clauses. Extending the model size as such allows a better scalability analysis on the large; in fact, a smaller model size would not be able to reap the benefits of the dissected query plan, as it becomes more likely that there will not be enough work to allocate; as more threads might be left idle in the pool, no speed-up can be achieved.

The results of our experiments are shown in Figure 12. Across all instances, the parallelisation pipeline (line with data-points) proves more performant than any single threaded executions (horizontal vertical bar). There also appears to be a great variation in speed-up for different scheduling policies; MONOTONIC DYNAMIC, TASK SIZE PREDICTION UNBALANCED DYNAMIC, and GUIDED DYNAMIC consistently perform worse than all others. In addition to this, the former schedules grant almost no gain with trace number, indicating that dynamic scheduling is not only less performant than static in our use case scenario, but also bears no potential gains by through thread scalability. This is especially true in the case of Alternate Precedence, where all static policies have improved performance by at least an order of magnitude. Schedules also show different degrees of speed-ups. For the dynamic and BLOCK-CYCLIC STATIC schedules, increasing the number of threads has little effect on performance. In fact, adding threads proves to be **detrimental** in some cases (BLOCK-CYCLIC STATIC & Chain Precedence). Conversely, the other static schedules (BLOCKED STATIC and TASK SIZE PREDICTION BLOCK STATIC) achieve a super-linear speed-up [51–53], as the thread count increases. The greatest gains in performance were found for Alternate Precedence and Alternate Response with thread sizes of eight; there are over two orders of magnitude improvement against a single threaded instance, and almost the same speed up compared with the static schedules. As our problem is heavily bounded on data access and on the size of it, reducing the task allocation size will create an overall increase of cache misses, while these are minimised by associating each thread with a greater amount of tasks.

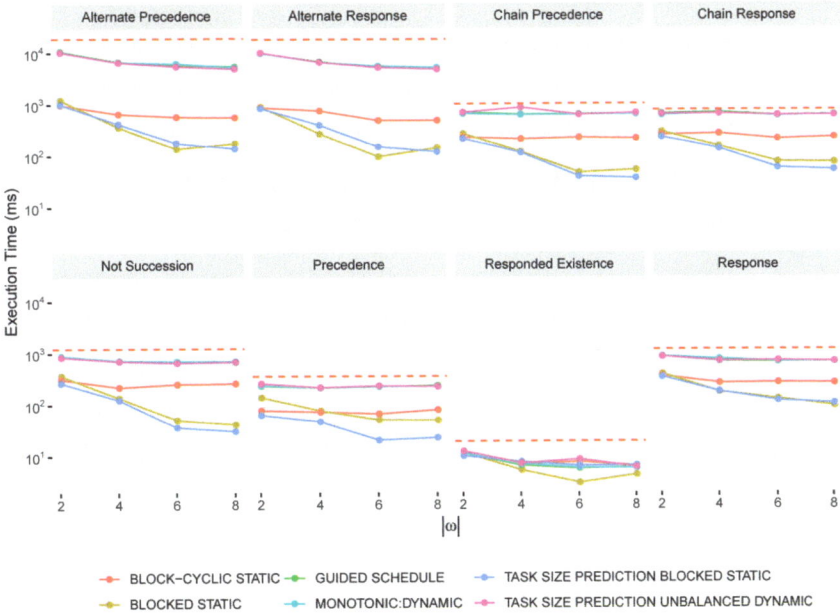

Figure 12. Results for parallelisation Section 7.3. ω indicates the set of threads in the thread pool, and the red dashed horizontal lines indicate running times for single threaded instances.

7.4. $\mathcal{D}_\varphi$-Encoding Atomisation Strategies

We now want to test how distinct query atomisation strategies affect the query run time. For this, we exploit a different dataset while we hardcoded some models suitable for highlighting such differences.

While the `AtomizeEverything` strategy guarantees that all activation and targets undergo the atomizaiton step if a clause is found that contains a data payload predicate, the `AtomizeOnlyOnDataPredicate` atomises only those conditions containing a data payload and considers the others as activity labels. As a consequence, the former is expected to have more weighted access to AttributeTable$_\mathcal{L}$, while the latter to ActivityTable$_\mathcal{L}$. We analyse the execution times over the same models M_1–M_5, where each model differs from the other in the number of clauses as well as in data conditions.

For these experiments, we exploited the full BPIC 2012 (Dutch loan company) dataset. This contained event/trace payload information and was comprised of activities occurring for a loan transaction. The models exploited are visualised in Supplement Table S1a. We define four models, increasing by five clauses, where each is a sub-model of the latter.

These clauses consisted of both data and data-less payload conditions, in order to adhere to our benchmarking hypothesis.

Results are shown in Figure 13 for both configurations, where there is a positive correlation between model size and execution time, with a constant increase with each additional set of clauses. For the smaller model size, `AtomizeEverything` outperforms `AtomizeOnlyOnDataPredicate`, though the former exhibits greater increases in running time as more clauses are added. This therefore suggests that accessing the ActivityTable$_\mathcal{L}$ becomes more expensive than the AttributeTable$_\mathcal{L}$ as the number of activation/target conditions increases. To explain this, the reader is encouraged to refer back to Supplement Table S1a, which defines the clauses that are added to each model, and therefore the new activities and atoms that may require decomposition. With increased model sizes, `AtomizeOnlyOnDataPredicate` suffers from duplicated memory access; as some events (e.g., A_SUBMITTED) are accessed in both tables: while returning the events satisfying an atom requires the access to the AttributeTable$_\mathcal{L}^k$ for any given attribute k of interest, returning all of the events having a given activity label requires accessing the ActivityTable$_\mathcal{L}$. The data access for the atomised queries may duplicate access to the ActivityTable$_\mathcal{L}$, which becomes more costly as our model size increases. Conversely, `AtomizeEverything` will atomize A_SUBMITTED from q_1, as clauses q_2 and q_3 contain payload conditions. Therefore, these queries only ever access the AttributeTable$_\mathcal{L}$, and the duplication of data access is removed. For the smaller model size M_1, this gain is less apparent as the duplicated data access becomes negligible.

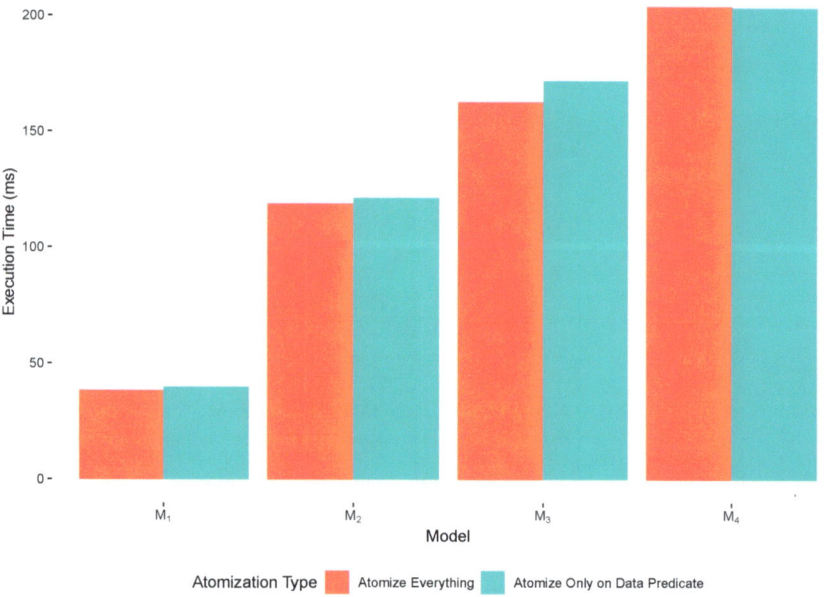

Figure 13. Running times over different models (Table S1a) for different atomisation strategies.

7.5. Data-Aware Conformance Checking

We now consider another state-of-the-art solution, Declare Analyzer [6] for conformance checking with payload information. This solution is tested against two different sets of models of increasing sizes, with each of them providing either the worst or the best case scenario for KnoBAB. These experiments exploit the same dataset as in the former experimental set-up, and also used in [6].

We represented the log for Declare Analyzer via MapDB (https://mapdb.org/), thus reflecting a relational model representation. The authors do not consider trace payloads, and therefore propose injecting trace payload as an extension of each event payload. On the other hand, KnoBAB injects the trace payload as a *unique* event at the beginning of the trace (Section 2), thus reducing the overhead of testing an activation/target condition *per event* while minimising data loading time. We wanted to investigate our solution's performance among the best/worst cases regarding the clauses of choice. Therefore, we provide two scenarios. The first scenario (SCENARIO 1), also described in our seminal paper [4], provides our *worst* case scenario models (Table S1a) where each additional set of clauses consist of entirely novel activity labels and clauses and, within each sub-model, each clause is distinguished by data payload conditions. Consequently, the query plan cannot exploit gains made from data access minimisation as every condition is considered a unique disjunction of atoms. Conversely, the second (SCENARIO 2) novel scenario describes our best case. We encourage the reader to refer to this, where activation and target conditions appear several times in different clauses (Table S2). Thus, there are many more instances where data access can be minimised; for example, the model $q_1 \wedge q_2 \wedge q_3 \wedge q_4 \wedge q_5$ considers the activity label A_SUBMITTED across five instances. Following strategies such as in [9], this can be reduced to one access. SCENARIO 1 (SCENARIO 2) results are shown from Figure 14a (Figure 14b). For either scenario, we average 2–3 orders of magnitude more performant than Declare Analyzer; even in the worst case (M_4), we are over an order of magnitude more performant. For both scenarios, we compute the following metrics: Conjunctive Query (CQ) and Support, to analyse any variations between the ensemble methods. KnoBAB + CQ outperforms KnoBAB + Support in all cases, where the cost increase is linear with model size.

SCENARIO 1. For Declare Analyzer, increases in model size results in a constant slope of 3.47×10^2 ms per model size, while our solution demonstrates an initial slope of 2×10^1 ms per model size, followed by a constant slope of 6×10^0 ms per model size. To explain this abrupt behaviour, the reader is encouraged to refer to Supplement Table S1a and the query plan from Figure 2. KnoBAB thrives when data access is minimised; if this cannot be achieved (due to the addition of novel activation/target conditions), potential gains cannot be exploited. **Every** clause from M_2 contains new activation/target labels/payload conditions compared to M_1. As a result, the number of atoms and leaves in the query plan is doubled. However, M_3 contains the activity label O_CANCELLED. This atom has already been considered in the previous model, and so data access is optimised. Therefore, the time increase from M_2 to M_3 is much less than that of the former. Subsequently, as M_3 is a sub-model of M_4, the same gains are seen here (M_4 contains entirely novel conditions). Overall, the results show that we are not bounded by model size unlike Declare Analyzer, which must perform an entire log scan per clause, while we can ignore irrelevant traces via bounding/indexing across our tabular representation available to the relational model. Still, our running times reflect the formal definition stated in Section 5.2.3, where queries still need to scan each model clause and therefore their expected running time is proportional to the model size.

SCENARIO 2. We now want to test whether clauses providing similar queries lead to lower running times. Here, the model sizes are smaller than the previous example, so as to demonstrate the potential optimisation from even small examples. The former contains only a single clause, while the latter consists of seven clauses. The slope between these models is 3.3×10^0 ms per model size, an order of magnitude less than the worst case scenario. To clarify the results, the reader is encouraged to compare the models q_1 vs. $q_1 \wedge q_2 \wedge q_3 \wedge q_4 \wedge q_5$. All atoms in the former are included in the latter, so we can have much greater data access minimisation, which these results confirm. Of course, a hand-made model is unlikely to contain such overlapping elements, but these results demonstrate the potential gains to be made, even for less bespoke scenarios such as data mining, where a huge amount of overlap might still occur while testing multiple clauses' combinations.

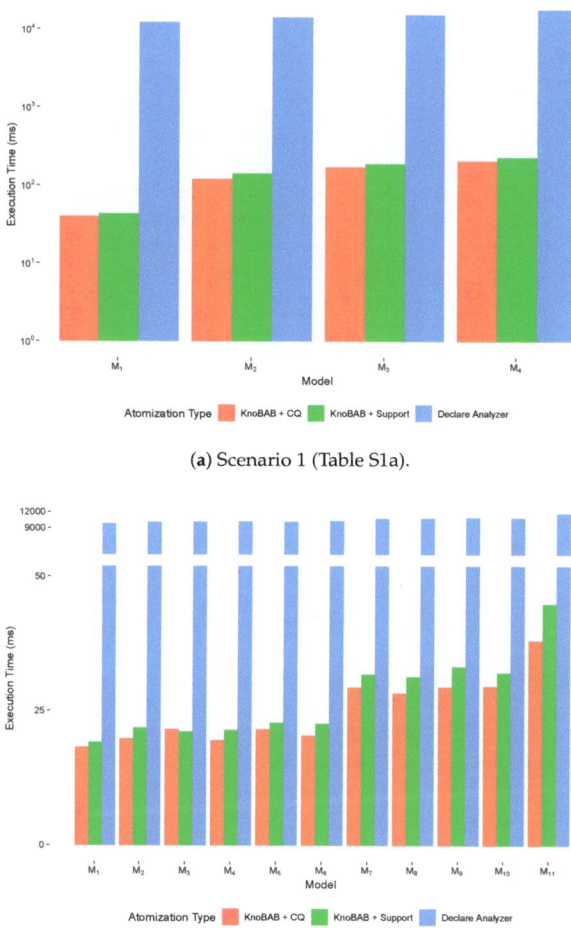

(**a**) Scenario 1 (Table S1a).

(**b**) Scenario 2 (Table S2).

Figure 14. Running times for data-aware conformance checking.

8. Conclusions

By summarizing the contributions of our paper, we showed how to express temporal logic through ad hoc temporal algebra ($xtLTL_f$) based on the relational model. The latter, defined both in its logical and physical model, has been suitably extended for log and operators' result representation. We showed how it is possible to load data on this model using suitable algorithms and how it is possible to represent a sequence of operations with a parallelisable query plan providing super-linear speed-up. As a new contribution to our previous work, we have also shown different implementations for the $xtLTL_f$ operators, thus showing how there is always a faster non-trivial implementation exploiting both the properties of the intermediate result representation as well as query rewriting. Our proposed solution, KnoBAB, leverages all of the aforementioned features, thus providing higher performance than current conformance checking and mining solutions, be it data or data-less.

This work encourages future KnoBAB developments and implementations, including more efficient data model mining algorithms and the use of views to reduce further the cost of allocating intermediate results. Furthermore, secondary memory representation of the log according to the percepts of Near Data Processing is in its infancy. Future developments will explore the possibility of using KnoBAB to learn temporal models from data and the ability to fully support trace repair operations in order to make deviant traces compliant to the given model. For this, we will consider the possibility of integrating our relational system with the BCDM relational model [54], thus fully supporting operations such as insertions, updates, and deletions required for trace repairs in conformance checking [25]. Finally, our future work will also consider vectorial data as a specific data representation [32,55]: this will enable KnoBAB to fully support spatial data representation, thus aiming for full spatio-temporal representation [56,57]. This, along with more advanced model mining algorithms, will enable us to efficiently mine spatio-temporal patterns from logs. Finally, we will also investigate the possibility of transferring the definition of such algebraic operators when logs are represented as graphs [58,59], thus further improving the efficiency of graph-based query languages.

Supplementary Materials: The following supporting information can be downloaded at: https://www.mdpi.com/article/10.3390/info14030173/s1.

Author Contributions: *Conceptualisation*, S.A. and G.B.; *methodology*, G.B.; *software*, S.A. and G.B.; *validation*, S.A. and G.B.; *formal analysis*, G.B.; *investigation*, S.A.; *resources*, G.B. and G.M.; *data curation*, S.A.; *writing—original draft preparation*, S.A. and G.B.; *writing—review and editing*, G.B. and G.M.; *visualisation*, S.A.; *supervision*, G.B. and G.M.; *project administration*, G.B.; *funding acquisition*, G.B. All authors have read and agreed to the published version of the manuscript.

Funding: Samuel Appleby's work is supported by Newcastle University.

Informed Consent Statement: Not applicable.

Data Availability Statement: The dataset associated with the presented experiments was available online the 5 March 2022: https://dx.doi.org/10.17605/OSF.IO/2CXR7.

Conflicts of Interest: The authors declare no conflict of interest.

Sample Availability: The most up-to-date version of KnoBAB is available on GitHub: https://github.com/datagram-db/knobab (5 March 2022).

Abbreviations

The following abbreviations are used in this manuscript:

DAG	Direct Acyclic Graph
KnoBAB	KNOwledge Base for Alignments and Business process modelling
LTL_f	Linear Temporal Logic over finite traces
RDBMS	Relational Database Management System
XES	eXtensible Event Stream
$xtLTL_f$	eXTended Linear Temporal Logic over finite traces

Appendix A

We now show some equivalence and correctness lemmas.

Appendix A.1

First, we want to show the equivalence of some unary operators as generalisations of some of the base operators. We now show that $\text{Init}_{A/T}^{\mathcal{L}}$ or $\text{Ends}_{A/T}^{\mathcal{L}}$ can be subsumed by appropriate combinations of Init or Ends with $\text{Activity}_{A/T}^{\mathcal{L},\tau}$. As the former set of operators cannot express data conditions for the events while the former can by replacing $\text{Activity}_{A/T}^{\mathcal{L},\tau}$ with an arbitrary sub-expression with $\text{Atom}_{A/T}^{\mathcal{L},\tau}$, we can trivially conclude that the former are less general than the latter.

Lemma A1.

$$\forall a \in \Sigma. \quad \text{Init}_{A/T}^{\mathcal{L}}(a) = \text{Init}(\text{Activity}_{A/T}^{\mathcal{L},\tau}(a))$$

Proof. We can expand the definition of the left-hand side of the equation for any a $\in \Sigma$ as follows:

$$\text{Init}^{\mathcal{L}}_{A/T}(a) = \{\, \langle i, 1, \{A/T(1)\}\rangle \mid \exists \phi.\, \langle \beta(a), i, 1, \bot, \phi \rangle \in \text{ActivityTable}_{\mathcal{L}} \,\}$$

The right-hand side of the equation can be rewritten as follows:

$$\text{Init}(\text{Activity}^{\mathcal{L},\tau}_{A/T}(a)) = \{\, \langle i, 1, \{A/T(1)\}\rangle \mid \exists \pi, \phi.\, \langle \beta(a), i, 1, \pi, \phi \rangle \in \text{ActivityTable}_{\mathcal{L}} \,\}$$

The goal is immediately closed by choosing $\pi = \bot$, as any first event will have always an empty Prev pointer. □

Lemma A2.

$$\forall a \in \Sigma. \quad \text{Ends}^{\mathcal{L}}_{A/T}(a) = \text{Ends}(\text{Activity}^{\mathcal{L},\tau}_{A/T}(a))$$

Proof. We can expand the definition of the left-hand side of the equation for any a $\in \Sigma$ as follows:

$$\text{Ends}^{\mathcal{L}}_{A/T}(a) = \{\, \langle i, 1, \{A/T(|\sigma^i|)\}\rangle \mid \exists \pi.\, \langle \beta(a), i, 1, \pi, \bot \rangle \in \text{ActivityTable}_{\mathcal{L}} \,\}$$

The right-hand side of the equation can be rewritten as follows:

$$\text{Ends}(\text{Activity}^{\mathcal{L},\tau}_{A/T}(a)) = \{\, \langle i, 1, L\rangle \mid \langle i, |\sigma^i|, L\rangle \in \text{ActivityTable}_{\mathcal{L}} \,\}$$
$$= \{\, \langle i, 1, \{A/T(|\sigma^i|)\}\rangle \mid \exists \pi.\, \langle \beta(a), i, |\sigma^i|, \pi, \bot \rangle \in \text{ActivityTable}_{\mathcal{L}} \,\}$$

The goal is immediately closed by choosing $\pi = \bot$, as any first event will always have an empty Prev pointer. □

On the other hand, as the $\text{Exists}^{\mathcal{L}}_{A/T}$ and $\text{Absence}^{\mathcal{L}}_{A/T}$ operators discard the activation and target marks of the associated events for the purposes of efficiency, we need to relax their notion of equivalence by ignoring the result being provided by the third component. Still, we can observe that they compute the same result trace-wise. Even in this scenario, as the former operators merely access the counting table for the purposes of efficiency, they cannot be generally exploited when the expression of data conditions is also required.

Lemma A3.

$$\forall a \in \Sigma. \forall \sigma^i \in \mathcal{L}. \exists L, L'.\, \langle i, 1, L\rangle \in \text{Exists}^{\mathcal{L}}_{A/T}(a, n) \Leftrightarrow \langle i, 1, L'\rangle \in \text{Exists}_n(\text{Activity}^{\mathcal{L},\tau}_{A/T}(a))$$

Proof.

$$\langle i, 1, L\rangle \in \text{Exists}^{\mathcal{L}}_{A/T}(a, n) \Leftrightarrow \langle i, 1, L'\rangle \in \text{Exists}_n(\text{Activity}^{\mathcal{L},\tau}_{A/T}(a))$$
$$\exists m \geq n.\, \langle \beta(a), i, m\rangle \in \text{CountingTable}_{\mathcal{L}} \Leftrightarrow n \leq |\{\, \langle i, j, L'\rangle \in \text{Activity}^{\mathcal{L},\tau}_{A/T}(a) \,\}|$$
$$\left|\{\, \sigma^i_j \in \sigma^i \mid \sigma^i_j = \langle a, p\rangle \,\}\right| \geq n \Leftrightarrow n \leq |\{\, \langle \beta(a), i, j, \pi, \phi\rangle \in \text{ActivityTable}_{\mathcal{L}} \,\}|$$

□

Lemma A4.

$$\forall a \in \Sigma. \forall \sigma^i \in \mathcal{L}. \exists L, L'.\, \langle i, 1, L\rangle \in \text{Absence}^{\mathcal{L}}_{A/T}(a, n) \Leftrightarrow \langle i, 1, L'\rangle \in \text{Absence}_n(\text{Activity}^{\mathcal{L},\tau}_{A/T}(a))$$

Proof. By simply replacing the $m \geq n$ and $n \leq |S|$ for any set S conditions in the former lemma to $m < n$ and $n > |S|$. This boils down to:

$$\langle i, 1, L \rangle \in \text{Absence}^{\mathcal{L}}_{A/T}(a, n) \Leftrightarrow \langle i, 1, L' \rangle \in \text{Absence}_n(\text{Activity}^{\mathcal{L},\tau}_{A/T}(a))$$

$$\exists m < n. \langle \beta(a), i, m \rangle \in \text{CountingTable}_{\mathcal{L}} \Leftrightarrow n > |\{ \langle i, j, L' \rangle \in \text{Activity}^{\mathcal{L},\tau}_{A/T}(a) \}|$$

$$|\{ \sigma^i_j \in \sigma^i \mid \sigma^i_j = \langle a, p \rangle \}| < n \Leftrightarrow n > |\{ \langle \beta(a), i, j, \pi, \phi \rangle \in \text{ActivityTable}_{\mathcal{L}} \}|$$

□

Appendix A.2

Next, we want to show that xtLTL_f is at least as expressive as LTL_f. To support this claim, we need to prove the two following lemmas where, as LTL_f does not support explicit activation and target conditions with Θ correlation conditions over the payload data, we are always going to assume $\Theta = \textbf{True}$ and that the atomic operators are never associated with an activation/target label, thus always returning an empty third component of the intermediate result. As we might observe, the following lemma entails that, differently from standard LTL_f semantics applied to each event trace at a time, xtLTL_f semantics returns all of the events for which the given temporal condition holds. This becomes very relevant for minimising the data access while scanning our relational representation of the log, as well as allowing better intermediate result reuse for any incoming sub-expression. The following lemma also entails a correspondence between timed xtLTL_f operators and LTL_f formulae.

Lemma A5. *For each LTL_f formula φ, a timed $xt\text{LTL}_f$ expression ψ^τ evaluated over an intended relational model representing a log $\mathcal{L}$ of finite and non-empty traces exists for which the latter returns $\langle i, j, L \rangle$ iff. $\sigma^i_j \vDash \varphi$. More formally:*

$$\forall \sigma^i_j \in \sigma^i, \sigma^i \in \mathcal{L}. \forall \varphi \in \text{LTL}_f. \exists \psi^\tau : \text{timed } xt\text{LTL}_f.(\langle i, j, L \rangle \in \psi^\tau \Leftrightarrow \sigma^i_j \vDash \varphi)$$

Proof. The constructive proof proceeds by structural induction over ψ^τ. We first need to consider a rewriting lemma stating that $\langle \beta(a), i, j, \pi, \chi \rangle \in \text{ActivityTable}_{\mathcal{L}}$ iff. a p exists such that $\sigma^i_j = \langle a, p \rangle$. Now, we can start the proof by induction.

$\varphi = \text{a}$: By applying the aforementioned rewriting lemma (from now on simply referred to as *by construction of ActivityTable*), we can immediately close the goal by choosing $\psi^\tau = \text{Activity}^\tau(a)$ as the model will only return data associated with the log of choice:

$$\langle i, j, L \rangle \in \text{Activity}^\tau(a) \Leftrightarrow \exists p. \sigma^i_j = \langle a, p \rangle \Leftrightarrow \sigma^i_j \vDash a$$

$\varphi = a \wedge q$: If the compound condition is also atomic for which q can be expressed as an interval query $low \leq \kappa \leq up$ for some payload key κ, we can follow a similar proof from the former case and choose the atom $\psi^\tau = \text{Compound}^\tau(a, \kappa, low, up)$, thus closing the goal as follows:

$$\langle i, j, L \rangle \in \text{Compound}^\tau(a, \kappa, [low, up]) \Leftrightarrow \exists p. \sigma^i_j = \langle a, p \rangle \wedge low \leq p(\kappa) \leq up$$
$$\Leftrightarrow \sigma^i_j \vDash a \wedge (low \leq \kappa \wedge \kappa \leq up)$$

$\varphi = \bigcirc \varphi'$: by inductive hypothesis, we know the ρ xtLTL_f expression returning ρ, which contains $\langle i, j+1, L \rangle$ when $\sigma^i_{j+1} \vDash \varphi'$. For this, we choose as $\psi^\tau = \text{Next}^\tau(\rho)$, which also guarantees that j never exceeds the trace's length ($j \leq |\sigma^i|$). We can therefore expand the definition of our proposed operator by obtaining:

$$\langle i, j, L \rangle \in \mathsf{Next}^\tau(\rho) \Leftrightarrow \langle i, j+1, L \rangle \in \rho \wedge 1 < j+1 \leq |\sigma^i|$$
$$\overset{IH}{\Leftrightarrow} \sigma^i_{j+1} \vDash \varphi \wedge 0 < j < |\sigma^i|$$
$$\Leftrightarrow \sigma^i_j \vDash \bigcirc, \varphi$$

$\varphi = \Box \varphi'$: The application of the induction is similar to the former and, similarly to the former case, we also proceed by expanding the definition of the relational operator. We can hereby choose $\psi^\tau = \mathsf{Globally}^\tau(\rho)$ where the induction is applied over ρ and φ'. We can close the goal as follows:

$$\langle i, j, L \rangle \in \mathsf{Globally}^\tau(\rho) \Leftrightarrow \langle i, j, L_j \rangle \in \rho \wedge |\sigma^i| - j + 1 = |\{\langle i, k, L_k \rangle \in \rho | j \leq k \leq |\sigma^i|\}|$$
$$\Leftrightarrow \forall j \leq k \leq |\sigma^i|. \langle i, k, L_k \rangle \in \rho$$
$$\overset{IH}{\Leftrightarrow} \forall j \leq k \leq |\sigma^i|.\sigma^i_k \vDash \varphi'$$
$$\Leftrightarrow \sigma^i_j \vDash \Box \varphi'$$

$\varphi = \Diamond \varphi'$: Similarly to globally, we obtain $\psi^\tau = \mathsf{Future}^\tau(\rho)$ for a ρ corresponding to φ' by inductive hypothesis.

$\varphi = \neg \varphi'$: Similarly to the previous unary operators, we choose as xtLTL$_f$ operator $\psi^\tau = \mathsf{Not}^\tau(\rho)$ where the inductive hypothesis links ρ to φ'. We can therefore close the goal as follows:

$$\langle i, j, L \rangle \in \mathsf{Not}^\tau(\rho) \Leftrightarrow \sigma^i_j \in \sigma^i \wedge \sigma^i \in \mathcal{L} \wedge \langle i, j, L \rangle \notin \rho$$
$$\overset{IH}{\Leftrightarrow} \sigma^i_j \in \sigma^i \wedge \sigma^i \in \mathcal{L} \wedge \sigma^i_j \vDash \neg \varphi$$
$$\Leftrightarrow \sigma^i_j \vDash \neg \varphi$$

This is doable as stating $\langle i, j, L \rangle \in \psi^\tau \Leftrightarrow \sigma^i_j \vDash \varphi$ is equivalent to $\langle i, j, L \rangle \notin \psi^\tau \Leftrightarrow \sigma^i_j \not\vDash \varphi$ where the latter can be rewritten as $\sigma^i_j \vDash \neg \varphi$.

$\varphi = \varphi' \wedge \varphi''$: As we have that two inductive hypotheses associate ρ' and ρ'' respectively to φ' and φ'', we choose the xtLTL$_f$ formula $\psi^\tau = \mathsf{And}^\tau_{\mathsf{True}}(\rho', \rho'')$ to be associated with $\varphi' \wedge \varphi''$. For this xtLTL$_f$ operator, we can state that a result $\langle i, j, \ell \rangle$ is returned by such an operator if and only if $\langle i, j, \varnothing \rangle \in \rho'$ and $\langle i, j, \varnothing \rangle \in \rho''$ per definition of operators never returning explicit activation or target condition. We close the goal as follows:

$$\langle i, j, L \rangle \in \mathsf{And}^\tau_{\mathsf{True}}(\rho, \rho') \Leftrightarrow \langle i, j, \varnothing \rangle \in \rho \wedge \langle i, j, \varnothing \rangle \in \rho'$$
$$\overset{IH}{\Leftrightarrow} \sigma^i_j \vDash \varphi \wedge \sigma^i_j \vDash \varphi'$$
$$\Leftrightarrow \sigma^i_j \vDash \varphi \wedge \varphi'$$

$\varphi = \varphi' \vee \varphi''$: We can firstly observe that $(A \wedge B) \vee (A \wedge \neg B) \vee (\neg A \wedge B)$ in the classical semantics is equivalent to $A \vee B$ for any possible proposition A and B (*OrRwLem*). After observing that the current operator is defined by extension of the previously proved one, we can exploit the previous one as a rewriting lemma. As we have that two inductive hypothesis associating ρ' and ρ'' respectively to φ' and φ'', we choose the xtLTL$_f$ formula $\psi^\tau = \mathsf{Or}^\tau_{\mathsf{True}}(\rho', \rho'')$ to be associated with $\varphi' \vee \varphi''$. We close the goal as follows:

$$\langle i,j,L \rangle \in \mathsf{Or}^\tau_{\mathsf{True}}(\rho,\rho') \overset{IH}{\Leftrightarrow} \langle i,j,L\rangle \in \mathsf{And}^\tau(\rho,\rho') \vee (\sigma^i_j \vDash \varphi \wedge \sigma^i_j \nvDash \varphi') \vee (\sigma^i_j \vDash \varphi' \wedge \sigma^i_j \nvDash \varphi)$$
$$\Leftrightarrow \sigma^i_j \vDash \varphi \wedge \varphi' \vee (\sigma^i_j \vDash \varphi \wedge \sigma^i_j \nvDash \varphi') \vee (\sigma^i_j \vDash \varphi' \wedge \sigma^i_j \nvDash \varphi)$$
$$\overset{OrRwLem}{\Leftrightarrow} \sigma^i_j \vDash \varphi \vee \varphi'$$

$\varphi = \varphi' \,\mathcal{U}\, \varphi''$: as both the results from the third element of the intermediate results are always empty by construction and preliminary assumption, and we have inductive hypothesis associating ρ and ρ' respectively to φ and φ', we can immediately close the goal after choosing the xtLTL$_f$ formula $\psi^\tau = \mathsf{Until}^\tau_{\mathsf{True}}(\rho',\rho'')$ to be associated with $\varphi = \varphi' \,\mathcal{U}\, \varphi''$.

□

The next lemma is required for closing the generic lemma stated at the beginning of this sub-section, as LTL$_f$ starts assessing the formulae from the beginning of each trace. We need to show that the former lemma applies to xtLTL$_f$ operators in a stricter version, which is the following one:

Lemma A6. *For each LTL$_f$ formula φ satisfied from the beginning of the trace, it exists an xtLTL$_f$ expression ψ returning a $\langle i, 1, L\rangle$, thus highlighting that the condition holds from the beginning of the trace. More formally:*

$$\forall \sigma^i \in \mathcal{L}. \forall \varphi \colon \mathsf{LTL}_f. \exists \psi \in \textit{xt}\mathsf{LTL}_f. (\sigma^i \vDash \varphi \Rightarrow \exists L.\, \langle i, 1, L\rangle \in \psi)$$

Proof. Similarly to the previous lemma, as LTL$_f$ cannot express activation and target conditions to be tested in Θ correlation conditions, we always choose $\Theta = \mathbf{True}$, and we decide to use base xtLTL$_f$ operators where none of these conditions is returned. Differently from the previous lemma, we now have to go by inductive structure over the LTL$_f$ formulae rather than on the xtLTL$_f$ ones. We can therefore consider the following inductive cases:

$\varphi = a$: By definition of the Init operator, it is sufficient to consider $\psi = \mathsf{Init}(a)$;

$\varphi = a \wedge p$: Under the assumption that the compound condition corresponds to an atomic query with $p := \mathit{low} \leq \kappa \leq \mathit{up}$, we can formulate the former as follows: $\psi = \mathsf{Init}(\mathsf{Compound}^{\mathcal{L},\tau}(a,\kappa,\mathit{low},\mathit{up}))$;

$\varphi = \bigcirc, \varphi'$: By rewriting this definition, this implies to prove that $\varphi^i_2 \vDash \varphi'$. As the Next$^\tau$ operator is a timed one and we cannot assess φ' from the beginning of the trace, we cannot exploit the inductive hypothesis for φ', but we need to apply the previously proven lemma for the conditions happening at any point in the trace. From the application of the previous lemma, we have that $\varphi^i_2 \vDash \varphi \Leftrightarrow \langle i, 2, L\rangle \in \rho$ for some xtLTL$_f$ expression returning ρ. From this, it follows that $\langle i, 1, L\rangle \in \mathsf{Next}^\tau(\rho)$. By its definition, Next$^\tau$ returns all events preceding the ones stated in ρ, while, for $\sigma^i \vDash \bigcirc, \varphi$, we are only interested in restricting all of the possible results of Next$^\tau$ to the ones also corresponding to the beginning of the trace. For this reason, we need to consider ψ as $\mathsf{And}^\tau(\mathsf{First}^\tau, \mathsf{Next}^\tau(\rho))$;

$\varphi = \Box\varphi'$: Similarly to the previous operator, φ' is timed and should be checked for all events σ^i_j of interest within the trace σ^i. Even in this case, we need to apply the previous lemma for φ', thus guaranteeing that an xtLTL$_f$ expression ρ exists containing $\langle i, j, L\rangle$ whenever $\sigma^i_j \vDash \varphi'$. As globally requires that all of the events satisfy φ', we have that $\mathsf{Globally}(\rho)$ responds by the intended semantics, and therefore we choose this as our ψ;

$\varphi = \Diamond\varphi'$: Similarly to the previous operator, we choose $\mathsf{Future}(\rho)$ when ρ is linked to the evaluation of φ' for any possible trace event by the previous lemma;

$\varphi = \neg\varphi'$: In this other scenario, we can directly apply the previous lemma, as the evaluation of φ' will always start from the beginning of the trace. After recalling that $\nexists x.P(x) \Leftrightarrow \forall x.\neg P(x)$, we rewrite the definition of φ while applying the inductive hypothesis for the present lemma over some ρ semantically linked to φ' as follows:

$$\sigma^i \vDash \neg\varphi \Leftrightarrow \sigma^i_1 \nvDash \varphi' \overset{IH}{\Leftrightarrow} \forall L. \langle i, 1, L\rangle \notin \rho$$

Per inductive hypothesis, ρ contains all of the records $\langle i, 1, L\rangle$ for which $\sigma^i_1 \vDash \varphi'$; as the untimed negation will return a record $\langle \iota, 1, \varnothing\rangle$ if and only if there is no event associated with the trace ι in the provided operand, we can choose $\psi = \mathsf{Not}(\rho)$ and close the goal as follows:

$$\langle i, 1, \varnothing\rangle \in \mathsf{Not}(\rho) \Leftrightarrow \forall j, L. \langle i, j, L\rangle \notin \rho \Leftrightarrow \forall L. \langle i, 1, L\rangle \notin \rho$$

$\varphi = \varphi' \, \mathcal{U} \, \varphi''$: Similarly to the former operators, both φ' and φ'' required a timed evaluation of the events along the trace of interest, for which we need to exploit the former lemma, thus obtaining timed xtLTL_f expressions ρ' and ρ''. We can immediately close the lemma by choosing $\psi = \mathsf{Until}_{\mathsf{True}}(\rho', \rho'')$;

$\varphi = \varphi' \wedge \varphi''$: Similarly to the negation operator, we can directly apply the inductive hypothesis on φ' and φ'', as these sub-operators will also be assessed from the beginning of a trace; these will be associated respectively to the xtLTL_f expressions ρ' and ρ'' having $\langle i, 1, \varnothing\rangle \in \rho'$ and $\langle i, 1, \varnothing\rangle \in \rho''$ as we exploit neither activation nor target conditions. As per construction ρ' and ρ'' will contain no record $\langle i, j+2, L\rangle$ for some natural number $j \geq 0$, we chose $\psi = \mathsf{And}_{\mathsf{True}}(\rho', \rho'')$;

$\varphi = \varphi' \vee \varphi''$: By exploiting similar consideration from the former operator, we chose $\psi = \mathsf{Or}_{\mathsf{True}}(\rho', \rho'')$ for some ρ' and ρ'' respectively associated by inductive hypothesis to φ' and φ''.

□

As a corollary of the two given lemmas, we have that xtLTL_f is at least as expressive as LTL_f, as any LTL_f formula can always be computed through an equivalent xtLTL_f formula. This validates the decision from our previous work [4] where we expressed the semantics of each template in Declare through a correspondent xtLTL_f expression. These were also checked through automated testing Appendix A.2. At this stage, we also want to ascertain that the untimed and timed operators work as expected, that is, that we can mimic the outcome of the timed operators over the timed ones if, for each event $\langle i, j, L\rangle$, we evaluate the corresponding untimed operator over the suffix $\sigma^i_j, \ldots, \sigma^i_{|\sigma^i|}$. This can be proven as follows:

Lemma A7. *For each timed xtLTL_f operator ψ^τ containing a result $\langle i, j, L\rangle$ over a relational representation of $\mathcal{L}$, generate a log of suffixes $\mathcal{L}' = \{\sigma^{i\oplus j}\}$, where $\sigma^{i\oplus j} := \sigma^i_j, \ldots, \sigma^i_{|\sigma^i|}$ of σ^i, and each event is defined as $\sigma^{i\oplus j}_k := \sigma^i_{j+k-1}$ for each $1 \leq k \leq |\sigma^i| - j + 1$. For this, an xtLTL_f expression ψ evaluated over the relational representation of $\mathcal{L}'$ always exists such that $\langle i \oplus j, 1, L\rangle \in \psi$.*

Proof. We prove the lemma by induction over ψ^τ by considering all of the timed operators having an untimed counterpart. Please observe that we discard the negation Not from our considerations, as we have previously mentioned that the timed and untimed versions of this serve different purposes. We also provide an implementation (https://github.com/datagram-db/knobab/blob/main/tests/ltlf_operators_test.cpp, 5 March 2023) of such proofs via automated testing.

$\psi^\tau = \mathbf{Activity}^{\mathcal{L},\tau}_{A/T}(a)$: This can be trivially closed by choosing $\mathsf{Init}^{\mathcal{L}'}_{A/T}(a)$;

$\psi^\tau = \text{Compound}_{A/T}^{\mathcal{L},\tau}(a, k, low, up)$: This can be trivially closed by choosing $\text{Init}(\text{Compound}_{A/T}^{\mathcal{L}'}(a, k, [low, up]))$;

$\psi^\tau = \text{Globally}^\tau(\rho)$: After observing that $|\sigma^{i \oplus j}| = |\sigma^i| - j + 1$, we obtain the following condition by operator's expansion, where ρ' is evaluated over $\mathcal{L}'$ as per inductive hypothesis:

$$\langle i, j, L \rangle \in \text{Globally}^\tau(\rho) \Leftrightarrow L := \cup_{\substack{j \leq k \leq |\sigma^i|, \\ \langle i,k,L_k \rangle \in \rho}} L_k \wedge |\sigma^i| - j + 1 = \left| \{ \langle i, k, L_k \rangle \in \rho | j \leq k \leq |\sigma^i| \} \right|$$

$$\Leftrightarrow L := \cup_{\substack{1 \leq k \leq |\sigma^{i \oplus j}|, \\ \langle i \oplus j,k,L_k \rangle \in \rho}} L_k \wedge |\sigma^{i \oplus j}| = \left| \{ \langle i \oplus j, k, L_k \rangle \in \rho | 1 \leq k \leq |\sigma^{i \oplus j}| \} \right|$$

$$\Leftrightarrow L := \cup_{\langle i \oplus j,k,L_k \rangle \in \rho} L_k \wedge |\sigma^{i \oplus j}| = \left| \{ \langle i \oplus j, k, L_k \rangle \in \rho \} \right|$$

$$\Leftrightarrow \langle i \oplus j, 1, L \rangle \in \text{Globally}(\rho');$$

$\psi^\tau = \text{Future}^\tau(\rho)$: By following similar consideration as per the former operator, we have:

$$\langle i, j, L \rangle \in \text{Future}^\tau(\rho) \Leftrightarrow L := \cup_{\substack{j \leq k \leq |\sigma^i| \\ \langle i,k,L_k \rangle \in \rho}} L_k \wedge \exists h \geq j, L. \langle i, h, L_h \rangle \in \rho$$

$$\Leftrightarrow L := \cup_{\substack{1 \leq k \leq |\sigma^{i \oplus j}| \\ \langle i \oplus j,k,L_k \rangle \in \rho}} L_k \wedge \exists h \geq 1, L. \langle i, h, L_h \rangle \in \rho$$

$$\Leftrightarrow L := \cup_{\langle i \oplus j,k,L_k \rangle \in \rho} L_k \wedge \exists h, L. \langle i, h, L_h \rangle \in \rho$$

$$\Leftrightarrow \langle i \oplus j, 1, L \rangle \in \text{Future}(\rho');$$

$\psi^\tau = \text{And}_\Theta^\tau(\rho_1, \rho_2)$: By rewriting the definition of the timed And operator, we obtain the following:

$$\langle i, j, L \rangle \in \text{And}_\Theta^\tau(\rho_1, \rho_2) \Leftrightarrow \exists L_1, L_2. \langle i, j, L_1 \rangle \in \rho_1 \wedge \langle i, j, L_2 \rangle \in \rho_2 \wedge$$
$$L := \mathcal{T}_\Theta^{E,i}([j \mapsto L_1], [j \mapsto L_2]) \wedge L \neq \textbf{False}$$

If And contains for both of its operands an event σ_j^i, it follows that there should be at least one match $\sigma_1^{i \oplus j}$ over the corresponding untimed operator $\text{And}_\Theta(\rho', \rho'')$ evaluated over $\mathcal{L}'$. For the latter operator, we can therefore ensure that a j exists and a j' being $j = j' = 1$ and L as well as L' for which the following condition holds:

$$\langle i, j, L \rangle \in \text{And}_\Theta^\tau(\rho_1, \rho_2) \Rightarrow \exists L_1, L_2. \langle i \oplus j, 1, L_1 \rangle \in \rho' \wedge \langle i \oplus j, 1, L_2 \rangle \in \rho'' \wedge$$
$$L := \mathcal{T}_\Theta^{E,i}([1 \mapsto \{L_j | \langle i, j, L_j \rangle \in \rho'\}], [1 \mapsto \{L_j | \langle i, j, L_j \rangle \in \rho''\}]) \wedge$$
$$L \neq \textbf{False}$$
$$\Leftrightarrow \langle i \oplus j, 1, L \rangle \in \text{And}_\Theta(\rho', \rho'');$$

$\psi^\tau = \text{Or}_\Theta^\tau(\rho_1, \rho_2)$: As this operator is derived from the definition of And_Θ^τ, we can directly close the goal by the previous inductive step if the result represents a match between the elements of the first and second operand. If there were no events that might have been matched, the data come either from the first or from the second operand. As the two cases are symmetric, we just provide proof for the former case. In this situation, we have a $\langle i, j, L \rangle \in \text{Or}_\Theta^\tau(\rho_1, \rho_2)$ corresponding to a $\langle i, j, L \rangle \in \rho_1$ for which there is no L' such that $\langle i, j, L' \rangle \in \rho_2$. If there still exists a j' and L' such that $\langle i, j', L' \rangle \in \rho_2$ for which there might be a match between L and L', then this case falls under the untimed And_Θ over $\mathcal{L}'$, and we still have some τ for which the latter returns $\langle i \oplus j, 1, \tau \rangle$; if match is never possible or no of such j' exists, then the untimed Or_Θ operator will return a $\langle i \oplus j, 1, \cup \{L | \exists k. \langle i \oplus j, k, L \rangle \in \rho_2 \} \rangle$ by definition;

$\psi^\tau = \mathbf{Until}^\tau(\rho_1, \rho_2)$: This is a mere rewriting exercise, as the untimed version of Until is a mere instantiation of the latter where only the case $k = 1$ is considered.

□

Appendix A.3

At this stage, we provide some rewriting lemmas motivating the introduction of derived operators. First, we want to show that the untimed $\mathsf{And}_\Theta(\rho, \rho')$ operator can also be exploited to compute $\mathsf{And}_\Theta(\mathsf{Future}(\rho), \mathsf{Future}(\rho'))$, thus motivating the peculiar definition of such operator with an existential interpretation over all of the possible matches in the future. We can formally prove this as follows:

Lemma A8.

$$\forall \rho, \rho'. \mathsf{And}_\Theta(\mathsf{Future}(\rho), \mathsf{Future}(\rho')) = \mathsf{And}_\Theta(\rho, \rho')$$

Proof. By expanding the definition of the operators, we obtain:

$\langle i, 1, L'' \rangle \in \mathsf{And}_\Theta(\mathsf{Future}(\rho), \mathsf{Future}(\rho')) \Leftrightarrow \exists L, L'. \langle i, 1, L \rangle \in \mathsf{Future}(\rho) \land \langle i, 1, L' \rangle \in \mathsf{Future}(\rho')$
$\quad L'' := \mathcal{T}_\Theta^{E,i}([1 \mapsto \cup \{L_j | \langle i, j, L_j \rangle \in \rho_1\}], [1 \mapsto \cup \{L_j | \langle i, j, L_j \rangle \in \rho_2\}]),$
$\quad L'' \neq \mathbf{False}$
$\Leftrightarrow \exists j, j', L, L'. \langle i, j, L \rangle \in \rho \land \langle i, j', L' \rangle \in \rho'$
$\quad L'' := \mathcal{T}_\Theta^{E,i}([1 \mapsto \cup \{L_j | \langle i, j, L_j \rangle \in \rho_1\}], [1 \mapsto \cup \{L_j | \langle i, j, L_j \rangle \in \rho_2\}]),$
$\quad L'' \neq \mathbf{False}$
$\Leftrightarrow \langle i, 1, L'' \rangle \in \mathsf{And}_\Theta(\rho, \rho')$

□

Please remember that the untimed And operator is also compliant with the LTL_f semantics as per our previous lemmas. We can therefore exploit the versatile definition of such operation to reduce the computational overhead provided by the additional and unrequired aggregation provided by Future. Given the previous lemma, we have as a Corollary that the semantics associated with the Choice Declare clause, i.e., $\mathsf{Or}_\Theta(\mathsf{Future}(\rho), \mathsf{Future}(\rho'))$, can equivalently be computed by $\mathsf{Or}_\Theta(\rho, \rho')$. The following proof motivates the choice of exploiting E_Θ^i as a correlation matching semantics for both And_Θ and Or_Θ.

Corollary A1.

$$\forall \rho, \rho'. \mathsf{Or}_\Theta(\mathsf{Future}(\rho), \mathsf{Future}(\rho')) = \mathsf{Or}_\Theta(\rho, \rho')$$

Proof. By expanding the definition of the untimed Or_Θ, we obtain:

$\mathsf{Or}_\Theta(\mathsf{Future}(\rho), \mathsf{Future}(\rho')) = \mathsf{And}_\Theta(\mathsf{Future}(\rho), \mathsf{Future}(\rho')) \cup \{ \langle i, 1, \cup\{L | \exists j. \langle i, j, L \rangle \in \mathsf{Future}(\rho)\} \rangle \mid \nexists j, L'. \langle i, j, L' \rangle \in \mathsf{Future}(\rho') \}$
$\cup \{ \langle i, 1, \cup\{L | \exists j. \langle i, j, L \rangle \in \mathsf{Future}(\rho')\} \rangle \mid \nexists j, L'. \langle i, j, L' \rangle \in \mathsf{Future}(\rho) \}$

For the previous lemma, this becomes:

$\mathsf{Or}_\Theta(\mathsf{Future}(\rho), \mathsf{Future}(\rho')) = \mathsf{And}_\Theta(\rho, \rho') \cup \{ \langle i, 1, \cup\{L | \exists j. \langle i, j, L \rangle \in \mathsf{Future}(\rho)\} \rangle \mid \nexists j, L'. \langle i, j, L' \rangle \in \mathsf{Future}(\rho') \}$
$\cup \{ \langle i, 1, \cup\{L | \exists j. \langle i, j, L \rangle \in \mathsf{Future}(\rho')\} \rangle \mid \nexists j, L'. \langle i, j, L' \rangle \in \mathsf{Future}(\rho) \}$

At this stage, we only need to test the contribution of the second component of the union, as the third one is symmetrical (ρ and ρ' are just inverted). As the elements of the second component of the union come from Future operators, we can rewrite such as follows:

$$\{ \langle i, 1, \cup\{L | \langle i, 1, L \rangle \in \mathsf{Future}(\rho)\} \rangle \mid \nexists L'. \langle i, 1, L' \rangle \in \mathsf{Future}(\rho') \}$$

We can also observe that $\langle i, 1, L \rangle \in \mathsf{Future}(\rho)$ for a given L if there exist a j and L'' for which $\langle i, j, L'' \rangle \in \rho$. Similar considerations come from the negated counterpart ($\langle i, 1, L \rangle \notin \mathsf{Future}(\rho)$). For this expansion, we can therefore close our goal. □

The remaining lemmas show the correctness of the logical formulation of the derived operators, thus motivating their adoption when possible. These lemmas were also tested in our implementation (See the end of https://github.com/datagram-db/knobab/blob/main/tests/until_test.cpp, 5 March 2023). The supplementary materials (Section II) show that it is possible to implement such derived operators so that they are faster than their corresponding LTL_f rewriting counterpart.

Lemma A9.
$$\forall \rho, \rho'. \text{And}_\Theta^\tau(\rho_1, \text{Future}^\tau(\rho_2)) = \text{AndFuture}_\Theta^\tau(\rho_1, \rho_2)$$

Proof.

$$\langle i,j,L \rangle \in \text{And}_\Theta^\tau(\rho_1, \text{Future}^\tau(\rho_2)) \Leftrightarrow \exists L_1, L_2. \langle i,j,L_1 \rangle \in \rho_1 \wedge \langle i,j,L_2 \rangle \in \text{Future}^\tau(\rho_2) \wedge$$
$$L := \mathcal{T}_\Theta^{E,i}([j \mapsto L_1],[j \mapsto L_2]) \wedge L \neq \textbf{False}$$
$$\Leftrightarrow \exists L_1, L_2. \langle i,j,L_1 \rangle \in \rho_1 \wedge \exists h \geq j, L. \langle i,h,L \rangle \in \rho_2 \wedge$$
$$L := \mathcal{T}_\Theta^{E,i}([j \mapsto L_1],[j \mapsto \cup_{\substack{j \leq k \leq |\sigma^i| \\ \langle i,k,L_k \rangle \in \rho}} L_k]) \wedge L \neq \textbf{False}$$
$$\Leftrightarrow \langle i,j,L \rangle \in \text{AndFuture}_\Theta^\tau(\rho_1, \rho_2)$$

□

Lemma A10.
$$\forall \rho, \rho'. \text{And}_\Theta^\tau(\rho_1, \text{Globally}^\tau(\rho_2)) = \text{AndGlobally}_\Theta^\tau(\rho_1, \rho_2)$$

Proof.

$$\langle i,j,L \rangle \in \text{And}_\Theta^\tau(\rho_1, \text{Globally}^\tau(\rho_2)) \Leftrightarrow \exists L_1, L_2. \langle i,j,L_1 \rangle \in \rho_1 \wedge \langle i,j,L_2 \rangle \in \text{Globally}^\tau(\rho_2) \wedge$$
$$L := \mathcal{T}_\Theta^{E,i}([j \mapsto L_1],[j \mapsto L_2]) \wedge L \neq \textbf{False}$$
$$\Leftrightarrow \exists L_1. \langle i,j,L_1 \rangle \in \rho_1 \wedge \langle i,j,L_j \rangle \in \rho_2 \wedge$$
$$|\sigma^i| - j + 1 = \left| \{\langle i,k,L_k \rangle \in \rho | j \leq k \leq |\sigma^i|\} \right| \wedge$$
$$L := \mathcal{T}_\Theta^{E,i}([j \mapsto L_1],[j \mapsto \cup_{\substack{j \leq k \leq |\sigma^i| \\ \langle i,k,L_k \rangle \in \rho}} L_k]) \wedge L \neq \textbf{False}$$
$$\Leftrightarrow \exists L_1. \langle i,j,L_1 \rangle \in \rho_1 \wedge \forall j \leq k \leq |\sigma^i|. \exists L'. \langle i,k,L_k \rangle \in \rho_2 \wedge$$
$$L := \mathcal{T}_\Theta^{E,i}([j \mapsto L_1],[j \mapsto \cup_{\substack{j \leq k \leq |\sigma^i| \\ \langle i,k,L_k \rangle \in \rho}} L_k]) \wedge L \neq \textbf{False}$$
$$\Leftrightarrow \langle i,j,L \rangle \in \text{AndGlobally}_\Theta^\tau(\rho_1, \rho_2)$$

□

References

1. Agrawal, R.; Imieliński, T.; Swami, A. Mining Association Rules between Sets of Items in Large Databases. *SIGMOD Rec.* **1993**, *22*, 207–216. [CrossRef]
2. Bergami, G.; Maggi, F.M.; Montali, M.; Peñaloza, R. Probabilistic Trace Alignment. In Proceedings of the 2021 3rd International Conference on Process Mining (ICPM), Eindhoven, The Netherlands, 31 October–4 November 2021; pp. 9–16. [CrossRef]
3. Schön, O.; van Huijgevoort, B.; Haesaert, S.; Soudjani, S. Correct-by-Design Control of Parametric Stochastic Systems. In Proceedings of the 2022 IEEE 61st Conference on Decision and Control, Cancun, Mexico, 6–9 December 2022.
4. Appleby, S.; Bergami, G.; Morgan, G. Running Temporal Logical Queries on the Relational Model. In Proceedings of the International Database Engineered Applications Symposium (IDEAS'22), Budapest, Hungary, 22–24 August 2022; pp. 222–231.
5. Schönig, S.; Rogge-Solti, A.; Cabanillas, C.; Jablonski, S.; Mendling, J. Efficient and Customisable Declarative Process Mining with SQL. In *Advanced Information Systems Engineering, Proceedings of the 28th International Conference, CAiSE 2016, Ljubljana, Slovenia, 13–17 June 2016*; Springer: Berlin/Heidelberg, Germany, 2016.
6. Burattin, A.; Maggi, F.M.; Sperduti, A. Conformance checking based on multi-perspective declarative process models. *Expert Syst. Appl.* **2016**, *65*, 194–211. [CrossRef]

7. Pesic, M.; Schonenberg, H.; van der Aalst, W.M.P. DECLARE: Full Support for Loosely-Structured Processes. In Proceedings of the 11th IEEE International Enterprise Distributed Object Computing Conference, Annapolis, MA, USA, 15–19 October 2007; pp. 287–300.
8. Musser, D.R. Introspective Sorting and Selection Algorithms. *Softw. Pract. Exp.* **1997**, *27*, 983–993. [CrossRef]
9. Bellatreche, L.; Kechar, M.; Bahloul, S.N. Bringing Common Subexpression Problem from the Dark to Light: Towards Large-Scale Workload Optimizations. In Proceedings of the 25th International Database Engineering & Applications Symposium, Montreal, QC, Canada, 14–16 July 2021.
10. Naldurg, P.; Sen, K.; Thati, P. A Temporal Logic Based Framework for Intrusion Detection. In Proceedings of the Formal Techniques for Networked and Distributed Systems—FORTE 2004: 24th IFIP WG 6.1 International Conference, Madrid, Spain, 27–30 September 2004; Núñez, M., Ed.; Springer: Berlin/Heidelberg, Germany, 2004; Volume 3235, pp. 359–376.
11. Ray, I. Security Vulnerabilities in Smart Contracts as Specifications in Linear Temporal Logic. Master's Thesis, University of Waterloo, Waterloo, ON, Canada, 2021.
12. Buschjäger, S.; Hess, S.; Morik, K. Shrub Ensembles for Online Classification. In Proceedings of the the AAAI Conference on Artificial Intelligence 2022, Virtual, 22 February–1 March 2022; pp. 6123–6131.
13. Huo, X.; Hao, K.; Chen, L.; song Tang, X.; Wang, T.; Cai, X. A dynamic soft sensor of industrial fuzzy time series with propositional linear temporal logic. *Expert Syst. Appl.* **2022**, *201*, 117176. [CrossRef]
14. Bergami, G.; Francescomarino, C.D.; Ghidini, C.; Maggi, F.M.; Puura, J. Exploring Business Process Deviance with Sequential and Declarative Patterns. *arXiv* **2021**, arXiv:2111.12454.
15. Zhou, H.; Milani Fard, A.; Makanju, A. The State of Ethereum Smart Contracts Security: Vulnerabilities, Countermeasures, and Tool Support. *J. Cybersecur. Priv.* **2022**, *2*, 358–378. [CrossRef]
16. Szabo, N. Smart contracts: Building blocks for digital markets. *Extropy J. Transhumanist Thought* **1996**, *18*, 28.
17. Fionda, V.; Greco, G.; Mastratisi, M.A. Reasoning About Smart Contracts Encoded in LTL. In Proceedings of the AIxIA 2021—Advances in Artificial Intelligence: 20th International Conference of the Italian Association for Artificial Intelligence, Virtual Event, 1–3 December 2021; Springer International Publishing: Cham, Switzerland, 2021; pp. 123–136.
18. Bank, H.S.; D'souza, S.; Rasam, A. Temporal Logic (TL)-Based Autonomy for Smart Manufacturing Systems. *Procedia Manuf.* **2018**, *26*, 1221–1229. [CrossRef]
19. Mao, X.; Li, X.; Huang, Y.; Shi, J.; Zhang, Y. Programmable Logic Controllers Past Linear Temporal Logic for Monitoring Applications in Industrial Control Systems. *IEEE Trans. Ind. Informatics* **2022**, *18*, 4393–4405. [CrossRef]
20. Boniol, P.; Linardi, M.; Roncallo, F.; Palpanas, T.; Meftah, M.; Remy, E. Unsupervised and scalable subsequence anomaly detection in large data series. *Vldb J.* **2021**, *30*, 909–931. [CrossRef]
21. Xu, H.; Pang, J.; Yang, X.; Yu, J.; Li, X.; Zhao, D. Modeling clinical activities based on multi-perspective declarative process mining with openEHR's characteristic. *BMC Med. Inform. Decis. Mak.* **2020**, *20-S*, 303. [CrossRef]
22. Rovani, M.; Maggi, F.M.; de Leoni, M.; van der Aalst, W.M.P. Declarative process mining in healthcare. *Expert Syst. Appl.* **2015**, *42*, 9236–9251. [CrossRef]
23. Bertini, F.; Bergami, G.; Montesi, D.; Veronese, G.; Marchesini, G.; Pandolfi, P. Predicting Frailty Condition in Elderly Using Multidimensional Socioclinical Databases. *Proc. IEEE* **2018**, *106*, 723–737. [CrossRef]
24. De Giacomo, G.; Maggi, F.M.; Marrella, A.; Patrizi, F. On the Disruptive Effectiveness of Automated Planning for LTLf-Based Trace Alignment. In Proceedings of the AAAI Conference on Artificial Intelligence 2017, San Francisco, CA, USA, 4–9 February 2017.
25. Bergami, G.; Maggi, F.M.; Marrella, A.; Montali, M. Aligning Data-Aware Declarative Process Models and Event Logs. In *Business Process Management*; Springer International Publishing: Berlin/Heidelberg, Germany, 2021; pp. 235–251.
26. Bergami, G. A Logical Model for joining Property Graphs. *arXiv* **2021**, arXiv:2106.14766.
27. Zhu, S.; Pu, G.; Vardi, M.Y. First-Order vs. Second-Order Encodings for LTLf-to-Automata Translation. *arXiv* **2019**, arXiv:1901.06108.
28. Ceri, S.; Gottlob, G. Translating SQL Into Relational Algebra: Optimization, Semantics, and Equivalence of SQL Queries. *IEEE Trans. Software Eng.* **1985**, *11*, 324–345. [CrossRef]
29. Calders, T.; Lakshmanan, L.V.S.; Ng, R.T.; Paredaens, J. Expressive power of an algebra for data mining. *ACM Trans. Database Syst.* **2006**, *31*, 1169–1214. [CrossRef]
30. Li, J.; Pu, G.; Zhang, Y.; Vardi, M.Y.; Rozier, K.Y. SAT-based explicit LTLf satisfiability checking. *Artif. Intell.* **2020**, *289*, 103369. [CrossRef]
31. Petermann, A.; Junghanns, M.; Müller, R.; Rahm, E. FoodBroker-Generating Synthetic Datasets for Graph-Based Business Analytics. In Proceedings of the 5th International Workshop, WBDB 2014, Potsdam, Germany, 5–6 August 2014.
32. Bergami, G. On Declare MAX-SAT and a finite Herbrand Base for data-aware logs. *arXiv* **2021**, arXiv:2106.07781.
33. Pichler, P.; Weber, B.; Zugal, S.; Pinggera, J.; Mendling, J.; Reijers, H.A. Imperative versus Declarative Process Modeling Languages: An Empirical Investigation. In Proceedings of the BPM 2011 International Workshops, Clermont-Ferrand, France, 29 August 2011; pp. 383–394.
34. Codd, E.F. A Relational Model of Data for Large Shared Data Banks. *Commun. ACM* **1970**, *13*, 377–387. [CrossRef]
35. Idreos, S.; Groffen, F.; Nes, N.; Manegold, S.; Mullender, K.S.; Kersten, M.L. MonetDB: Two Decades of Research in Column-oriented Database Architectures. *IEEE Data Eng. Bull.* **2012**, *35*, 40–45.

36. Boncz, P.A.; Manegold, S.; Kersten, M.L. Database Architecture Evolution: Mammals Flourished long before Dinosaurs became Extinct. *Proc. VLDB Endow.* **2009**, *2*, 1648–1653.
37. Roth, M.A.; Korth, H.F.; Silberschatz, A. Extended Algebra and Calculus for Nested Relational Databases. *ACM Trans. Database Syst.* **1988**, *13*, 389–417. [CrossRef]
38. Wang, J.; Ntarmos, N.; Triantafillou, P. GraphCache: A Caching System for Graph Queries. In Proceedings of the International Conference on Extending Database Technology (EDBT) 2017, Venice, Italy, 21–24 March 2017; pp. 13–24.
39. Keller, A.M.; Basu, J. A Predicate-based Caching Scheme for Client-Server Database Architectures. *VLDB J.* **1996**, *5*, 35–47. [CrossRef]
40. Davey, B.A.; Priestley, H.A. *Introduction to Lattices and Order*, 2nd ed. ; Cambridge University Press: Cambridge, UK, 2002.
41. de Berg, M.; Cheong, O.; van Kreveld, M.J.; Overmars, M.H. *Computational Geometry: Algorithms and Applications*, 3rd ed.; Springer: Berlin/Heidelberg, Germany, 2008.
42. Elmasri, R.; Navathe, S.B. *Fundamentals of Database Systems*, 7th ed.; Pearson: Upper Saddle River, NJ, USA, 2015.
43. Polyvyanyy, A.; ter Hofstede, A.H.M.; Rosa, M.L.; Ouyang, C.; Pika, A. Process Query Language: Design, Implementation, and Evaluation. *arXiv* **2019**, arXiv:1909.09543.
44. Coffman, E.G.; Graham, R.L. Optimal Scheduling for Two-Processor Systems. *Acta Inform.* **1972**, *1*, 200–213. [CrossRef]
45. Sugiyama, K.; Tagawa, S.; Toda, M. Methods for Visual Understanding of Hierarchical System Structures. *IEEE Trans. Syst. Man. Cybern.* **1981**, *11*, 109–125. [CrossRef]
46. Bergami, G. On Efficiently Equi-Joining Graphs. In Proceedings of the 25th International Database Engineering & Applications Symposium 2021, Montreal, QC, Canada, 14–16 July 2021.
47. Dittrich, J. *Patterns in Data Management: A Flipped Textbook*; CreateSpace Independent Publishing Platform: Charleston, SC, USA, 2016.
48. Schönig, S. SQL Queries for Declarative Process Mining on Event Logs of Relational Databases. *arXiv* **2015**, arXiv:1512.00196.
49. Shoshany, B. A C++17 Thread Pool for High-Performance Scientific Computing. *arXiv* **2021**, arXiv:2105.00613.
50. Klemm, M.; Cownie, J. 8 Scheduling parallel loops. In *High Performance Parallel Runtimes*; De Gruyter Oldenbourg: Berlin, Germany; Boston, MA, USA, 2021; pp. 228–258.
51. Ristov, S.; Prodan, R.; Gusev, M.; Skala, K. Superlinear speedup in HPC systems: Why and when? In Proceedings of the 2016 Federated Conference on Computer Science and Information Systems (FedCSIS), Gdańsk, Poland, 11–14 September 2016; pp. 889–898.
52. Yan, B.; Regueiro, R.A. Superlinear speedup phenomenon in parallel 3D Discrete Element Method (DEM) simulations of complex-shaped particles. *Parallel Comput.* **2018**, *75*, 61–87. [CrossRef]
53. Nagashima, U.; Hyugaji, S.; Sekiguchi, S.; Sato, M.; Hosoya, H. An experience with super-linear speedup achieved by parallel computing on a workstation cluster: Parallel calculation of density of states of large scale cyclic polyacenes. *Parallel Comput.* **1995**, *21*, 1491–1504. [CrossRef]
54. Anselma, L.; Bottrighi, A.; Montani, S.; Terenziani, P. Extending BCDM to Cope with Proposals and Evaluations of Updates. *IEEE Trans. Knowl. Data Eng.* **2013**, *25*, 556–570. [CrossRef]
55. Bergami, G.; Bertini, F.; Montesi, D. Hierarchical embedding for DAG reachability queries. In Proceedings of the IDEAS 2020: 24th International Database Engineering & Applications Symposium, Seoul, Republic of Korea, 12–14 August 2020; Desai, B.C., Cho, W., Eds.; ACM: New York, NY, USA, 2020; pp. 24:1–24:10.
56. Revesz, P.Z. *Introduction to Databases—From Biological to Spatio-Temporal*; Texts in Computer Science; Springer: Berlin/Heidelberg, Germany, 2010.
57. Revesz, P. Geographic Databases. In *Introduction to Databases: From Biological to Spatio-Temporal*; Springer: London, UK, 2010; pp. 81–109.
58. Zaki, N.M.; Helal, I.M.A.; Awad, A.; Hassanein, E.E. Efficient Checking of Timed Order Compliance Rules over Graph-encoded Event Logs. *arXiv* **2022**, arXiv:2206.09336.
59. Rost, C.; Gómez, K.; Täschner, M.; Fritzsche, P.; Schons, L.; Christ, L.; Adameit, T.; Junghanns, M.; Rahm, E. Distributed temporal graph analytics with GRADOOP. *VLDB J.* **2022**, *31*, 375–401. [CrossRef]

Disclaimer/Publisher's Note: The statements, opinions and data contained in all publications are solely those of the individual author(s) and contributor(s) and not of MDPI and/or the editor(s). MDPI and/or the editor(s) disclaim responsibility for any injury to people or property resulting from any ideas, methods, instructions or products referred to in the content.

Article

Streamlining Temporal Formal Verification over Columnar Databases

Giacomo Bergami

School of Computing, Faculty of Science, Agriculture and Engineering, Newcastle University, Newcastle upon Tyne NE4 5TG, UK; giacomo.bergami@newcastle.ac.uk

Abstract: Recent findings demonstrate how database technology enhances the computation of formal verification tasks expressible in linear time logic for finite traces (LTL_f). Human-readable declarative languages also help the common practitioner to express temporal constraints in a straightforward and accessible language. Notwithstanding the former, this technology is in its infancy, and therefore, few optimization algorithms are known for dealing with massive amounts of information audited from real systems. We, therefore, present four novel algorithms subsuming entire LTL_f expressions while outperforming previous state-of-the-art implementations on top of KnoBAB, thus postulating the need for the corresponding, leading to the formulation of novel $xtLTL_f$-derived algebraic operators.

Keywords: temporal formal verification; columnar databases; verified artificial intelligence; linear time logic for finite traces

1. Introduction

Grounded in formal methods, verified artificial intelligence [1] is concerned with defining, designing, and verifying systems represented mathematically. In context-free data, this focuses on a system $\mathfrak{S}$ to be verified through properties described in Φ, while the model of the environment $\mathfrak{E}$ is neglected. In this regard, a *formal verification* task ascertains whether a given system complies with a specification $\mathfrak{S} \vDash \Phi$. In the context of business process management, we can consider *model* [2], *conformance* [3], or *compliance* [4,5] *checking* as all synonyms of the former. Concerning temporal data, we focus our attention on systems described as logs, a collection of temporally ordered records (i.e., *traces*) of observed and completed (or aborted) labelled activities unravelling one possible run of a process. These real-world processes might include the auditing of malware in terms of system calls being invoked [6,7], records describing patients' hospitalization procedures [8–10], as well as transactions between producers and retailers through a brokerage system [11]. As an example, each trace of a log can describe three distinct patient registration events at an emergency department (ED) [12] as given by the following log expressed in terms of the activity labels associated to our events:

$$\mathfrak{S} = \{ \langle \texttt{registration}, \texttt{examination}, \texttt{discharge} \rangle, \\ \langle \texttt{registration}, \texttt{redirection}, \texttt{clinical test}, \\ \texttt{examination}, \texttt{discharge} \rangle, \\ \langle \texttt{registration}, \texttt{redirection}, \texttt{examination}, \\ \texttt{discharge} \rangle \} \quad (1)$$

In all these contexts, a formal verification task returns whether the current instances of the processes being collected as traces of a log $\mathfrak{S}$ abide by specific temporal quality requirements Φ while determining which temporal constraints in Φ are explicitly violated. Linear Temporal Logic over Finite traces (LTL_f, Section 2.1) [13] can be used to express these temporal specifications Φ. This logic is defined as linear since it assumes there is only

Citation: Bergami, G. Streamlining Temporal Formal Verification over Columnar Databases. *Information* **2024**, *15*, 34. https://doi.org/10.3390/info15010034

Academic Editor: Peter Revesz

Received: 8 December 2023
Revised: 4 January 2024
Accepted: 5 January 2024
Published: 8 January 2024

Copyright: © 2024 by the author. Licensee MDPI, Basel, Switzerland. This article is an open access article distributed under the terms and conditions of the Creative Commons Attribution (CC BY) license (https://creativecommons.org/licenses/by/4.0/).

one future possible event immediately following a given event in a sequence of events of interest. Such low-level semantics are then exploited to give the semantics of temporal templates, expressing occurring temporal correlations of interest; the present paper will discuss Declare [14].

The emerging area of temporal big data analytics, having data with time as a first-class citizen, makes the need to efficiently process the aforementioned tasks more pressing [15,16]. In such real scenarios, adopting relational databases provides an ideal setting for dealing with such temporal data [17]. This also includes the storage and querying of numerical time series [18], or considering different versions in time of entities and relationships represented in the relational model [19–22]. In recent years, researchers have demonstrated that time series can be represented as traces via time series segmentation by discretizing the variation in time series into discrete, observable, linear events that are distinct from each other, enabling identification of a system's transitional states [23] as well as variations in the values associated with time series [24]. As a result of such segmentation, pattern searches can now be run using streamlined approaches. LTL_f has now been applied to a widespread set of applications in real use case scenario contexts, such as controlling actuation upon sensing the environment in Industry 4.0 settings [25] as well as for the verification of smart contracts [26], for which this technology proved to be effective for verified artificial intelligence. The large adoption of such formal language pushes us to focus on this well-known and consolidated language [13,27].

In the context of formal specification tasks expressed in LTL_f, recent research clearly remarked on the inadequacy of off-the-shelf row-based relational databases and SQL as a query language for expressing LTL_f temporal constraints, as it clearly showed that a customized relational algebra for expressing formal specification (eXTended LTL_f, $xtLTL_f$ [28]) and query plan minimizing the running of sub-queries [29] running on customized column-based storage (KnoBAB [28,30]) outperformed the previous solution. The main benefit of this approach is that any LTL_f can be directly expressed in terms of $xtLTL_f$, while high-level and human-readable temporal constraints expressed through temporal clauses can be directly specified in a semantics query at warm-up, thus allowing the support of any declarative temporal language (queryplan in Figure 1). As this line of research is in its infancy, very few algorithms for efficiently running $xtLTL_f$ are known. We now remark on two use cases addressed for the first time in the present paper.

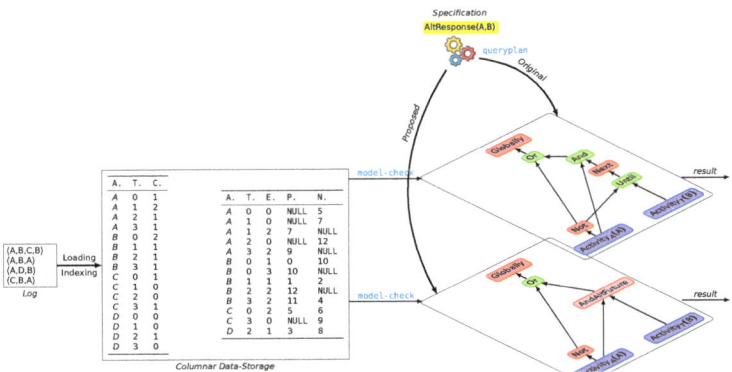

Figure 1. High-level representation of the KnoBAB query plan for running a AltResponse(A, B) for different specifications over a pre-loaded log within a columnar data-storage. After loading and indexing some traces stored as a log, we obtain a columnar data storage. At warm-up time, we can specify a queryplan which, at formal verification (model-check) time, converts a Declare specification into a $xtLTL_f$ query plan. As KnoBAB supports multiple queryplans at once, we can run the same formal verification task over different resulting query plans.

First, due to their formulation, some of the logical operators such as the timed until operator $\text{U{\scriptsize NTIL}}_{\text{True}}^{\tau}(\varphi, \varphi')$ ($\varphi \mathcal{U} \varphi'$ in LTL_f) are associated with very high computational complexity, as it prescribes that the occurrence of at least one future event matching a φ' condition per trace shall always be preceded by events matching φ. Under the occasions that this temporal post-condition shall be considered only after determining the occurrence of a first event φ'', this could drastically reduce the amount of computation associated with the overall task. This is not taken into account in our previous implementation in KnoBAB, as it computed a union between the cases where φ'' does not occur and the ones where φ'' occurs, for which the evaluation of $\text{U{\scriptsize NTIL}}_{\text{True}}^{\tau}(\varphi, \varphi')$ is extended to any event occurring of the trace. Walking in the footsteps of relational algebra, where θ-joins are expressed as the combination of natural joins [31] or cross-products [32] with θ-selections and join operations can be streamlined through cogrouping [33], we then propose similarly derived operators, combining the matching of a given pre-condition with the subsequent requirement that all the intermediate events should meet the alternance requirements dictated by $\text{U{\scriptsize NTIL}}_{\ominus}^{\tau}$. This paper will then contextualize the need for such derived operators for two specific Declare temporal templates, AltPrecedence and AltResponse, thus substantiating the interest in these temporal patterns from the current literature (Table 1).

Table 1. Declare templates as exemplifying clauses. A (B) represents the *activation* (*target*) condition as an activity label.

	Exemplifying Clause (c_l)	Natural Language Specification for Traces	LTL_f Semantics ($[\![c_l]\!]$)
In this paper	ChainPrecedence (A, B)	The activation is immediately preceded by the target.	$\Box(\bigcirc A \Rightarrow B)$
	ChainResponse (A, B)	The activation is immediately followed by the target.	$\Box(A \Rightarrow \bigcirc B)$
	AltResponse (A, B)	If activation occurs, no other activations must happen until the target occurs.	$\Box(A \Rightarrow \bigcirc(\neg A \, \mathcal{U} \, B))$
	AltPrecedence (A, B)	Every activation must be preceded by a target without any other activation in between	$\neg B \, \mathcal{W} \, A \land \Box(A \Rightarrow \bigcirc(\neg A \, \mathcal{W} \, B))$
Not subject to optimization in this paper	Init (A)	The trace should start with an activation	A
	Exists (A, n)	Activations should occur at least n times	$\Diamond(A \land \bigcirc([\![\text{Exists}(A, n-1)]\!])_{n>0})$
	Absence $(A, n+1)$	Activations should occur at most n times	$\neg[\![\text{Exists}(A, n+1)]\!]$
	Precedence (A, B)	Events preceding the activations should not satisfy the target	$\neg B \, \mathcal{W} \, A$
	Choice (A, A')	One of the two activation conditions must appear.	$\Diamond A \lor \Diamond A'$
	Response (A, B)	The activation is either followed by or simultaneous to the target.	$\Box(A \Rightarrow \Diamond B)$
	RespExistence (A, B)	The activation requires the existence of the target.	$\Diamond A \Rightarrow \Diamond B$
	ExlChoice (A, A')	Only one activation condition must happen.	$[\![\text{Choice}(A, A')]\!] \land [\![\text{NotCoExistence}(A, A')]\!]$
	CoExistence (A, B)	RespExistence, and vice versa.	$[\![\text{RespExistence}(A, B)]\!] \land [\![\text{RespExistence}(B, A)]\!]$
	Succession (A, B)	The target should only follow the activation.	$[\![\text{Precedence}(A, B)]\!] \land [\![\text{Response}(A, B)]\!]$
	ChainSuccession (A, B)	Activation immediately follows the target, and the target immediately preceeds the activation.	$\Box(A \Leftrightarrow \bigcirc B)$
	NotCoExistence (A, B)	The activation nand the target happen.	$\neg(\Diamond A \land \Diamond B)$
	NotSuccession (A, B)	The activation requires that no target condition should follow.	$\Box(A \Rightarrow \neg \Diamond B)$

Legend: Globally: $\Box \phi$, Next: $\bigcirc \phi$, Implication: $\phi \Rightarrow \phi'$, Until: $\phi' \, \mathcal{U} \, \phi$, Weak Until: $\phi \, \mathcal{W} \, \phi'$, Future: $\Diamond \phi$.

Example 1. *AltResponse*(A, B) *requires that, when A occurs, B shall occur anytime in the future while no other A shall occur in between. In xtLTL$_f$, this can be expressed as* $\Box(\neg A \lor (A \land \bigcirc(\neg A \, \mathcal{U} \, B)))$ *(Original in Figure 1). On the other hand, the present paper shows that, by replacing $A \land \bigcirc(\neg A \, \mathcal{U} \, B)$ with a single operator, we obtain a significant reduction in running time by reducing the amount of result scans and data allocations. This is possible by providing a different (Proposed) xtLTL$_f$ queryplan while implementing AndAltFuture as a novel operator. This difference is remarked in the two resulting query plans in Figure 1.*

Second, temporal constraints requiring that events abiding by a φ specification shall always precede (or follow) other events abiding by φ' are currently implemented in KnoBAB by equi-joining all the events matching φ with the ones matching φ', while the predicate is $i = i' \land j = j - 1$ (or $i = i' \land j = j' + 1$), where i (or i') and j (or j') are, respectively, referring to the trace id and event id associated to a record coming from the first (or second) operand (see $\text{And}_{\ominus}^{\tau}$ xtLTL$_f$ in Section 2.2.2). Even this implementation can be further boosted by minimizing the data table access to just one operator (e.g., φ) for directly accessing the immediately preceding or following events within the relational database and checking whether they abide by φ'. Even this second observation is motivated by the existence of ChainResponse and ChainPrecedence Declare templates, thus requiring the definition of novel derived operators for performance purposes.

To support our research claims, we extend (https://github.com/datagram-db/knobab/releases/tag/v2.3, accessed on 3 January 2024) the current implementation of KnoBAB [34], a column-oriented main memory DBMS supporting formal verification and specification mining tasks by defining relational operations for temporal logic and customary mining algorithms. Despite this being a main memory engine, it currently supports intra-query parallelism and hybrid algorithms (Section 2.2.1). To our knowledge, no other database management system for temporal formal verification over LTL$_f$ provides these features, for which we choose to extend such a system. Furthermore, KnoBAB already proved to consistently outperform previous state-of-the-art algorithms on both tasks [35], thus including competing approaches interpreting the same temporal constraints over SQL and row-oriented relational database architecture [17]. After providing a brief literature overview on the landscape of formal verification for temporal data (Section 2), we outline the following main contributions leading to the our performance analysis result for our newly proposed xtLTL$_f$ operators:

- We formally introduce the novel temporal operators optimizing the aforementioned scenarios in the context of Declare as a declarative language for formal verification (Section 3).
- We describe the implementation of the aforementioned operators over the KnoBAB architecture leveraging columnar-oriented main memory storage (Section 4).
- We present experimental results to evaluate the effectiveness of such newly introduced operators in the context of formal verification in Declare (Section 5).

2. Related Works

2.1. Languages for Temporal Formal Specifications

2.1.1. LTL$_f$

Taking the possible worlds as finite traces, LTL$_f$ is a well-established extension of modal logic with modalities referring to time; it assumes that all the events of interest are fully observable and therefore deterministic and that, for each occurring event, they should be immediately followed by at most one event. This entails that the i-th trace σ^i in a log $\mathfrak{S}$ can be considered as a sequence of n totally ordered events $\sigma_0^i \ldots \sigma_{n-1}^i$, where each event σ_j^i is associated to a single activity label $\lambda(\sigma_j^i) \in \Sigma$ [3]. When events are associated to a payload represented as a key-value association $\varsigma(\sigma_j^i)$, we refer to such logs as *dataful* and as *dataless* otherwise. In the eventuality of the former, such payloads can be represented as finite functions V^K, where K is the set of the keys and V is the overall set of non-NULL values. Concerning our datasets of interest, we only consider ones where trace events are not associated with a data payload, and therefore even such logs can be considered as *dataless*. On the other hand, with reference to Equation (1), event payloads can store patient information, thus registering the recorded medical condition being observed [28]; in the context of good brokerage, such payload might contain the relevant contract information between the supplier and the customer which are required to be respected (e.g., delivery times), as well as the location of the goods, their number, and quality [11].

LTL$_f$ semantics is usually defined in terms of First-Order Logic [36]; more informally, Next ($\bigcirc \phi$) requires ϕ to occur from the subsequent temporal step, Globally ($\square \phi$) that ϕ always holds from the current instant of time, Future ($\lozenge \phi$) that ϕ must eventually hold, and Until $\phi \mathcal{U} \phi'$ that ϕ must hold until the first occurrence of ϕ' does. Weak Until is a *derived operator* for $\varphi \mathcal{W} \varphi' := \varphi \mathcal{U} \varphi' \vee \square \varphi$, while the logical implication can be rewritten as $\varphi \Rightarrow \varphi' := (\neg \varphi) \vee (\varphi \wedge \varphi')$. Please observe that LTL_f does not provide full support for handing data correlation conditions between operands of binary operations, as it only supports the declaration of data conditions that can be applied to one single event [3]. To the best of our knowledge, $xtLTL_f$ (Section 2.2.2) is the only extension of this language supporting data payload correlation across events matched by both arguments of the binary operator, thus providing a complete *dataful* support.

2.1.2. Declare

Declare [14,37] provides a human-readable declarative language on top of LTL_f (first column of Table 1), where each template is associated with a specific LTLf formula (third column), which can be instantiated with arbitrary activity labels. We refer to the instantiation of such templates via activity labels in a finite set Σ as *(declarative) clauses*. Declare circumscribes the set of all the possible behaviors expressible in LTL_f to the ones of interest over a set of possible Σ; Table 1 recalls some of the most used templates while remarking on the four templates of interest optimized in the present paper.

At the time of writing, Declare expresses specifications Φ as a set of clauses c_l being usually associated with an LTL_f semantics $[\![c_l]\!]$; in this context, a trace $\sigma \in \mathfrak{S}$ satisfies a Declare specification Φ if it jointly satisfies all the clauses associated to the specification. If these clauses can be characterized by a precondition which, if satisfied by some event, imposes the occurrence of a post-condition, then we refer to these as *activation* and *target* conditions, respectively. Please observe that post-conditions are considered as such merely in terms of causal implication (i.e., $\Rightarrow$) and not necessarily in temporal terms, e.g., while ChainResponse requires the target to immediately follow any existing activation, ChainPrecedence requires that the targeted event shall instead precede the activation. Please consider that Declare clauses do not necessarily reflect association rules, as the latter do not provide temporal constraints correlating the activation of activation and target conditions. In this paper, we focus on Declare clauses only predicating over the events' activity labels, which are then referred to as *dataless*; on the other hand, *dataful* Declare clauses can express data payload conditions over both activation and target conditions, as well as representing Θ payload correlation conditions between activating and targeted conditions [28]. Thus, both clauses and logs are referred to *dataful* otherwise.

Despite the fact that the four clauses of interest in Table 1 might appear to express similar behavior, they express substantially different concepts. Table 2 provides four traces distinguishing the behavior of such four templates, the validity of which can be easily controlled by transforming the associated LTL_f formulæ into a DFA (http://ltlf2dfa.diag.uniroma1.it/dfa, accessed on 3 January 2024).

Table 2. Traces from the Log in Figure 1 distinguishing the temporal behavior of the Declare clauses of interest in this paper, where each trace $\sigma_0^i \ldots \sigma_{n-1}^i$ is expressed in terms of their associated activity labels, $\langle \lambda(\sigma_0^i), \ldots, \lambda(\sigma_{n-1}^i) \rangle$. ✓(and ✗) remarks a trace satisfying (violating) a corresponding clause.

Traces	ChainResponse(A,B)	ChainPrecedence(B,A)	AltResponse(A,B)	AltPrecedence(B,A)
⟨A,B,C,B⟩	✓	✗	✓	✗
⟨A,B,A⟩	✗	✓	✗	✗
⟨A,D,B⟩	✗	✗	✓	✗
⟨C,B,A⟩	✗	✗	✗	✓

2.2. KnoBAB and xtLTL$_f$

We now summarize our previous contributions on temporal formal verification tasks run over our proposed main memory columnar database, KnoBAB.

2.2.1. KnoBAB

KnoBAB [28,34] is a column database store tailored for both loading *dataful* logs being represented in XES [38] and *dataless* ones described as a tab-separated file. This outperformed the previous state of the art in terms of both specification mining [39] and formal verification [35] tasks on tailored non-database solutions.

Logical and Physical Model

The resulting column-based relational database is then represented through some tables having fixed schema independently from its data representation. As the present paper focuses on dataless datasets, we describe in this paper just two of those; Table 3 describes the relational representation of the log presented in Equation (1). The ActivityTable (Table 3a) lists each trace event of a given log, where records are sorted in ascending order for activity label, trace id, and event id. Cells under the Prev (and Next) column store a pointer to the record representing the immediately preceding (and following) event in the same trace if any. After mapping each existing activity label in the log a to a unique natural number $\beta(a)$, we can define a primary dense and clustered index that can be accessed in $O(1)$ time as it is an array of offset pointers. We also define a secondary index structured as a block of two records, associating each trace in the log to the first and last trace event; given that all the traces are associated with a unique natural number, this index can also be accessed on $O(1)$ time by trace id. The CountTable (Table 3b), also created at loading time like the previous, merely lists the number of occurrences of each activity label per trace and can be used to determine the absence or presence of an event with a given activity label per trace.

Formal Verification Tasks over Query Plans

In spite of the ActivityTable also appearing in SQLMiner's log representation [17] (except for the Prev and Next columns), this still used an off-the-shelf relational database engine and a translation of Declare specification into SQL for carrying out formal verification tasks over a dataless log. KnoBAB showed a new pathway for enhancing temporal queries over customary main memory relational database through the combined provision of both ad hoc relational operators expressing LTL_f over relational tables ($xtLTL_f$) and the definition of a query plan represented as a rooted DAG where shared subqueries are computed only once [29]. This was sensibly different from competing approaches [40,41] also relying on main memory engines where, instead, the query plan associated to a formal verification task is always expressed in terms of trees, thus not allowing the detection of shared sub-expressions to be merged to avoid wasteful recomputations. As vertices for a DAG can be sorted topologically, we can obtain for free the scheduling order in which the operators must be executed and, by associating each node a maximum distance value from the root, we can safely run in parallel all the operators laying at the same depth level, as all the previously called operators will pertain their information in an intermediate cache, thus achieving intraquery parallelism as a free meal [28]. This parallelization approach greatly differs from straightforward parallelization algorithms known in the Business Process Management area, where they simply run each declarative clause occurring in the specification in a separate thread [35]. In addition to the former, KnoBAB guarantees efficient access to the tables through the provision of specific indexing data structures such as primary indices for directly accessing the blocks of the table concerning a specific activity label as well as the provision of secondary indices mapping a specific trace id i and event id j for σ_j^i into a table offset. KnoBAB outperformed SQLMiner run over PostgreSQL within two to five orders of magnitude, thus demonstrating the inadequacy of using customary relational operators for computing temporal tasks over relational databases.

Table 3. KnoBAB representation for the dataless log in Equation (1). (**a**) ActivityTable; (**b**) CountTable.

(a)

ActivityLabel	TraceId	EventId	Prev	Next
Clinical Test	1	2	7	5
Discharge	0	2	4	NULL
Discharge	1	4	5	NULL
Discharge	2	3	6	NULL
Examination	0	1	9	1
Examination	1	3	0	2
Examination	2	2	8	3
Redirection	1	1	10	0
Redirection	2	1	11	6
Registration	0	0	NULL	4
Registration	1	0	NULL	7
Registration	2	0	NULL	8

(b)

ActivityLabel	TraceId	Count
Clinical Test	0	0
Clinical Test	1	1
Clinical Test	1	0
Discharge	0	1
Discharge	1	1
Discharge	2	1
Examination	0	1
Examination	1	1
Examination	2	1
Redirection	0	0
Redirection	1	1
Redirection	2	1
Registration	0	1
Registration	1	1
Registration	2	1

KnoBAB enables the specification of user-defined template names in terms of xtLTL$_f$ operators through a `queryplan "semanticsname" {...}` query, thus allowing the co-presence of multiple possible definitions of declarative clauses. Then, we can select the most appropriate semantics while carrying out the formal verification task by specifying such a name, e.g., `model-check...plan "semanticsname"...`. This then enables us in this paper to test multiple possible specifications of Declare clauses without necessarily recompiling the database's source code.

Walking in the footsteps of the BAT algebra for columnar databases [42], each of the novel temporal operands for xtLTL$_f$ (Section 2.2.2) not requiring direct data access to the aforementioned KnoBAB tables both accepts as an input and returns a uniform data representation ρ with schema:

$$\texttt{IntermediateRepresentation(TraceId, EventId, Witnesses(Tag))} \qquad (2)$$

where the first (and second) argument refers to the trace (and event) id matching a specific temporal condition of choice, while `witnesses` represents the relevant activated or targeted conditions occurring from the position `EventId` in a given `TraceId` trace onwards via a tagged extension of semiring provenance [43]; such tags mainly refer to the distinction between activated and targeted events, respectively A and T. Dataful matching occurring between witnessed activated $A(i)$ and targeted events $T(j)$ certified via a Θ binary predicate are represented as $M(i,j)$. Matches can be represented as semiring products, while the listing of all the activated, targeted, and matched events can be represented as a semiring sum; the latter is simply rendered as a list. As the table is sorted by trace id and event id by design for any given activity label, such intermediate representation also returns trace entities sorted by ascending trace id and event id.

2.2.2. xtLTL$_f$

We now discuss some *xtLTL$_f$* operators of relevance for the current paper. By using KnoBAB as a computational model, we can also discuss the time complexity associated with such operators. While LTL$_f$ operators can mainly be used to establish a yes/no question about whether a single trace abides by some temporal specification, an xtLTL$_f$ expression returns all the traces in the log conforming to a temporal specification by composing the trace events as records through temporal operations. Furthermore, the latter can also be directly exploited to express confidence, maximum satisfiability, and support metrics similar to association rules. So to better support future explainable temporal AI tasks, xtLTL$_f$ also carries out information concerning activated/targeted events justifying the algorithmics' outcome, while the cache associated to the leaves can be analyzed so as to check which events were activated/targeted without necessarily satisfying the temporal requirements computed through xtLTL$_f$.

Table Access ("Leaf") Operators

We determine all the events being associated with a specific activity label through the ActivityLabel's primary block index and express the outcome of this retrieval in terms of intermediate representation:

$$\text{Activity}_{A/T}^{\mathfrak{S},\tau}(a) = \{\langle i,j, \{A/T(j)\}\rangle \,|\, \exists \pi, \phi.\, \langle a, i, j, \pi, \phi\rangle \in \text{ActivityTable}\}$$

where A/T provides the optional tags for remarking the matching event of interest as being part of an activation/target condition. By associating each activity label a with a unique natural number $\beta(a)$, we can now seek the presence of events with label a in $O(1)$ time and retrieve all the events $\#a \ll |\mathfrak{S}|$ associated to such a label. If, on the other hand, we are interested in events matching a specific data predicate q, we define the following operator:

$$\text{Atom}_{A/T}^{\mathfrak{S},\tau}(B, q) = \{\langle i,j, A/T(j)\rangle \,|\, q(\sigma_j^i) \wedge \lambda(\sigma_j^i) = B\}$$

Despite the fact that this might appear as a simple selection operation, the atomization of a predicate into mutually exclusive data conditions required for both minimizing the data access to the tables holding the key-value payload associations within the dataful events and merging multiple equivalent sub-expressions into one makes both its associated query plan and its actual formal definition quite convoluted. As describing this is not the major purpose of the paper, we refer to [28] for any further information. By accessing the secondary index of the ActivityTable, we can collect the last events for each trace in linear time over the log's size $O(|\mathfrak{S}|)$ using the following operator:

$$\text{Last}_A^{\mathfrak{S},\tau} = \{\langle i, |\sigma^i|, \{A(|\sigma^i|)\}\rangle \,|\, \exists a, \pi.\, \langle \beta(a), i, |\sigma^i|, \pi, \text{NULL}\rangle \in \text{ActivityTable}\}$$

Unary Operators

We discuss the main difference between operators' execution in xtLTL$_f$ from corresponding ones in LTL$_f$; the latter computes semantics from the first occurring operator appearing in the formula towards the leaves, whereas the former assumes intermediate results coming from the leaves. In this, the downstream operator is completely agnostic about the semantics associated with the upstream operator, so it must combine the intermediate results appropriately. Therefore, the Next(ρ) (timed) xtLTL$_f$ unary operator returns all the events σ_j^i witnessing the satisfaction of an activation, target, or correlation condition being returned by a downstream operator as an intermediate result ρ, while $\bigcirc \varphi$ will simply increment the internal time counter over φ, thus determining the time from which to assess the specification in φ.

Due to this structural discrepancy in the order of computation, xtLTL_f must distinguish *timed* operators (assessing the occurrence of a specification sub-expression anytime in the trace) from the *untimed* operators (determining the properties holding from the beginning of the trace). The aforementioned xtLTL_f operator can therefore be expressed as follows:

$$\text{Next}^\tau(\rho) = \{\, \langle i, j-1, L\rangle \mid \langle i,j,L\rangle \in \rho, j > 0 \,\}$$

This operator can then be computed in linear time over the size of the input, i.e., $O(|\rho|)$. On the other hand, the timed negation operator $\text{Not}^\tau(\rho)$ subtracts from the universal relation, being all the events occurring in any trace, the ActivityTables events appearing in ρ while still guaranteeing the return of the records in ascending order for trace and event id. Given ϵ, the maximum trace length, this operator takes at most $O(|\mathfrak{S}|\epsilon)$ time by assuming $|\rho| \ll |\mathfrak{S}|\epsilon$. The globally timed operator prescribes to return a $\langle i,j,L\rangle \in \rho$ if also all the subsequent events within the same trace are in ρ, and can be computed in $O(|\rho|\log|\rho|)$ time by starting scanning the events from the last occurring in the trace.

Binary Operators

We now stress further differences between xtLTL_f and LTL_f in terms of binary operators. While xtLTL_f can express dataful matching conditions between activation and target conditions, LTL_f can only express properties associated with one single event at a time through atoms. In these regards, timed logical conjunction ($\text{And}^\tau_\Theta(\rho, \rho')$) extending the logical conjunction in LTL_f with a binary match condition Θ over the event's payloads can be expressed as a nested Θ-join returning the records from both operands having the same trace id and event id, while all the pairs of witnessed events satisfying an activation $A(i)$ and target $T(j)$ conditions from the matching record shall satisfy the Θ matching condition when provided; the matching is then registered with $M(i,j)$. Timed logical disjunction ($\text{Or}^\tau_\Theta(\rho, \rho')$) can be similarly expressed through a full outer Θ-join. Given that the ActivityTable is pre-sorted at indexing time, we can efficiently implement such algorithms through sorted joins. As these can be computed with a joint linear scan of both operands, both operators have at most a time complexity in $O(|\rho|+|\rho'|)$. The timed until operator ($\text{Until}^\tau_{\text{True}}(\rho, \rho')$) for $\Theta = \text{True}$ is defined similarly to the corresponding LTL_f operator; it returns all the events within a given log trace in the second operand and the events from the first operand if all the immediately following events until the first occurrence of an event in the second operand also belong to the first:

$$\text{Until}^\tau_{\text{True}}(\rho, \rho') = \rho' \cup \{\, \langle i, k, L \cup L'\rangle \mid \exists j > k. \langle i,j,L\rangle \in \rho', (\forall k \leq h < j. \langle i,h,L'\rangle \in \rho) \,\}$$

This can be computed in $O(|\rho|^2|\rho'|)$ time in its worst-case scenario. The in-depth discussion concerning the formal definition of such an operator when matching a non-trivially true matching condition Θ is deferred due to its technicalities and can be retrieved from the original paper [28].

2.3. Algebraic Specification for Queries

We now compare $xtLTL_f$ with other long-standing definitions of temporal operators regarding database temporal representations.

Current research [17] outlined the possibility of loading logs composed of multiple traces within row-based relational databases while providing a direct translation of *dataless* Declare-driven formal verification and specification mining tasks into SQL [44]. Our previous research remarked on the inefficiency of directly expressing temporal formal verification tasks on top of off-the-shelf relational databases, thus motivating the definition of a novel query plan specification directly exploiting temporal algebra operators, xtLTL_f [28]. As SQL queries are translated into query plans where each operator expresses an implementation of a relational algebra operator, this demonstrates the overall inefficiency of exploiting traditional relational algebra for representing temporal queries. Please observe that LTL$_f$ temporal requirements cannot be expressed in traditional relational algebra without aggregation operators while not naturally assuming a columnar database storage. Therefore, traditional relational algebra cannot be directly exploited to predicate about the necessity or the eventuality of a given event to occur without any further extension.

For all these considerations, our proposed algebra more resembles BAT from MonetDB [42,45], where the intermediate result output for each operator records the table's record being selected, without necessarily carrying out values stored within the specific row. Given the specificity of our scenario, our intermediate results carry the trace id and the event id as unique record identifiers. We further had to extend this representation to possibly carry out the activated and targeted events as witnesses of the computation's correctness, providing explainable justifications for the computation, and correctly expressing Θ predicates over dataful logs. xtLTL$_f$ then provides a required extension of such a representation for new computation needs.

Concerning Allen's algebra for temporal intervals [46], we can *first* see that such algebra considers events as temporal intervals that might also be overlapping, while xtLTL$_f$ inherits the same assumptions from LTL$_f$ and considers events as pointwise and non-overlapping activities. *Secondly*, while the former only supports conditions on the activity labels, xtLTL$_f$ also supports predicating on the conditions for the payload values (expressed as key-value pairs) associated with the specific events [28], as well as supporting binary predicates to be tested across activated and targeted conditions similarly to θ-joins. Recent extensions of Allen's algebra aimed at supporting single data conditions over single events [40]. *Thirdly*, such algebra only expresses temporal correlations between two single events, albeit expressed with a duration and a termination time, and can predicate natively neither the eventuality nor the necessity of some properties to occur in a trace (e.g., globally and future) from a given instant in time.

Concerning the temporal relational algebra [22] defined over temporal relational databases [47] (also referred to as *temporal modules* [21]), it mainly proposes timestamp transformation operations currently supported by Oracle Cloud [48] as well as windowing functions, thus retaining the entities and relationships occurring within a window time frame. This allows the slicing of a temporal module into a finite sequence of finite database states, where such a snapshot sequence can be ascribed to a single trace and each event can be mapped to a single database state [49]. Despite time being considered as a first citizen within these operators, no operator of such an algebra temporally correlates entities at different timestamps while also requiring the eventuality or the necessity for a specific condition within a given lapse of time. An orthogonal contemporary approach attempted at mapping LTL$_f$ to TSQL2 [50], a de facto extension of SQL for querying temporal modules [51]. Differently from the approach mentioned above, this preserved LTL$_f$ temporal operators such as Until ($\mathcal{U}$); as the authors preceded the definition of LTL$_f$ extensions considering data payload conditions [3,28], these are not considered in their transformation. Furthermore, as these temporal modules represent one single distinct trace as a result of temporal snapshotting of a single database into multiple distinct states, they cannot be effectively used to run a single formal verification task over numerous traces as per our proposed approach, as this would require running a single TSQL2 query over multiple databases, one for each log trace. In fact, our solution can assess multiple traces simultaneously by leveraging an extended relational representation to the one initially described in [17].

3. Proposed Derived Operators

Similarly to the definition of the derived operators in relational algebra, we now provide the definition of our proposed operators extending xtLTL$_f$ by expressing those in terms of the ones already known in such a temporal algebra. These are then defined in Equations (3), (5), (7) and (9).

3.1. AndAltFuture

We want this operator to seek all the instants of time when an event activates the Declare clause while the target follows anytime in the future, while requiring that no further activation occurs between these two events. This operator aims to optimize the AltResponse(A,B) clause and can be then expressed in terms of basic $xtLTL_f$ operators as follows:

$$\text{AndAltFuture}_{\ominus}^{\tau}(\rho, \rho') \overset{\text{def}}{=} \text{And}_{\ominus}^{\tau}\left(\rho, \text{Next}\left(\text{Until}_{\text{True}}^{\tau}(\text{Not}^{\tau}(\rho), \rho')\right)\right) \quad (3)$$

By implementing this operator from scratch, we want to avoid running the costly computation of the timed Until^τ unless the activation condition associated with the intermediate result returned as ρ is satisfied. Furthermore, we want to avoid explicitly computing the negation of the activation condition and express this by explicitly checking that, given any activating event in σ_j^i in ρ with an immediately following targeting one σ_k^i in ρ' with $|\sigma^i| > k > j$, no other events σ_{j+h}^i in ρ with $j + h < k$ shall occur. We can then express the aforementioned Declare clause in terms of the recently defined operator as follows:

$$\text{Globally}^\tau\left(\text{Or}_{\text{True}}^{\tau}(\text{Not}^\tau(\rho), \text{AndAltFuture}_{\text{True}}^{\tau}(\rho, \rho'))\right) \quad (4)$$

where $\rho = \text{Activity}_A^{\ominus,\tau}(A)$ and $\rho' = \text{Activity}_T^{\ominus,\tau}(B)$ under the dataless assumption.

Example 2. *With reference to the log in Equation (1), AltResponse(redirect, examine) requires that a patient redirected to a given department shall be examined before being further redirected. This constraint satisfies all the traces within that equation. By considering only the events from the second trace, in our previous $xtLTL_f$ solution we have intermediate results $\rho = \text{Activity}_A(\text{redirect}) = \{\langle 1, 1, [A(1)]\rangle\}$ for the activation condition and $\rho' = \text{Activity}_T(\text{examine}) = \{\langle 1, 3, [T(3)]\rangle\}$ for the target one. The timed Until $\rho'' = \text{Until}_{\text{True}}^{\tau}(\neg\rho, \rho')$ returns:*

$$\{\langle 1, 0, [T(1)]\rangle, \ldots, \langle 1, 3, [T(3)]\rangle, \langle 1, 4, []\rangle\}$$

as each event in $xtLTL_f$ can only witness a future event, and $\rho''' = \text{Next}^\tau(\rho'')$ returns:

$$\{\langle 1, 1, [T(3)]\rangle, \ldots, \langle 1, 2, [T(3)]\rangle, \langle 1, 3, []\rangle\}$$

Hence, $\text{And}_{\text{True}}^{\tau}(\rho, \rho''')$ returns just $f = \{\langle 1, 1, [M(1,3)]\rangle\}$, while witnessing that, from that time onwards, both activation $A(1)$ and target $T(3)$ condition will occur from the same event 1. The rest of the events will be returned via $\neg\rho$, which are finally grouped-by temporally via untimed Globally. Before running it, we previously ran the timed Until operator independently from the occurrence of ρ'' in a trace.

On the other hand, our new AndAltFuture operator directly returns f after taking as an argument ρ and ρ'; this scans the events in ρ' occurring after each occurrence of events in ρ while immediately discarding the events in ρ containing another redirect event in between. This reduces the memory footprint and the number of scans from our previous query plan.

3.2. AndAltWFuture

Reflecting upon the definition of AltPrecedence(A,B) which this operator is aiming to optimize, we can observe that implementing an ad hoc operator AndAltWFuture for this might provide even greater optimization, as we might as well avoid checking the global absence of A-labelled events if no B occurs in a trace after an A. Therefore, this operator acts as an extension of the former by either requiring an alternate occurrence between activation and target condition, as previously, or requiring the absence of any future activation if no targeting event is expected to occur. $\text{AndAltWFuture}_{\Theta}^{\tau}(\rho,\rho')$ can be then defined as follows:

$$\text{And}_{\Theta}^{\tau}\left(\rho, \text{Next}\left(\text{Or}_{\text{True}}^{\tau}\left(\text{Until}_{\text{True}}^{\tau}(\text{Not}^{\tau}(\rho),\rho'), \text{Globally}^{\tau}(\text{Not}^{\tau}(\rho))\right)\right)\right) \quad (5)$$

We can now express AltPrecedence(A,B) by replacing, in the original xtLTL$_f$ Declare semantics, the previous equation with the currently introduced operator, thus obtaining:

$$\text{Or}_{\text{True}}^{\tau}\left(\text{Until}^{\tau}\left(\text{Not}^{\tau}(\rho'),\rho\right), \text{Globally}^{\tau}\left(\text{Or}_{\text{True}}^{\tau}\left(\text{Not}^{\tau}(\rho), \text{AndAltWFuture}_{\text{True}}^{\tau}(\rho,\rho')\right)\right)\right) \quad (6)$$

3.3. AndNext

This operator aims to optimize the ChainResponse operator by reducing the data access by accessing the `ActivityTable` just for the activation condition. This makes this operator intrinsically unary, as the target condition, both in terms of data predicate and activity label, has to be provided as additional arguments for the operator alongside the Θ correlation condition for dataful scenarios. To check whether the target condition occurs immediately after the operand's current event, we need to check whether it is associated with an activity table and whether it satisfies a predicate q. This can be then expressed in xtLTL$_f$ in terms of the following derived operator:

$$\text{AndNext}_{B,q,\Theta}^{\tau}(\rho) \stackrel{\text{def}}{=} \text{And}_{\Theta}^{\tau}\left(\rho, \text{Next}^{\tau}(\text{Atom}_{T}^{\mathfrak{S},\tau}(B,q))\right) \quad (7)$$

At this stage, we can then express the semantics associated to the Declare template ChainResponse(A,B) as follows:

$$\text{Globally}^{\tau}\left(\text{Or}_{\text{True}}^{\tau}\left(\text{Not}^{\tau}(\rho), \text{AndNext}_{B,\text{True},\text{True}}^{\mathfrak{S},\tau}(\rho)\right)\right) \quad (8)$$

where $\rho = \text{Activity}_{A}^{\mathfrak{S},\tau}(A)$ in a dataless scenario.

3.4. NextAnd

The second operator aims at optimizing ChainPrecedence(A,B) similarly to the previous one, but with a swapped temporal occurrence. Please observe that negating the fact that an event shall occur after another can be expressed in terms of all the events occurring at the end of a trace and all of the events not matching the activation condition a when occurring in a non-first position. So, ChainPrecedence is usually represented as:

$$\text{Globally}^{\tau}\left(\text{Or}_{\text{True}}^{\tau}\left(\text{Or}_{\text{True}}^{\tau}\left(\text{Last}^{\mathfrak{S},\tau}, \text{Next}^{\tau}(\text{Not}^{\tau}(\rho))\right), \text{And}_{\text{True}}^{\tau}\left(\text{Next}^{\tau}(\rho),\rho'\right)\right)\right)$$

where $\rho = \text{Activity}_{A}^{\mathfrak{S},\tau}(A)$ and $\rho' = \text{Activity}_{T}^{\mathfrak{S},\tau}(B)$ in a dataless scenario. After compactly representing the subexpression in the second row of the previous definition, as follows:

$$\text{NextAnd}_{B,q,\Theta}^{\tau}(\rho) \stackrel{\text{def}}{=} \text{And}_{\Theta}^{\tau}\left(\text{Next}^{\tau}(\rho), \text{Atom}_{T}^{\mathfrak{S},\tau}(B,q)\right) \quad (9)$$

we aim to optimize this last declarative clause by using this last introduced operator by rewriting the semantics associated to ChainPrecedence(A,B) as such:

$$\text{Globally}^{\tau}\left(\text{Or}_{\text{True}}^{\tau}\left(\text{Or}_{\text{True}}^{\tau}\left(\text{Last}^{\mathfrak{S},\tau}, \text{Next}^{\tau}(\text{Not}^{\tau}(\rho))\right), \text{NextAnd}_{B,\text{True},\text{True}}^{\tau}(\rho)\right)\right) \quad (10)$$

Please observe that the intended optimization induced by these operators can be considered as non-trivial, as these do not directly subsume the entire xtLTL$_f$ semantics associated to a template, rather than optimizing a specific part.

4. Algorithmic Implementation

We discuss the implementation of the previously introduced operators outlined in Algorithm 1, thus justifying their definition as novel derived operators. For each of them, we briefly discuss their computational complexity and compare it to the expected theoretical speed-up not considering the cost of memory allocation and page-faults.

Algorithm 1 Newly proposed xtLTL$_f$ operators.

1: **function** ANDALTFUTURE$_\Theta^\tau(\rho, \rho')$
2:   **for all** $\langle i,j,L \rangle, \langle i,k,L' \rangle \in (\rho \times \rho')$ s.t. $j < k$ **do**
3:     **if** $\nexists h > 0. \langle i, j+h, L \rangle \in \rho$ s.t. $j + h < k$ **then**
4:       **if** $L' \neq \emptyset$ and $L \neq \emptyset$ and $\Theta \neq$ True **then**
5:         $L'' \leftarrow \{M(j',k') | \Theta(\sigma_{j'}^i, \sigma_{k'}^i), A(j') \in L, T(k') \in L'\}$
6:         **if** $L'' \neq \emptyset$ **then yield** $\langle i,j,L'' \rangle$
7:       **else yield** $\langle i,j,L \cup L'' \rangle$
8:       **end if**
9:     **end if**
10:   **end for**

11: **function** ANDALTWFUTURE$_\Theta^\tau(\rho, \rho')$
12:   **for all** $\langle i,j,L \rangle \in \rho$ **do**
13:     **for all** $\langle i,k,L' \rangle \in \rho'$ s.t. $j \leq k$ **do**
14:       **if** $\nexists h > 0. \langle i, j+h, L \rangle \in \rho$ s.t. $j + h < k$ **then**
15:         **if** $j = |\sigma^i| - 1$ **continue**;
16:         **if** $L \neq \emptyset$ and $L' \neq \emptyset$ and $\Theta \neq$ True **then**
17:           $L'' \leftarrow \{M(j',k') | \Theta(\sigma_{j'}^i, \sigma_{k'}^i), A(j') \in L, T(k') \in L'\}$
18:           **if** $L'' \neq \emptyset$ **then yield** $\langle i,j,L'' \rangle$
19:         **else yield** $\langle i,j,L \cup L'' \rangle$
20:         **end if**
21:       **end if**
22:     **end for**
23:     **if** $\nexists k, h. \langle i,k,L' \rangle \in \rho' \wedge \langle i,h,L'' \rangle \in \rho \wedge j < k, j < h$ **then**
24:       **yield** $\langle i,j,L \rangle$
25:     **end if**
26:   **end for**

27: **function** ANDNEXT$_{B,q,\Theta}^\tau(\rho)$
28:   **if** $\nexists \sigma^i \in \mathfrak{S}, \sigma_j^i \in \sigma^i.\lambda(\sigma_j^i) = B$ **then return** $\emptyset$
29:   **for all** $\langle i,j,L \rangle \in \rho$ s.t. $j < |\sigma^i| - 1$ and $\lambda(\sigma_{j+1}^i) = B$ **do**
30:     $L' \leftarrow L \cup \{T(j+1)\}$
31:     **if** $\Theta \neq$ True **then**
32:       **if** $L \neq \emptyset$ and $\nexists A(k) \in L.\theta(\sigma_k^i, \sigma_{j+1}^i)$ **then continue**
33:       **else** $L' \leftarrow \{M(k, j+1) | A(k) \in L\}$
34:     **end if**
35:     **if** $q \neq$ True $\vee q(\sigma_{j+1}^i)$ **then yield** $\langle i,j,L' \rangle$
36:   **end for**

37: **function** NEXTAND$_{B,q,\Theta}^\tau(\rho)$
38:   **for all** $\langle i, j+1, L \rangle \in \rho$ s.t. $j \geq 0$ and $\lambda(\sigma_j^i) = B$ **do**
39:     $L' \leftarrow L \cup \{T(j)\}$
40:     **if** $\Theta \neq$ True **then**
41:       **if** $L \neq \emptyset$ and $\nexists A(k) \in L.\theta(\sigma_k^i, \sigma_j^i)$ **then continue**
42:       **else** $L' \leftarrow \{M(k,j) | A(k) \in L\}$
43:     **end if**
44:     **if** $q \neq$ True $\vee q(\sigma_j^i)$ **then yield** $\langle i, j+1, L' \rangle$
45:   **end for**

4.1. AndAltFuture

As all the intermediate results in the KnoBAB pipeline are always sorted by ascending trace and event id, we can scan all the events within the same trace where the targets follow the activations in linear time similarly to the timed *and* operator, despite this being expressed in pseudocode with a cross product for simplifying the overall notation (Line 2). We then consider all the events in the same trace having no immediate subsequent event in ρ prior to the occurrence of the next event in ρ'; this can be simply checked in ρ by determining that the next record appearing in ρ after $\langle i, j, L \rangle$ has an event id less than k (Line 3). If there is a non-trivially true Θ predicate, we also impose that at least one activation occurring after or at σ_j^i and at least one target occurring after or at σ_k^i matches with Θ (Line 6). Otherwise, we compute no match, and we straightforwardly collect the activation and target conditions from both events (Line 7). In the code, we explicitly injected an early-stopping condition avoiding testing subsequent events in ρ' within the same trace as soon as we detect one event in ρ, invalidating the condition at Line 3. By considering the time complexities for each xtLTL$_f$ operator in Section 2.3, we can argue that the time complexity associated with computing this operator as in the previous section without the aforementioned computation is totalled to $O(|\rho| + (||\mathfrak{S}|| - |\rho|)^2|\rho'| + 2((||\mathfrak{S}|| - |\rho|) + |\rho'|))$, where $||\mathfrak{S}|| = |\mathfrak{S}|\epsilon$. On the other hand, by assuming to always scan each trace quadratically of length ϵ for each event in ρ, we obtain the time complexity of $O(|\rho|\epsilon^2/2 + |\rho'|)$ for the derived operator when implemented as per the previous discussion. If we assume that ρ and ρ' are associated with a single activity label, as per the scenario in Declare, where the number of events and the activity labels are uniformly distributed such that #a $\approx |\mathfrak{S}|\epsilon/|\Sigma|$ for each a $\in \Sigma$, we can derive that the provided algorithm always provides a positive speed-up if compared to the original formulation in Equation (3).

4.2. AndAltWFuture

This algorithm works similarly to the previous, where we relax the until condition with a weak until, thus also admitting an absence of activation conditions after the first occurrence (of an activation) if no further target events are present (Line 23). Even in this scenario, we have a similar time complexity to the previous, while the original formulation in Equation (5) introduced an additional overhead to the previous by computing an additional timed disjunction and the global computation over the negation of the possibly activating events. Therefore, we expect an even greater speed up for this latest operator.

4.3. AndNext

As previously observed in the formal definition of this operator, we transformed this into an unary operator where, instead of retrieving two sets of events associated with two activity labels, we just scan one of the two. Before starting any form of scan, we immediately return if, after a $O(|\mathfrak{S}|)$ scan of the CountTable, we detect that no event is associated with the target condition (Line 28). Otherwise, we consider only events both coming from traces containing an event with activity label B and not being at the end of the trace, and for which the immediately next event is associated to an activity label B as a target condition ($T(j+1)$, Line 29); we implementationally further enhanced this by completely skipping any test whether the event resides in a trace where no B event resides. If $\Theta \neq$ **True**, then we also have to guarantee that each activation condition appearing in ρ should match with the target event at time $j+1$ (Line 33) and, upon provision of q, the target condition should also match with this (Line 35). The computational complexity of this operator is in $O(|\rho| + |\mathfrak{S}|)$ and, if we are taking into account the accessing time to the immediately following event, if any, we obtain a time in $2|\rho| + |\mathfrak{S}|$. If compared to the time complexity of Equation (7) of $|\rho| + 2|\rho'|$, we then obtain a positive speed up, i.e., $\frac{|\rho|+2|\rho'|}{2|\rho|+|\mathfrak{S}|} > 1$, for $|\rho'| > |\rho|/2$ and $0 < |\mathfrak{S}| < 2|\rho'| + |\rho|$.

4.4. NextAnd

This other operator works similarly to the previous, where we are checking instead the immediately preceding event instead of looking at the immediately following one, thus requiring that each element of interest in ρ shall never be at the beginning of the trace. The same considerations over speed-up and time complexity follow from the previous algorithm.

After associating each of the novel operators in the aforementioned algorithmic implementation, Equations (4), (6), (8) and (10) will then provide the semantics generating the query plan as Proposed in this current paper, while the direct translation of the LTL$_f$ expressions in Table 1 to the operators outlined in Section 2.2.2 provides the Original formulation of the query plan also in [28], where none of the previous algorithms are used.

5. Empirical Evaluation

Given that the aim of our derived operators is to enhance formal verification tasks conducted over temporal clauses expressed in Declare, we compare the different running times of carrying out formal verification tasks over our previous set of operators as well as by replacing those with our currently proposed derived ones, while focusing on benchmarking formal verification tasks over specifications written in Declare. We discard from our evaluation the benchmark of the single operator, as this is insufficient to remark on their adequacy in enhancing formal verification tasks in Declare. Thus, we compare different query plans being generated from different Declare semantics being specified at runtime through the queryplan "name" {...} query. With this, achieving a positive speed-up in Declare formal verification tasks as in our previous work [28] by using the proposed operators will tell us that, under specific data conditions, the original xtLTLf query plan associated with the declarative clauses available in KnoBAB constitutes the major computational bottleneck. Having a negligible speed-up will likely remark other components in the query plan dominating the overall running time, while having a negative speed-up only on specific data conditions will motivate some future work on hybrid algorithms, thus allowing us to choose between different algorithms for specific temporal operators depending on the data distribution within the loaded dataset [52].

Our benchmarks exploited a Dell Mobile Precision Workstation 5760 on Ubuntu 22.04: Intel® Xeon(R) W-11955M CPU @ 2.60 GHz × 16, 64 GB DDR4 3200 MHz RAM. We took two real-world datasets and a synthetic one for our experiments, both being dataless. The first real dataset (Hospital) monitors the patient flow and different medical procedures to which the patients in question were subjected; each trace tracks a single patient from his hospitalization to dismissal, and each activity label describes the name associated to such phases [10]. The second one (Cybersecurity) provides the auditing step of different malware, where each trace represents a single malware being audited, while each activity label identifies one single system call event being audited as invoked by the malware [6,7]. The synthetic dataset was derived from temporal graphs generated by FoodBroker [11] while describing trades and shipments of goods mediated by a brokerage company. For each GraphTransaction, we sort all the vertices describing an event occurring at a specific date, thus also including creation timestamps. For vertices describing a ticket being filed by a client raising a complaint, we return an activity label associated with the type of complaint (problem); otherwise, we keep the original vertex label. We then collect the set of temporally ordered activity labels and represent those as log traces. The updated FoodBroker codebase for generating event logs is also available online (https://github.com/jackbergus/foodbroker/, accessed on 3 January 2024).

For each dataset, we then obtain the sampled trace length distribution, and we sample sub-logs of various sizes while trying to abide by the trace distribution from the original dataset, notwithstanding their skewness. For the first and third (or second) datasets, we sample the logs so that their sizes are powers of ten (or nine) while always guaranteeing that each sub-log $|\mathfrak{S}_h| = 10^h$ (or $|\mathfrak{S}_h| = 9^h$) is always a subset of any larger sub-log. We also keep the original log as the last sample dataset. This random sampling mechanism is required to better assess the scalability of the proposed operator's implementation while guaranteeing an approximation of the original trace length distribution across the board to guarantee similar running time conditions. Figure 2 reports the sample PDF trace length for each of the sampled logs alongside the size of each sample. The FoodBroker synthetic dataset contains the shorter traces (Figure 2a); all the sampled logs except the first one have a maximum trace length of 24, while the first sublog has a maximum trace length of 21. On the other hand, the first two smaller log samples of the real-world Hospital dataset (Figure 2b) have traces with a maximum length of 1200, while the remaining two have a maximum trace length of 1814. The Cybersecurity dataset (Figure 2c) contains the longest traces, having a maximum trace length of 1.23×10^6 for the smaller two sub-logs and of 1.76×10^6 for the remaining ones. This information will soon become relevant while conducting our following analysis of the algorithmic speed-ups given by our proposed derived operators while performing formal verification over the models described in the following paragraph.

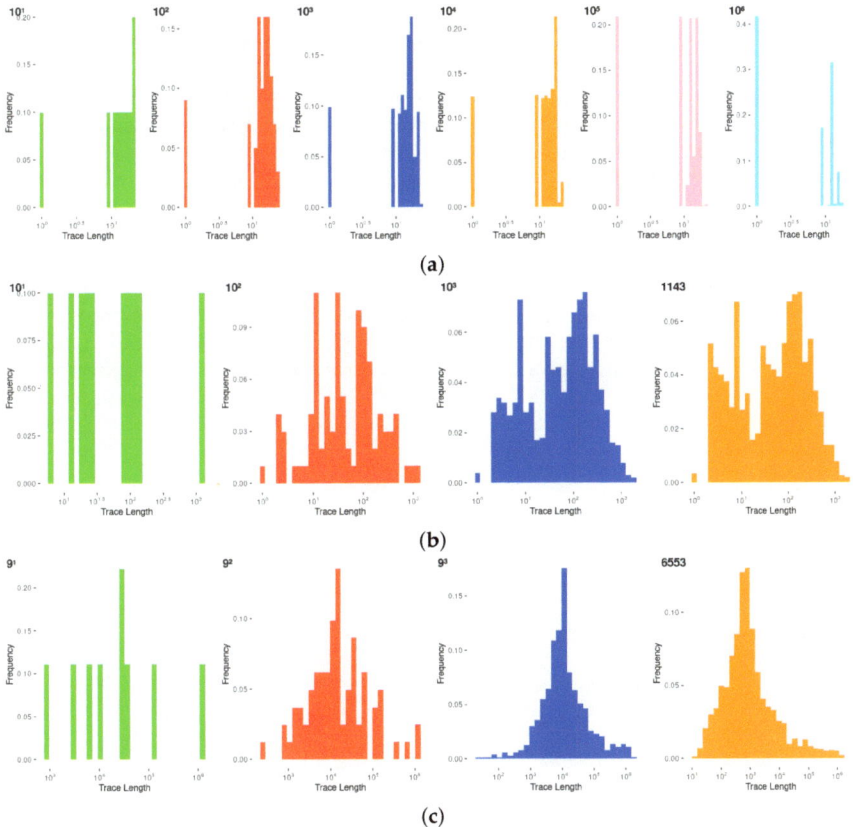

Figure 2. Sampled probability density function associated with the length of the traces for each sub-log extracted from each original dataset: (**a**) FoodBroker, (**b**) Hospital, and (**c**) Cybersecurity.

Given that we aim to test these newly introduced xtLTL$_f$ operators in the context of a Declare-based formal verification task when xtLTL$_f$ is used to represent its semantics, we generate four specifications $\Phi_1^c, \ldots, \Phi_4^c$ for each declarative clause of interest c, AltPrecedence, AltResponse, ChainPrecedence, and ChainResponse, where each Φ_i^c contains exactly i binary clauses determined by instantiating an activity label among the most frequently occurring ones within the smaller sub-log. We then use the same specifications generated for the smaller log and the greater sub logs, thus comparing the running times for each sub-log over the same Declare specifications. We then use the specifications to conduct a formal verification task via a `model-check...` query. The resulting logs and specifications are freely available online (https://osf.io/6y8cv/, accessed on 3 January 2024).

Last, as our previous work already showed that computing such queries on top of relational databases such as PostgreSQL with shorter traces leads to a greater running time than running similar queries over KnoBAB, we just focus on comparing the results from our previous implementation with the ones after applying the changes discussed in this paper.

With reference to Figure 3, AndAlt✶ operators can be deemed responsible for evidently outperforming the proposed query plan if compared to the one from the previous implementation, as they lead to an associated speed-up always strictly greater than one. Our previous definition of the Declare operators is greatly affected by the number of clauses within the model, which becomes even more apparent when the maximum and average trace length ϵ per sampled log increases. On the other hand, running our former formal verification query plan for AndAlt✶ clauses over the Cybersecurity dataset always took more than one 1H (3.6×10^6 ms), thus demonstrating an increased running time for the original query plan strategy when longer traces occur. We stopped recording the running time, as the overhead introduced by the intermediate operators for carrying out the actual matching between activation and target conditions was strikingly evident, while our proposed operators could instead carry out the formal verification task within one minute. Although no out-of-memory exceptions were observed before the timeout, these were clearly observed in larger specifications and log sizes, thus clearly demonstrating KnoBAB's limits as a main memory engine by not maintaining the query intermediate results in secondary memory. Despite the code allowing the clearing of intermediate caches to be run to free extra memory, this only partially addresses the out-of-memory failure for larger specifications. Overall, this demonstrates that this proposed extension for AndAlt✶ operators outperforms our previous query plan definition, as also expected from our previous analysis concerning the overall theoretical time complexity.

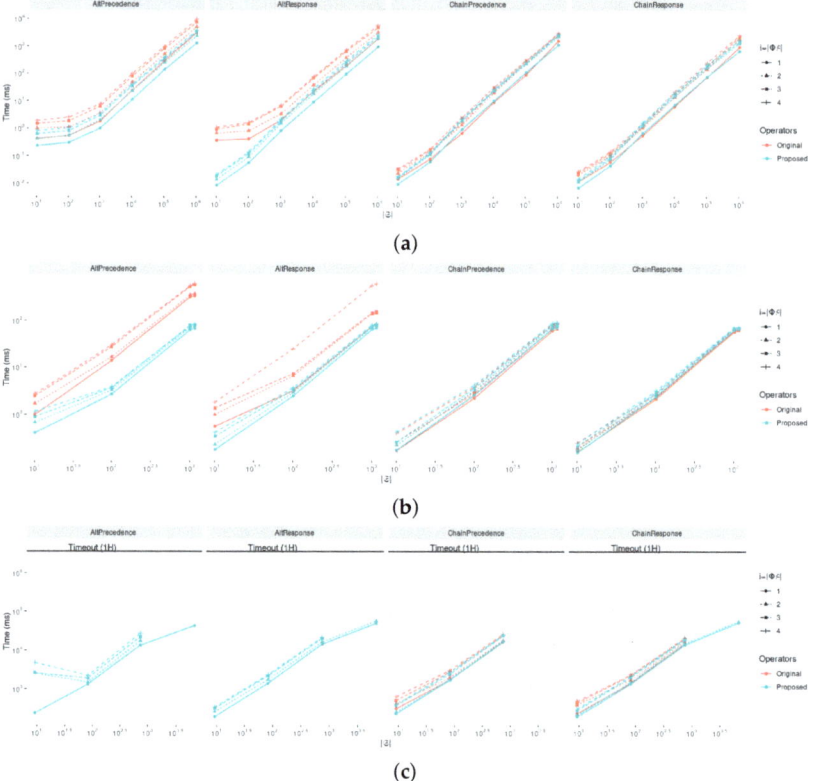

Figure 3. *Cont.*

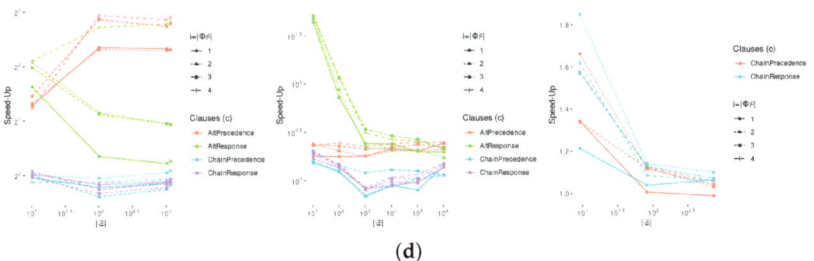

(d)

Figure 3. Comparing the proposed implementation of the derived operators with the previous implementation given in KnoBAB. (**a**) FoodBroker dataset; (**b**) Hospital dataset; (**c**) Cybersecurity dataset. (**d**) Datasets' speed-up: (**left**) FoodBroker, (**center**) Hospital, and (**right**) Cybersecurity.

Chain∗ operators provide a more convoluted scenario to be examined carefully. First, we observe a clear trend correlating datasets with longer traces with an overall increase in speed-up. In fact, the Hospital datasets exhibit more speed-ups compared to the FoodBroker one, where the recently proposed operators yield comparable or underperforming running times. Notwithstanding the former, we can clearly observe that the recently proposed operators consistently outperform our previous solution over the Cybersecurity dataset. Differently from our previous set-up, we can now observe that the original formulation of the declarative clauses without the currently presented operators now runs out of memory before hitting the 1H timeout for the larger sample, being the full dataset, while our solution still manages to carry out some temporal formal verification tasks over specifications containing fewer clauses. Last, we consistently observe that such operators still provide greater speed-ups over datasets with smaller log sizes, thus providing theoretical validation to our speed-up equations for such operators. This postulates the need for such operators while dealing with massive datasets, while advocating the usage of hybrid algorithms for switching between the previous solution and the currently proposed one.

6. Conclusions and Future Works

This paper proposes an extension to our previous work on KnoBAB by optimizing our previously proposed query plan by introducing novel algebraic temporal operators expressing formal verification tasks on column database storages in main memory. As a consequence, we extended our temporal algebra xtLTL$_f$ with four novel operators, subsuming entire xtLTL$_f$ expressions which before could only be represented in terms of combinations of costly basic operators. Preliminary results over such operators provide non-negligible speed-up to the formal verification tasks over realistic datasets, where several events are audited and collected in a larger collection of traces.

Despite these experiments demonstrating the efficiency of carrying out formal verification computations on columnar databases implemented as a main memory engine, the consistent out-of-memory faults that we experienced over larger collections of data containing more events (i.e., longer traces) encourage us to store the intermediate query results in secondary memory, as customary for off-the-shelf databases such as PostgreSQL. We see this as the last required step for fully supporting real data alongside the orthogonal operator optimization, as discussed in the present paper. Despite the fact that putting this solution in place will come at the detriment of overall performance, this will guarantee the carriage of the entire formal verification computation. This drawback might be alleviated by determining at runtime whether to represent intermediate results in primary or secondary memory depending on the log and trace size. Another possible way to alleviate such a problem is to re-implement the overall pipeline using a pull-based strategy [33] when operations are not run concurrently. Another way to challenge this primary memory limitation would be migrating the proposed architecture over Oracle Cloud [48], which already supports traditional time-transactional database operations compatible with the aforementioned *temporal modules*. While doing so, we will be walking in the footsteps of previous literature [50] by attempting to rewrite dataful xtLTL$_f$ specifications into the supported temporal extension of SQL.

The current experiment noted the optimality of the proposed operators when dealing with datasets with longer traces (i.e., greater ϵ). Future work will consider the possibility of defining hybrid algorithms [27] over the operators, optimizing Chain∗ clauses by empirically determining the table size threshold over which we prefer the derived operators over the original. As an orthogonal approach, we will also define the "dual" operators for AndNext and NextAnd so as to start scanning from the target condition while moving backwards towards any existing activation condition when the number of targets is deemed to be fewer than the activations. Our future works will also aim to further benchmark these operators in the context of *dataful* logs, where events are also associated with a payload expressed as a key-value pair as in customary semi-structured data formats. These works will then outline the overhead required to compute a Θ correlation condition between activation and target event.

Previous research on temporal modules demonstrated the possibility of expressing LTL$_f$ specifications when traces have multiple events occurring at one specific point in time [50]; the current theoretical literature on conformance checking suggests that this is actually possible by representing each single event as a conjunction of multiple mutually exclusive events, thus obtaining the characterization of composite events. However, realizing this in practice for events with distinct labels would require a drastic overhaul of KnoBAB's relational representation, as the current architecture is focused on the linear representation of each individual trace. Future work will therefore contemplate the possibility of extending the current relational model with an object-oriented one [53], better supporting the nesting and composition of objects, a feature also required for coalescing multiple events in a single instant in time.

Finally, an interesting outcome of these observations on relational databases would be the application of such an algebra in the context of temporal graphs [54], thus enabling the efficient temporal verification under this different data representation. Despite the recent attempt at representing logs as temporal graphs [55], the aforementioned is still a desideratum, as no graph temporal operator for expressing formal verification tasks is currently known. Differently from the previously pursued approach [56], this will then require us to define tailored temporal operators for graph query languages similarly to xtLTL$_f$.

Funding: This research received no external funding.

Institutional Review Board Statement: Not applicable.

Informed Consent Statement: Not applicable.

Data Availability Statement: The dataset is available at https://osf.io/6y8cv/, accessed on 3 January 2024.

Conflicts of Interest: The author declares no conflict of interest.

References

1. Seshia, S.A.; Sadigh, D.; Sastry, S.S. Toward verified artificial intelligence. *Commun. ACM* **2022**, *65*, 46–55. [CrossRef]
2. Baier, C.; Katoen, J. *Principles of Model Checking*; MIT Press: Cambridge, MA, USA, 2008.
3. Bergami, G.; Maggi, F.M.; Marrella, A.; Montali, M. Aligning Data-Aware Declarative Process Models and Event Logs. In Proceedings of the Business Process Management-19th International Conference, BPM 2021, Rome, Italy, 6–10 September 2021; Lecture Notes in Computer Science; Polyvyanyy, A., Wynn, M.T., Looy, A.V., Reichert, M., Eds.; Springer: Berlin/Heidelberg, Germany, 2021; Volume 12875, pp. 235–251.
4. Awad, A.; Decker, G.; Weske, M. Efficient Compliance Checking Using BPMN-Q and Temporal Logic. In Proceedings of the Business Process Management, 6th International Conference, BPM 2008, Milan, Italy, 2–4 September 2008; Lecture Notes in Computer, Science; Dumas, M., Reichert, M., Shan, M., Eds.; Springer: Berlin/Heidelberg, Germany, 2008; Volume 5240, pp. 326–341.

5. Weidlich, M.; Polyvyanyy, A.; Desai, N.; Mendling, J.; Weske, M. Process compliance analysis based on behavioural profiles. *Inf. Syst.* **2011**, *36*, 1009–1025. [CrossRef]
6. Catak, F.O.; Ahmed, J.; Sahinbas, K.; Khand, Z.H. Data augmentation based malware detection using convolutional neural networks. *PeerJ Comput. Sci.* **2021**, *7*, e346. [CrossRef] [PubMed]
7. Yazi, A.F.; Çatak, F.Ö.; Gül, E. Classification of Methamorphic Malware with Deep Learning (LSTM). In Proceedings of the 27th Signal Processing and Communications Applications Conference, SIU 2019, Sivas, Turkey, 24–26 April 2019; IEEE: Piscataway, NJ, USA, 2019; pp. 1–4.
8. Zheng, W.; Du, Y.; Wang, S.; Qi, L. Repair Process Models Containing Non-Free-Choice Structures Based on Logic Petri Nets. *IEEE Access* **2019**, *7*, 105132–105145. [CrossRef]
9. Xu, H.; Pang, J.; Yang, X.; Yu, J.; Li, X.; Zhao, D. Modeling clinical activities based on multi-perspective declarative process mining with openEHR's characteristic. *BMC Med. Inform. Decis. Mak.* **2020**, *20-S*, 303. [CrossRef] [PubMed]
10. van Dongen, B. Real-Life Event Logs-Hospital Log. 2011. Available online: https://data.4tu.nl/articles/_/12716513/1 (accessed on 3 January 2024).
11. Petermann, A.; Junghanns, M.; Müller, R.; Rahm, E. FoodBroker-Generating Synthetic Datasets for Graph-Based Business Analytics. In Proceedings of the Big Data Benchmarking-5th International Workshop, WBDB 2014, Potsdam, Germany, 5–6 August 2014; Revised Selected Papers; Lecture Notes in Computer Science; Rabl, T., Sachs, K., Poess, M., Baru, C.K., Jacobsen, H., Eds.; Springer: Berlin/Heidelberg, Germany, 2014; Volume 8991, pp. 145–155.
12. Petsis, S.; Karamanou, A.; Kalampokis, E.; Tarabanis, K. Forecasting and explaining emergency department visits in a public hospital. *J. Intell. Inf. Syst.* **2022**, *59*, 479–500. [CrossRef]
13. Giacomo, G.D.; Vardi, M.Y. Linear Temporal Logic and Linear Dynamic Logic on Finite Traces. In Proceedings of the IJCAI 2013, Proceedings of the 23rd International Joint Conference on Artificial Intelligence, Beijing, China, 3–9 August 2013; Rossi, F., Ed.; AAAI Press: Menlo Park, CA, USA, 2013; pp. 854–860.
14. Pesić, M.; Schonenberg, H.; van der Aalst, W.M. DECLARE: Full Support for Loosely-Structured Processes. In Proceedings of the 11th IEEE International Enterprise Distributed Object Computing Conference (EDOC 2007), Annapolis, MD, USA, 15–19 October 2007; p. 287.
15. Cuzzocrea, A. Temporal Big Data Analytics: New Frontiers for Big Data Analytics Research. In Proceedings of the 28th International Symposium on Temporal Representation and Reasoning (TIME 2021), Klagenfurt, Austria, 27–29 September 2021; Combi, C., Eder, J., Reynolds, M., Eds.; Leibniz International Proceedings in Informatics (LIPIcs): Dagstuhl, Germany, 2021; Volume 206, pp. 4:1–4:7.
16. Amer-Yahia, S.; Palpanas, T.; Tsytsarau, M.; Kleisarchaki, S.; Douzal, A.; Christophides, V. Temporal Analytics in Social Media. In *Encyclopedia of Database Systems*, 2nd ed.; Liu, L., Özsu, M.T., Eds.; Springer: Berlin/Heidelberg, Germany, 2018. [CrossRef]
17. Schönig, S.; Rogge-Solti, A.; Cabanillas, C.; Jablonski, S.; Mendling, J. Efficient and Customisable Declarative Process Mining with SQL. In *Advanced Information Systems Engineering, Proceedings of the 28th International Conference, CAiSE 2016, Ljubljana, Slovenia, 13–17 June 2016*; Lecture Notes in Computer, Science; Nurcan, S., Soffer, P., Bajec, M., Eder, J., Eds.; Springer: Berlin/Heidelberg, Germany, 2016; Volume 9694, pp. 290–305.
18. Huang, S.; Zhu, E.; Chaudhuri, S.; Spiegelberg, L. T-Rex: Optimizing Pattern Search on Time Series. *Proc. ACM Manag. Data* **2023**, *1*, 130:1–130:26. [CrossRef]
19. Anselma, L.; Bottrighi, A.; Montani, S.; Terenziani, P. Extending BCDM to Cope with Proposals and Evaluations of Updates. *IEEE Trans. Knowl. Data Eng.* **2013**, *25*, 556–570. [CrossRef]
20. Kaufmann, M.; Vagenas, P.; Fischer, P.M.; Kossmann, D.; Färber, F. Comprehensive and Interactive Temporal Query Processing with SAP HANA. *Proc. VLDB Endow.* **2013**, *6*, 1210–1213. [CrossRef]
21. Wang, X.S.; Jajodia, S.; Subrahmanian, V.S. Temporal Modules: An Approach Toward Federated Temporal Databases. *Inf. Sci.* **1995**, *82*, 103–128. [CrossRef]
22. Wang, X.S. Algebraic Query Languages on Temporal Databases with Multiple Time Granularities. In Proceedings of the CIKM '95, 1995 International Conference on Information and Knowledge Management, Baltimore, MD, USA, 28 November–2 December 1995; Pissinou, N., Silberschatz, A., Park, E.K., Makki, K., Eds.; ACM: New York, NY, USA, 1995; pp. 304–311.
23. Wang, C.; Yu, K.; Zhou, T.; Cai, Z. Time2State: An Unsupervised Framework for Inferring the Latent States in Time Series Data. *Proc. ACM Manag. Data* **2023**, *1*, 17:1–17:18. [CrossRef]
24. Huo, X.; Hao, K.; Chen, L.; Tang, X.; Wang, T.; Cai, X. A dynamic soft sensor of industrial fuzzy time series with propositional linear temporal logic. *Expert Syst. Appl.* **2022**, *201*, 117176. [CrossRef]
25. Mao, X.; Li, X.; Huang, Y.; Shi, J.; Zhang, Y. Programmable Logic Controllers Past Linear Temporal Logic for Monitoring Applications in Industrial Control Systems. *IEEE Trans. Ind. Inform.* **2022**, *18*, 4393–4405. [CrossRef]
26. Fionda, V.; Greco, G.; Mastratisi, M.A. Reasoning about Smart Contracts Encoded in LTL. In Proceedings of the AIxIA, Milan, Italy, 1–3 December 2021; pp. 123–136.
27. Pnueli, A. The temporal logic of programs. In Proceedings of the 18th Annual Symposium on Foundations of Computer Science (sfcs 1977), Providence, RI, USA, 31 October–2 November 1977; pp. 46–57. [CrossRef]

28. Bergami, G.; Appleby, S.; Morgan, G. Quickening Data-Aware Conformance Checking through Temporal Algebras. *Information* **2023**, *14*, 173. [CrossRef]
29. Bellatreche, L.; Kechar, M.; Bahloul, S.N. Bringing Common Subexpression Problem from the Dark to Light: Towards Large-Scale Workload Optimizations. In Proceedings of the IDEAS, Montreal, QC, Canada, 14–16 July 2021; ACM: New York, NY, USA, 2021.
30. Appleby, S.; Bergami, G.; Morgan, G. Running Temporal Logical Queries on the Relational Model. In Proceedings of the 26th International Database Engineered Applications Symposium, Budapest, Hungary, 22–24 August 2022.
31. Atzeni, P.; Ceri, S.; Paraboschi, S.; Torlone, R. *Database Systems—Concepts, Languages and Architectures*; McGraw-Hill Book Company: New York, NY, USA, 1999.
32. Elmasri, R.; Navathe, S.B. *Fundamentals of Database Systems*, 7th ed.; Pearson: London, UK, 2015.
33. Dittrich, J. *Patterns in Data Management: A Flipped Textbook*; CreateSpace Independent Publishing Platform: Scotts Valley, CA, USA, 2016.
34. Bergami, G.; Appleby, S.; Morgan, G. Specification Mining over Temporal Data. *Computers* **2023**, *12*, 185. [CrossRef]
35. Burattin, A.; Maggi, F.M.; Sperduti, A. Conformance checking based on multi-perspective declarative process models. *Expert Syst. Appl.* **2016**, *65*, 194–211. [CrossRef]
36. Zhu, S.; Pu, G.; Vardi, M.Y. First-Order vs. Second-Order Encodings for \textsc ltl_f -to-Automata Translation. In *Theory and Applications of Models of Computation, Proceedings of the 15th Annual Conference, TAMC 2019, Kitakyushu, Japan, 13–16 April 2019*; Lecture Notes in Computer, Science; Gopal, T.V., Watada, J., Eds.; Springer: Berlin/Heidelberg, Germany, 2019; Volume 11436, pp. 684–705.
37. Li, J.; Pu, G.; Zhang, Y.; Vardi, M.Y.; Rozier, K.Y. SAT-based explicit LTLf satisfiability checking. *Artif. Intell.* **2020**, *289*, 103369. [CrossRef]
38. Acampora, G.; Vitiello, A.; Stefano, B.N.D.; van der Aalst, W.M.P.; Günther, C.W.; Verbeek, E. IEEE 1849: The XES Standard: The Second IEEE Standard Sponsored by IEEE Computational Intelligence Society [Society Briefs]. *IEEE Comput. Intell. Mag.* **2017**, *12*, 4–8. [CrossRef]
39. Maggi, F.M.; Bose, R.P.J.C.; van der Aalst, W.M.P. Efficient Discovery of Understandable Declarative Process Models from Event Logs. In *Advanced Information Systems Engineering*; Springer: Berlin/Heidelberg, Germany, 2012; pp. 270–285.
40. de Murillas, E.G.L.; Reijers, H.A.; van der Aalst, W.M.P. Data-Aware Process Oriented Query Language. In *Process Querying Methods*; Polyvyanyy, A., Ed.; Springer: Berlin/Heidelberg, Germany, 2022; pp. 49–83. [CrossRef]
41. Kammerer, K.; Pryss, R.; Reichert, M. Retrieving, Abstracting, and Changing Business Process Models with PQL. In *Process Querying Methods*; Polyvyanyy, A., Ed.; Springer: Berlin/Heidelberg, Germany, 2022; pp. 219–254. [CrossRef]
42. Idreos, S.; Groffen, F.; Nes, N.; Manegold, S.; Mullender, K.S.; Kersten, M.L. MonetDB: Two Decades of Research in Column-oriented Database Architectures. *IEEE Data Eng. Bull.* **2012**, *35*, 40–45.
43. Green, T.J.; Karvounarakis, G.; Tannen, V. Provenance Semirings. In Proceedings of the Twenty-Sixth ACM SIGMOD-SIGACT-SIGART Symposium on Principles of Database Systems, New York, NY, USA, 11–13 June 2007; pp. 31–40. [CrossRef]
44. Schönig, S. SQL Queries for Declarative Process Mining on Event Logs of Relational Databases. *arXiv* **2015**, arXiv:1512.00196.
45. Boncz, P.A.; Manegold, S.; Kersten, M.L. Database Architecture Evolution: Mammals Flourished long before Dinosaurs became Extinct. *Proc. VLDB Endow.* **2009**, *2*, 1648–1653. [CrossRef]
46. Allen, J.F. Maintaining Knowledge about Temporal Intervals. *Commun. ACM* **1983**, *26*, 832–843. [CrossRef]
47. Revesz, P.Z. *Introduction to Databases—From Biological to Spatio-Temporal*; Texts in Computer Science; Springer: Berlin/Heidelberg, Germany, 2010.
48. Kvet, M. *Developing Robust Date and Time Oriented Applications in Oracle Cloud: A Comprehensive Guide to Efficient Date and Time Management in Oracle Cloud*; Packt Publishing: Birmingham, UK, 2023.
49. Tuzhilin, A.; Kedem, Z. *Using Temporal Logic and Datalog to Query Databases Evolving in Time*; New York University: New York, NY, USA, 1989.
50. Böhlen, M.H.; Chomicki, J.; Snodgrass, R.T.; Toman, D. Querying TSQL2 databases with temporal logic. In *Advances in Database Technology—EDBT '96, Proceedings of the 5th International Conference on Extending Database Technology, Avignon, France, 25–29 March 1996*; Apers, P., Bouzeghoub, M., Gardarin, G., Eds.; Springer: Berlin/Heidelberg, Germany, 1996; pp. 325–341.
51. Snodgrass, R.T. TSQL2. In *Encyclopedia of Database Systems*; Liu, L., Özsu, M.T., Eds.; Springer: Boston, MA, USA, 2009; pp. 3192–3197. [CrossRef]
52. Musser, D.R. Introspective Sorting and Selection Algorithms. *Softw. Pract. Exp.* **1997**, *27*, 983–993. [CrossRef]
53. van der Aalst, W.M.P. Object-Centric Process Mining: Unraveling the Fabric of Real Processes. *Mathematics* **2023**, *11*, 2691. [CrossRef]
54. Rost, C.; Gómez, K.; Täschner, M.; Fritzsche, P.; Schons, L.; Christ, L.; Adameit, T.; Junghanns, M.; Rahm, E. Distributed temporal graph analytics with GRADOOP. *VLDB J.* **2022**, *31*, 375–401. [CrossRef]

55. Khayatbashi, S.; Hartig, O.; Jalali, A. Transforming Event Knowledge Graph to Object-Centric Event Logs: A Comparative Study for Multi-dimensional Process Analysis. In Proceedings of the 42nd International Conference on Conceptual Modeling, Lisbon, Portugal, 6–9 November 2023.
56. Zaki, N.M.; Helal, I.M.A.; Hassanein, E.E.; Awad, A. Efficient Checking of Timed Ordered Anti-patterns over Graph-Encoded Event Logs. In *Model and Data Engineering: Proceedings of the 11th International Conference, MEDI 2022, Cairo, Egypt, 21–24 November 2022*; Lecture Notes in Computer Science; Fournier-Viger, P., Yousef, A.H., Bellatreche, L., Eds.; Springer: Berlin/Heidelberg, Germany, 2022; Volume 13761, pp. 147–161.

Disclaimer/Publisher's Note: The statements, opinions and data contained in all publications are solely those of the individual author(s) and contributor(s) and not of MDPI and/or the editor(s). MDPI and/or the editor(s) disclaim responsibility for any injury to people or property resulting from any ideas, methods, instructions or products referred to in the content.

Enhancing Flight Delay Predictions Using Network Centrality Measures

Joseph Ajayi, Yao Xu *, Lixin Li and Kai Wang

Department of Computer Science, Georgia Southern University, Statesboro, GA 30458, USA; ja20859@georgiasouthern.edu (J.A.); lli@georgiasouthern.edu (L.L.); kwang@georgiasouthern.edu (K.W.)
* Correspondence: yxu@georgiasouthern.edu

Abstract: Accurately predicting flight delays remains a significant challenge in the aviation industry due to the complexity and interconnectivity of its operations. The traditional prediction methods often rely on meteorological conditions, such as temperature, humidity, and dew point, as well as flight-specific data like departure and arrival times. However, these predictors frequently fail to capture the nuanced dynamics that lead to delays. This paper introduces network centrality measures as novel predictors to enhance the binary classification of flight arrival delays. Additionally, it emphasizes the use of tree-based ensemble models, specifically random forest, gradient boosting, and CatBoost, which are recognized for their superior ability to model complex relationships compared to single classifiers. Empirical testing shows that incorporating centrality measures improves the models' average performance, with random forest being the most effective, achieving an accuracy rate of 86.2%, surpassing the baseline by 1.7%.

Keywords: flight delay prediction; network centrality; machine learning; random forest; gradient boosting; CatBoost

1. Introduction

In the realm of aviation, the efficiency of flight operations significantly hinges on the ability to anticipate and mitigate delays. As the Federal Aviation Administration (FAA) reports, its Air Traffic Organization (ATO) orchestrates the movement of over 45,000 flights daily, servicing 2.9 million passengers across an expansive airspace exceeding 29 million square miles [1]. This volume is projected to swell by 4.9% annually over the next two decades, underscoring a pressing need for robust predictive models that can adeptly forecast flight delays, thereby enabling airlines to optimize scheduling and resource allocation [2]. Despite the proliferation of predictive methodologies ranging from traditional statistical techniques to advanced machine learning algorithms like decision trees (DTs), random forests (RFs), Bayesian networks (BNs), and linear regression (LR), the quest for high-accuracy predictions remains largely unfulfilled. This challenge is compounded by the unpredictable nature of many delay-inducing factors, such as adverse weather conditions, and the computational demands posed by the voluminous and growing datasets of airline operations [3].

In recent years, deep learning methods have shown promise in various prediction tasks due to their ability to model complex non-linear relationships in large datasets [4–6]. However, these models often require extensive computational resources and large amounts of data for training, which can limit their applicability in certain scenarios. In contrast, the traditional machine learning techniques, such as support vector machine (SVM), DT, RF, and gradient boosting (GB), offer robust performance while being less resource-intensive and more interpretable [7–9]. Given these advantages, this paper introduces a novel approach to predict whether a flight will be delayed or not, leveraging network centrality measures within a binary classification framework.

By constructing a network model wherein airports serve as nodes and flight routes as edges, this study integrates centrality metrics to enhance the predictive capabilities of tree-based ensemble models. These models are renowned for their efficacy in capturing complex non-linear relationships that elude the traditional base classifiers. This integration aims to shed light on how the structural properties of the airport network can influence delay propagation and, by extension, the overall network performance.

The motivation for this research is twofold: Firstly, flight delays are a pervasive issue that undermines operational efficiency and diminishes passenger satisfaction, with a notable 20% of flights in 2023 experiencing delays across the United States alone [10]. Secondly, the existing predictive models often fall short of the accuracy needed for effective planning and resource management, partly due to their reliance on a limited set of predictors that may not fully encapsulate the intricacies of the aviation system [3]. By incorporating network centrality measures into the predictive models, this study aspires to bridge this gap, offering a more comprehensive and nuanced understanding of the factors that contribute to flight delays.

The primary research question addressed in this study is whether the inclusion of network centrality measures can improve the accuracy of flight delay predictions. The innovation of this study lies in the novel integration of network centrality measures into machine learning models for flight delay prediction, which, to the best of our knowledge, has not been explored in the literature. This approach provides new insights and improves the predictive accuracy beyond the traditional methods. Specifically, the study examines how these centrality measures affect the performance of traditional machine learning models, including RF, GB, and CatBoost (CB). It also explores which network centrality measures, such as degree, betweenness, and closeness centrality, contribute most significantly to enhancing the predictive accuracy of these models.

The rest of the paper is organized as follows: It begins with a literature review of the current landscape of flight delay prediction methodologies. The subsequent sections describe the methodology employed in constructing the network model and integrating centrality measures into the ensemble predictive models. The Results section presents a comparative analysis of the model performances, highlighting the enhanced accuracy achieved through the inclusion of centrality measures. Finally, the conclusion reflects on the implications of these findings for airline operations and potential future research directions.

2. Literature Review

Flight delay prediction has been extensively studied due to its critical implications for airline operations and passenger satisfaction. The early approaches primarily relied on traditional statistical models such as linear regression and time series analysis. For instance, Hsiao and Hansen [11] utilized econometric models to assess the impact of morning queuing delays, while Zou et al. [12] explored the relationship between flight delays, capacity investment, and social welfare, underscoring the importance of strategic investments.

The advent of machine learning (ML) technologies introduced more sophisticated methods for delay prediction, capturing complex non-linear relationships in data. Re-bollo and Balakrishnan [13] applied RF algorithms to integrate temporal and spatial delay states, improving the prediction accuracy. Kim et al. [4] leveraged convolutional neural networks (CNNs) with historical flight and weather data, achieving higher accuracy. Choi et al. [8] emphasized the importance of weather data in improving predictions through various ML algorithms, while Nigam et al. [14] showcased the efficiency of cloud-based logistic regression in real-time delay prediction.

Later work has further integrated deep learning models with traditional ML techniques. Yin et al. [15] utilized reinforcement learning for predicting taxi-out times, optimizing airport operations. Pamplona, et al. [16] introduced a supervised neural network incorporating multiple factors, and Yu et al. [5] combined deep belief networks with support vector regression, demonstrating effective delay prediction at Beijing International

Airport. Gui et al. [17] and Liu et al. [18] also explored big data analytics, using DT, RF, and GB for large-scale delay prediction.

Network centrality measures have increasingly been recognized for their potential in delay prediction. However, the prior studies primarily utilized these measures for structural analysis rather than as direct input features for prediction models. Cai et al. [6] and Wu et al. [19] applied deep learning models to time-evolving graphs and spatiotemporal data, respectively, focusing on network dynamics without directly integrating centrality measures as predictive features. Li et al. [20] advanced this area by combining CNNs and LSTM networks to capture spatial and temporal dependencies, although they did not use centrality metrics as input variables.

Comparative and cluster-based methodologies have also been explored. Güvercin, Ferhatosmanoglu, and Gedik [21] proposed the Clustered Airport Modeling (CAM) approach using network features and REG-ARIMA models to predict delays. Paramita et al. [22] demonstrated the effectiveness of RF algorithms in a cluster computing environment, while Wei et al. [23] introduced a BiLSTM-Attention network to predict delays across airport clusters.

Finally, studies on the structural properties of air transportation networks have offered key insights into delay prediction. Cheung and Gunes [24] used complex network metrics to reveal small-world characteristics and assess the network's resiliency to disruptions. Anderson and Revesz [25] developed algorithms for MaxCount and threshold operators on moving objects, applicable in monitoring airplane congestion, a factor in delay prediction.

While significant advancements have been made in flight delay prediction, a notable gap remains in the use of network centrality measures as direct input features in predictive models. Previous studies have largely focused on structural analysis without fully leveraging these measures to improve prediction accuracy. This study addresses that gap by directly integrating network centrality measures into machine learning models, offering a novel approach that enhances the accuracy of flight delay predictions.

3. Preliminaries

3.1. Network Centrality Measures

Network centrality measures are crucial for identifying the most influential nodes within a network, such as airports in an air transportation network. In this study, we focus on three key centrality measures: degree centrality, betweenness centrality, and closeness centrality. Given a network of N nodes, representing airports in this study, the definitions of the three centrality measures are as follows.

Degree Centrality: This centrality measure quantifies the number of direct connections a node has [26]. It is calculated as

$$C_d(v) = \frac{\deg(v)}{N-1}, \qquad (1)$$

where $\deg(v)$ is the degree of node v, that is, the total number of nodes directedly connected to v. High degree centrality indicates that an airport is a major hub with numerous direct flights, making it a critical point for delay propagation.

Betweenness Centrality: This metric reflects the number of times a node acts as a bridge along the shortest path between two other nodes [26]. It is calculated as

$$C_b(v) = \sum_{s \neq v \neq t} \frac{\sigma_{st}(v)}{\sigma_{st}}, \qquad (2)$$

where σ_{st} represents the total number of shortest paths from node s to node t, and $\sigma_{st}(v)$ is the number of those paths that pass through node v. Airports with high betweenness centrality are crucial in the flow of air traffic and are more likely to influence delays across the network.

Closeness Centrality: This is a measure of the average shortest distance between a node and all other reachable nodes, indicating how close a node is to all other nodes in the network [26,27]. It is calculated as

$$C_c(v) = \frac{r(v)}{N-1} \cdot \frac{r(v)}{\sum_t d(v,\ t)},\qquad(3)$$

where $r(v)$ is the total number of nodes v can reach, and $d(v,\ t)$ is the shortest distance between nodes v and t. Airports with high closeness centrality can quickly disseminate delays throughout the network, affecting overall network performance.

3.2. Machine Learning Models

The machine learning methods employed in this study include RF, GB, and CB. These models were selected for their ability to handle complex non-linear relationships and large datasets typical of air transportation networks.

Random Forest (RF): This ensemble learning method constructs multiple decision trees during training and combines their outputs, either by taking the mode for classification tasks or the mean for regression tasks [28]. By using multiple trees, RF effectively reduces overfitting and enhances the model's generalization, making it robust for various predictive tasks.

Gradient Boosting (GB): An iterative ensemble technique that builds models sequentially, where each new model corrects errors made by the previous ones [29,30]. GB is particularly effective in handling high-dimensional data and capturing complex interactions between features, making it a powerful tool for improving predictive accuracy.

CatBoost (CB): A high-performing variant of GB, which is specifically designed to handle categorical data with minimal preprocessing [31]. It addresses overfitting through ordered boosting, which prevents information leakage by using a permutation of the training data, making it particularly effective for datasets with categorical features.

4. Data and Methodology

4.1. Data Collection and Preparation

The data used in this research were obtained from the US Bureau of Transportation Statistics (BTS) TranStats database [32], which is publicly available. The dataset we used in this study is from the database named "Airline On-Time Performance Data", which contains detailed records of on-time arrivals and departures for non-stop domestic flights. For this study, the initial dataset included 7,107,203 flights connecting 370 airports from July 2022 to June 2023.

Data preprocessing involved handling missing data by excluding records with null or missing values to maintain dataset integrity. The final dataset comprised 6,955,805 flights. Key features selected for the analysis included flight information such as ORIGIN_AIRPORT_ID, DEST_AIRPORT_ID, DEP_TIME, and ARR_TIME and delay information like DEP_DELAY and ARR_DELAY, along with DISTANCE as an operational factor. Among all the features, ORIGIN_AIRPORT_ID and DEST_AIRPORT_ID were converted to categorical features for model training. Table 1 displays the key attributes of the dataset after data preprocessing.

Table 1. Key attributes of the dataset.

Attribute Name	Description	Type
ORIGIN_AIRPORT_ID	Origin airport	Categorical
DEST_AIRPORT_ID	Destination airport	Categorical
DEP_TIME	Scheduled departure time	Numerical
DEP_DELAY	Flight delay (in minutes)	Numerical
ARR_DELAY	Arrival delay (in minutes)	Numerical

4.2. Methodology

This study focuses on integrating network centrality measures into machine learning models to improve the accuracy of flight delay predictions. The methodology encompasses constructing the airport network, calculating centrality measures, and applying machine learning models for prediction.

4.2.1. Airport Network Construction and Centrality Integration

A directed graph representing the airport network was constructed to compute the network centrality measures. Airports were represented as vertices, with flights between them forming the edges, and edge weights were determined by the distances between airports. This graph enabled the calculation of degree centrality, betweenness centrality, and closeness centrality, as defined by Equations (1)–(3).

Figure 1 presents the top 20 airports ranked by their degree, betweenness, and closeness centrality scores, highlighting their structural importance within the US air transportation network. DFW (Dallas/Fort Worth), DEN (Denver), and ATL (Atlanta) top the list for degree centrality, indicating their extensive connectivity as major hubs. For betweenness centrality, DFW, DEN, and ORD (Chicago O'Hare) rank highest, reflecting their critical roles as key transfer points in air traffic flow. Closeness centrality is also led by DFW, DEN, and ORD, demonstrating their central positions within the network. These centrality measures align with the real-world functions of these airports, confirming their effectiveness in understanding network dynamics and predicting flight delays.

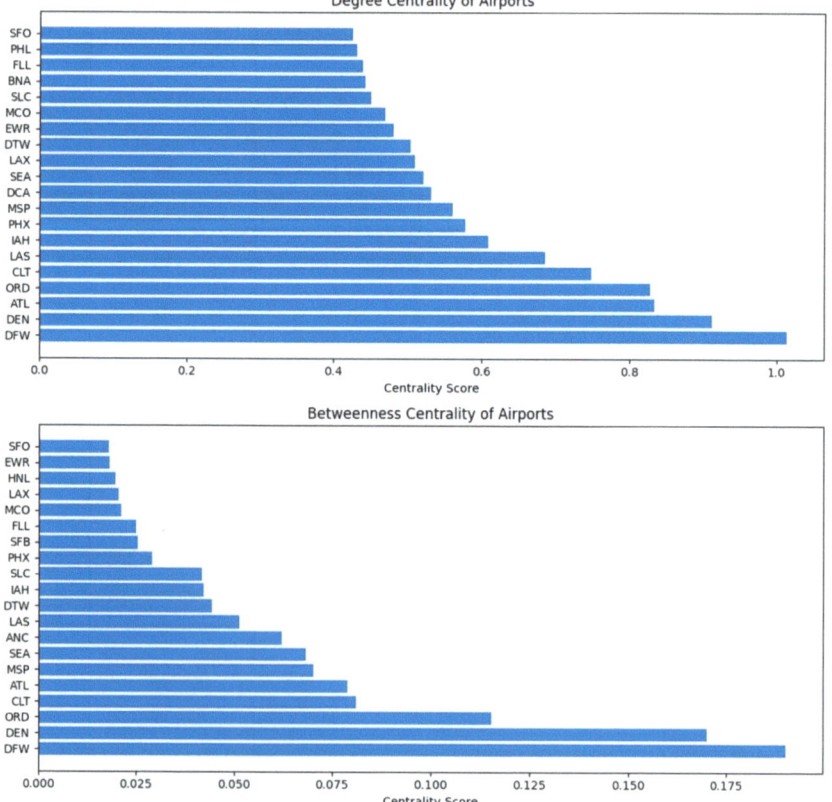

Figure 1. *Cont.*

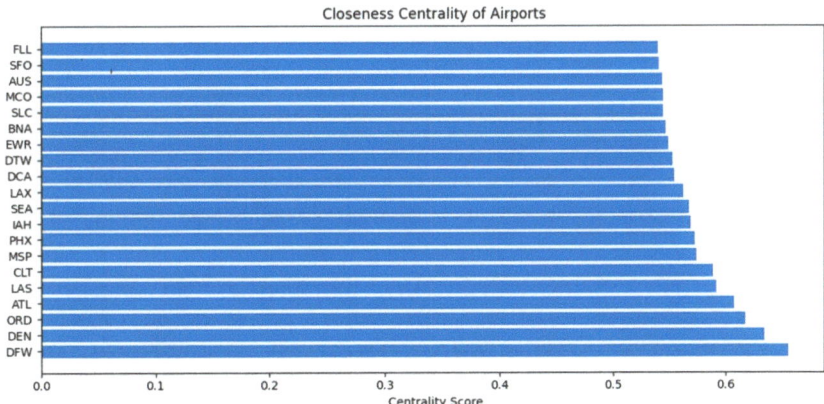

Figure 1. Top 20 airports by degree, betweenness, and closeness centrality scores.

The centrality values were incorporated into the flight dataset as additional features corresponding to the origin and destination airports of each flight. Since each airport has its own computed degree, betweenness, and closeness centrality scores, six additional features were added to the dataset: three for the origin airport and three for the destination airport. Table 2 displays the key attributes of the updated dataset.

Table 2. Key attributes of the dataset after integrating network centrality measures.

Attribute Name	Description	Type
ORIGIN_AIRPORT_ID	Origin airport	Categorical
DEST_AIRPORT_ID	Destination airport	Categorical
DEP_TIME	Scheduled departure time	Numerical
DEP_DELAY	Flight delay (in minutes)	Numerical
ARR_DELAY	Arrival delay (in minutes)	Numerical
Origin_Degree_Centrality	Degree centrality of origin airport	Numerical
Dest_Degree_Centrality	Degree centrality of destination airport	Numerical
Origin_Betweenness_Centrality	Betweenness centrality of origin airport	Numerical
Dest_Betweenness_Centrality	Betweenness centrality of destination airport	Numerical
Origin_Closeness_Centrality	Closeness centrality of origin airport	Numerical
Dest_Closeness_Centrality	Closeness centrality of destination airport	Numerical

4.2.2. Machine Learning Model Training

This study implements three machine learning models, specifically RF, GB, and CB, to predict flight delays. These models were selected for their ability to effectively manage complex relationships within large datasets.

The target variable for prediction was arrival delay, which originally comprised both positive values (indicating delays) and negative or zero values (indicating on-time or early arrivals). To facilitate binary classification, these values were transformed into categorical variables: delays exceeding 15 min were coded as 1, while delays within 15 min, on-time, or early arrivals were coded as 0.

For model training, two distinct datasets were prepared. The first dataset included only baseline features, such as origin and destination airports, scheduled departure time, and departure delay. The second dataset extended the baseline features by incorporating network centrality measures to evaluate their additional predictive value. Departure delay is included in both datasets to assess whether centrality measures provide better predictions of delay propagation within the network, even when traditional features like departure delay are used. Both datasets were split into training and testing sets using an 80/20 ratio, ensuring a robust evaluation of the models' performance.

5. Results

This section presents the evaluation of the three machine learning models applied, emphasizing the impact of integrating network centrality measures into the prediction framework. The analysis covers feature importance assessments and model performance comparisons.

5.1. Permutation Feature Importance

Permutation feature importance (PFI) was employed to evaluate the contribution of each feature to the predictive performance of the models. The importance is determined by measuring the decline in model performance when the values of a particular feature are randomly shuffled, which disrupts its relationship with the target variable.

Figure 2 shows the PFI for the RF model. It ranks destination and origin betweenness centrality as the most important features, followed by degree centrality for both origin and destination airports. These centrality measures outperformed DEP_DELAY, highlighting the value of network-based features.

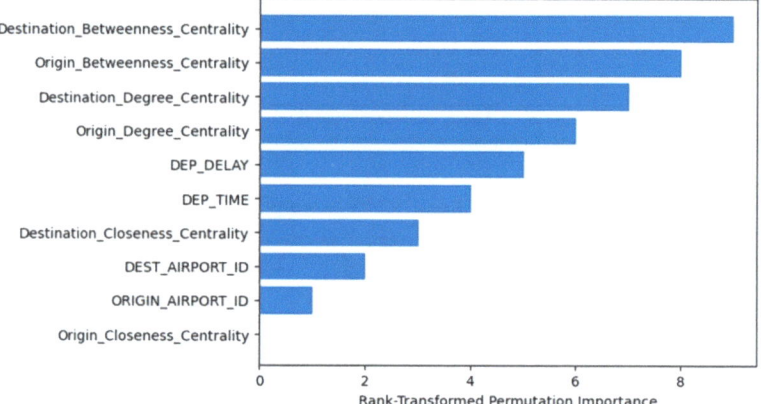

Figure 2. Permutation feature importance for RF.

Figure 3 shows the PFI for the GB model. It ranks ORIGIN_AIRPORT_ID as the top feature, with destination betweenness centrality also showing high importance. While traditional features like airport IDs and DEP_TIME are dominant, centrality measures still play a significant role.

Figure 4 shows the PFI for the CB model. It ranks DEST_AIRPORT_ID and origin degree centrality as the most critical features, with centrality measures consistently proving influential. DEP_DELAY is less impactful, further emphasizing the importance of network centrality in predictions.

Although DEP_DELAY might seem like an obvious predictor, since a delayed departure often leads to a delayed arrival, it remains in our feature set for the following reasons. Not all delayed departures result in delayed arrivals; factors like air traffic control, weather, and efficient operations can mitigate delays. While DEP_DELAY captures immediate operational delays, centrality measures offer a broader understanding of how the network structure influences delays. Including DEP_DELAY allows us to assess whether centrality metrics provide unique predictive insights beyond simple delay variables. This comparison helps to determine if centrality measures can more effectively predict the delay propagation within the network, even when traditional features like DEP_DELAY are considered.

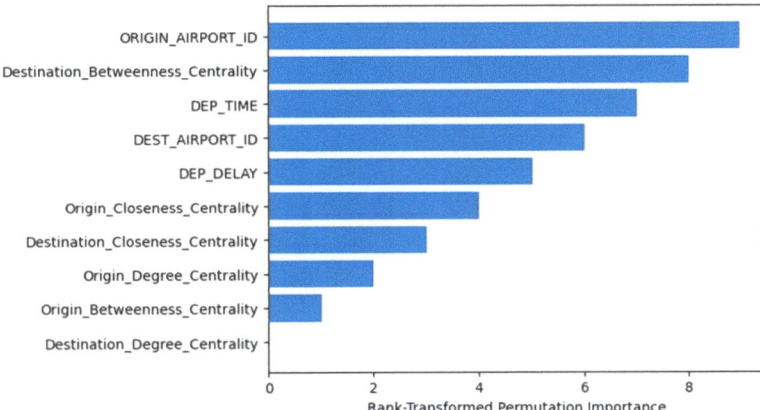

Figure 3. Permutation feature importance for GB.

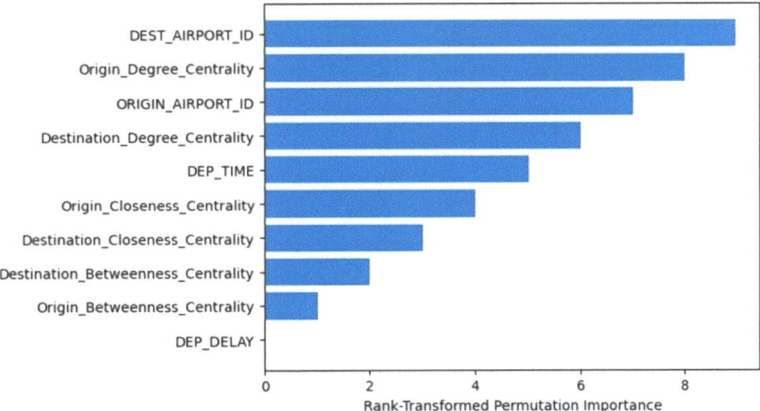

Figure 4. Permutation feature importance for CB.

Across all the models, the network centrality measures, particularly betweenness and degree centrality, are consistently ranked among the top features, indicating their critical role in improving flight delay prediction. These results highlight the importance of integrating network structure insights into machine learning models for more accurate predictions.

5.2. Comparison with Baseline Models

The comparison between those models trained with and without network centrality measures reveals notable improvements in performance across all the metrics: accuracy, precision, recall, and F1-score, as shown in Figure 5.

Accuracy measures the proportion of correctly predicted instances out of the total predictions. As shown in the figure, the inclusion of centrality measures increases the accuracy for all the models: RF improves from 84.5% to 86.2%, GB from 85.1% to 85.8%, and CB from 85% to 85.6%.

Precision indicates the proportion of true positive predictions among all the positive predictions. Precision also improves with the addition of centrality features: RF's precision increases from 86.3% to 86.9%, GB from 87.6% to 88.8%, and CB maintains a high precision with a slight improvement from 88.5% to 88.5%.

Recall (also known as sensitivity) measures the proportion of actual positive instances correctly identified by the model. The figure shows an increase in recall for all the models: RF from 72.9% to 74.2%, GB from 74% to 74.6%, and CB from 87% to 88%.

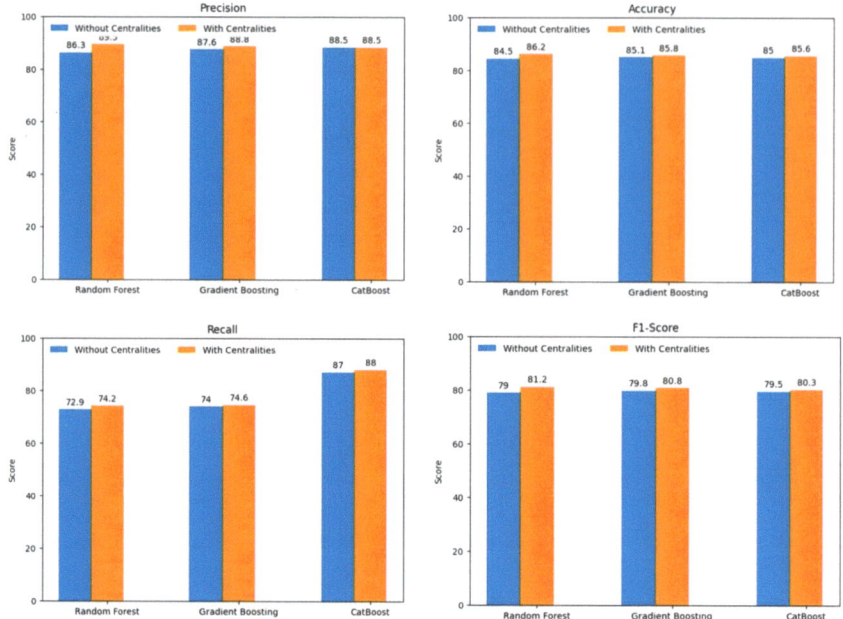

Figure 5. Performance comparison of RF, GB, and CB models trained with and without network centrality measures, evaluated using accuracy, precision, recall, and F1-score.

F1-score is the harmonic mean of precision and recall, providing a balance between the two metrics. The F1-scores reflect an overall improvement, with RF rising from 79 to 81.2, GB from 79.8 to 80.8, and CB from 79.5 to 80.3.

The results clearly demonstrate that incorporating network centrality measures into the models enhances their predictive performance. This improvement is evident across all the evaluated metrics, confirming that the integration of network structure insights contributes to more accurate and reliable flight delay predictions.

6. Discussion

The results from the permutation feature importance analyses indicate that network centrality measures significantly enhance the performance of flight delay prediction models. These measures, particularly betweenness and degree centrality, consistently ranked among the most important features across all the models. This finding underscores the relevance of network structure in understanding and predicting delays within the complex air transportation system.

While traditional features such as scheduled departure and arrival time, departure delay, and airport IDs remain crucial, the inclusion of centrality measures adds a valuable layer of predictive insight. This suggests that the structural properties of the airport network, including the connectivity and centrality of the airports within the network, play a critical role in the propagation of delays.

Previous studies have also applied RF or GB for flight delay prediction with varying degrees of success. For example, Choi et al. [8] implemented DT, RF, AdaBoost, and k-Nearest-Neighbors, achieving the highest accuracy of 83.4% with RF using data from BTS. Our model, also using RF on BTS data, achieved a higher accuracy of 86.2%. Another study [17] applied RF for large-scale delay prediction, achieving a 90.2% accuracy in binary classification. However, this study used a completely different dataset from China, which was created with a proprietary big data platform and included weather information. Similarly, Liu et al. [18], using the same big data platform, applied GB and obtained an accuracy of 87.72%. In comparison, our accuracies of 86.2% with RF and 85.8% with GB are slightly lower, but this difference can be attributed to the use of different data sources.

Despite the promising results, there are limitations to this study. The current model does not predict the duration of delays, which is crucial for practical applications. Future research should explore the use of alternative predictors, including real-time data, and consider the impact of other factors such as weather conditions.

Finally, while the results are encouraging, further validation on different datasets and with more complex models, such as deep learning models, is necessary.

7. Conclusions

This study integrated network centrality measures into machine learning models to enhance the accuracy of flight delay predictions. The models, including RF, GB, and CB, showed improved performance with the inclusion of the centrality measures. The accuracy increased from an average of 84.5% to 86.2% for RF, 85.1% to 85.8% for GB, and 85.0% to 85.6% for CB. The precision, recall, and F1-scores also improved, highlighting the value of the centrality features. The importance of these measures, especially betweenness and degree centrality, was confirmed through feature importance analysis.

The innovation of this study lies in the novel integration of network centrality measures into machine learning models for flight delay prediction, which, to the best of our knowledge, has not been explored in the literature. This approach provides new insights and improves the predictive accuracy beyond the traditional methods.

However, the study's limitations, including the focus on binary classification and the use of departure delay as a predictor, suggest several directions for future research. The future work should explore predicting the duration of delays, considering additional features like weather conditions, incorporating real-time data, and comparing the results with those of published studies using the same datasets. Expanding the methodology to include other ensemble learning methods and applying it to different transportation networks could further enhance the prediction accuracy and robustness.

In conclusion, while this study makes significant strides in improving flight delay predictions, ongoing research and refinement are necessary to fully realize the potential of network centrality measures in this domain.

Author Contributions: Methodology, Y.X.; Formal analysis, J.A.; Resources, J.A.; Writing – original draft, J.A.; Writing – review & editing, Y.X., L.L. and K.W. All authors have read and agreed to the published version of the manuscript.

Funding: This research received no external funding.

Institutional Review Board Statement: Not applicable.

Informed Consent Statement: Not applicable.

Data Availability Statement: The data presented in this study are available on request from the corresponding author.

Conflicts of Interest: The authors declare no conflict of interest.

References

1. Federal Aviation Administration (FAA). Air Traffic by the Numbers. Available online: https://www.faa.gov/airtraffic/air-traffic-numbers (accessed on 22 August 2024).
2. Boeing. Boeing Forecasts Demand for Nearly 44,000 New Airplanes Through 2043 as Air Travel Surpasses Pre-Pandemic Levels. Available online: https://investors.boeing.com/investors/news/press-release-details/2024/Boeing-Forecasts-Demand-for-Nearly-44000-New-Airplanes-Through-2043-as-Air-Travel-Surpasses-Pre-Pandemic-Levels/default.aspx (accessed on 22 August 2024).
3. Dai, M. A hybrid machine learning-based model for predicting flight delay through aviation big data. *Sci. Rep.* **2024**, *14*, 4603. [CrossRef] [PubMed]
4. Kim, Y.J.; Choi, S.; Briceno, S.; Mavris, D. A deep learning approach to flight delay prediction. In Proceedings of the 2016 IEEE/AIAA 35th Digital Avionics Systems Conference (DASC), Sacramento, CA, USA, 25–29 September 2016; pp. 1–6.

5. Yu, B.; Guo, Z.; Asian, S.; Wang, H.; Chen, G. Flight delay prediction for commercial air transport: A deep learning approach. *Transp. Res. Part E Logist. Transp. Rev.* **2019**, *125*, 203–221. [CrossRef]
6. Cai, K.; Li, Y.; Fang, Y.P.; Zhu, Y. A deep learning approach for flight delay prediction through time-evolving graphs. *IEEE Trans. Intell. Transp. Syst.* **2021**, *23*, 11397–11407. [CrossRef]
7. Esmaeilzadeh, E.; Mokhtarimousavi, S. Machine learning approach for flight departure delay prediction and analysis. *Transp. Res. Rec.* **2020**, *2674*, 145–159. [CrossRef]
8. Choi, S.; Kim, Y.J.; Briceno, S.; Mavris, D. Prediction of weather-induced airline delays based on machine learning algorithms. In Proceedings of the 2016 IEEE/AIAA 35th Digital Avionics Systems Conference (DASC), Sacramento, CA, USA, 25–29 September 2016; pp. 1–6.
9. Khan, R.; Akbar, S.; Zahed, T.A. Flight delay prediction based on gradient boosting ensemble techniques. In Proceedings of the 2022 16th International Conference on Open Source Systems and Technologies (ICOSST), Lahore, Pakistan, 14–15 December 2022; pp. 1–5.
10. KXAN. Which Airports Had the Most Delays and Cancellations in 2023? Available online: https://www.kxan.com/news/national-news/which-airports-had-the-most-delays-and-cancellations-in-2023/ (accessed on 22 August 2024).
11. Hsiao, C.Y.; Hansen, M. Econometric analysis of US airline flight delays with time-of-day effects. *Transp. Res. Rec.* **2006**, *1951*, 104–112. [CrossRef]
12. Zou, B.; Hansen, M. Flight delays, capacity investment and social welfare under air transport supply-demand equilibrium. *Transp. Res. Part A Policy Pract.* **2012**, *46*, 965–980. [CrossRef]
13. Rebollo, J.J.; Balakrishnan, H. Characterization and prediction of air traffic delays. *Transp. Res. Part C Emerg. Technol.* **2014**, *44*, 231–241. [CrossRef]
14. Nigam, R.; Govinda, K. Cloud based flight delay prediction using logistic regression. In Proceedings of the 2017 International Conference on Intelligent Sustainable Systems (ICISS), Palladam, India, 7–8 December 2017; pp. 662–667.
15. Yin, J.; Hu, Y.; Ma, Y.; Xu, Y.; Han, K.; Chen, D. Machine learning techniques for taxi-out time prediction with a macroscopic network topology. In Proceedings of the 2018 IEEE/AIAA 37th Digital Avionics Systems Conference (DASC), London, UK, 23–27 September 2018; pp. 1–8.
16. Pamplona, D.A.; Weigang, L.; De Barros, A.G.; Shiguemori, E.H.; Alves, C.J.P. Supervised neural network with multilevel input layers for predicting of air traffic delays. In Proceedings of the 2018 International Joint Conference on Neural Networks (IJCNN), Rio de Janeiro, Brazil, 8–13 July 2018; pp. 1–6.
17. Gui, G.; Liu, F.; Sun, J.; Yang, J.; Zhou, Z.; Zhao, D. Flight delay prediction based on aviation big data and machine learning. *IEEE Trans. Veh. Technol.* **2019**, *69*, 140–150. [CrossRef]
18. Liu, F.; Sun, J.; Liu, M.; Yang, J.; Gui, G. Generalized flight delay prediction method using gradient boosting decision tree. In Proceedings of the 2020 IEEE 91st Vehicular Technology Conference (VTC2020-Spring), Antwerp, Belgium, 25–28 May 2020; pp. 1–5.
19. Wu, Y.; Yang, H.; Lin, Y.; Liu, H. Spatiotemporal propagation learning for network-wide flight delay prediction. *IEEE Trans. Knowl. Data Eng.* **2023**, *36*, 386–400. [CrossRef]
20. Li, Q.; Guan, X.; Liu, J. A CNN-LSTM framework for flight delay prediction. *Expert Syst. Appl.* **2023**, *227*, 120287. [CrossRef]
21. Güvercin, M.; Ferhatosmanoglu, N.; Gedik, B. Forecasting flight delays using clustered models based on airport networks. *IEEE Trans. Intell. Transp. Syst.* **2020**, *22*, 3179–3189. [CrossRef]
22. Paramita, C.; Supriyanto, C.; Syarifuddin, L.A.; Rafrastara, F.A. The Use of Cluster Computing and Random Forest Algoritm for Flight Delay Prediction. *Int. J. Comput. Sci. Inf. Secur. (IJCSIS)* **2022**, *20*, 19–22.
23. Wei, X.; Li, Y.; Shang, R.; Ruan, C.; Xing, J. Airport Cluster Delay Prediction Based on TS-BiLSTM-Attention. *Aerospace* **2023**, *10*, 580. [CrossRef]
24. Cheung, D.P.; Gunes, M.H. A complex network analysis of the United States air transportation. In Proceedings of the 2012 IEEE/ACM International Conference on Advances in Social Networks Analysis and Mining, Istanbul, Turkey, 26–29 August 2012; pp. 699–701.
25. Anderson, S.; Revesz, P. Efficient MaxCount and threshold operators of moving objects. *Geoinformatica* **2009**, *13*, 355–396. [CrossRef]
26. Freeman, L.C. Centrality in social networks conceptual clarification. *Soc. Netw.* **1978**, *1*, 215–239. [CrossRef]
27. Wasserman, S. *Social Network Analysis: Methods and Applications*; The Press Syndicate of the University of Cambridge: Cambridge, UK, 1994.
28. Breiman, L. Random forests. *Mach. Learn.* **2001**, *45*, 5–32. [CrossRef]
29. Friedman, J.H. Greedy function approximation: A gradient boosting machine. *Ann. Stat.* **2001**, *29*, 1189–1232. [CrossRef]
30. Friedman, J.H. Stochastic gradient boosting. *Comput. Stat. Data Anal.* **2002**, *38*, 367–378. [CrossRef]
31. Prokhorenkova, L.; Gusev, G.; Vorobev, A.; Dorogush, A.V.; Gulin, A. CatBoost: Unbiased boosting with categorical features. In Proceedings of the 32nd International Conference on Neural Information Processing Systems, Montréal, Canada, 2–8 December 2018; pp. 6639–6649.
32. Bureau of Transportation Statistics (BTS). TranStats Database. Available online: https://www.transtats.bts.gov/ (accessed on 22 August 2024).

Disclaimer/Publisher's Note: The statements, opinions and data contained in all publications are solely those of the individual author(s) and contributor(s) and not of MDPI and/or the editor(s). MDPI and/or the editor(s) disclaim responsibility for any injury to people or property resulting from any ideas, methods, instructions or products referred to in the content.

Article

Correction of Threshold Determination in Rapid-Guessing Behaviour Detection

Muhammad Alfian [1], Umi Laili Yuhana [1], Eric Pardede [2,*] and Akbar Noto Ponco Bimantoro [1]

1. Institut Teknologi Sepuluh Nopember, Surabaya 60111, Indonesia; ini.muhalfian@gmail.com (M.A.); yuhana@if.its.ac.id (U.L.Y.); akbarnotopb@gmail.com (A.N.P.B.)
2. Department of Computer Science and Information Technology, La Trobe University, Melbourne, VIC 3000, Australia
* Correspondence: e.pardede@latrobe.edu.au

Abstract: Assessment is one benchmark in measuring students' abilities. However, assessment results cannot necessarily be trusted, because students sometimes cheat or even guess in answering the questions. Therefore, to obtain valid results, it is necessary to separate valid and invalid answers by considering rapid-guessing behaviour. We conducted a test to record exam log data from undergraduate and postgraduate students to model rapid-guessing behaviour by determining the threshold response time. Rapid-guessing behaviour detection is inspired by the common k-second method. However, the method flattens the application of the threshold, thus allowing misclassification. The modified method considers item difficulty in determining the threshold. The evaluation results show that the system can identify students' rapid-guessing behaviour with a success rate of 71%, which is superior to the previous method. We also analysed various aggregation techniques of response time and compared them to see the effect of selecting the aggregation technique.

Keywords: rapid-guessing behaviour; threshold determination; response time

Citation: Alfian, M.; Yuhana, U.L.; Pardede, E.; Bimantoro, A.N.P. Correction of Threshold Determination in Rapid-Guessing Behaviour Detection. *Information* **2023**, *14*, 422. https://doi.org/10.3390/info14070422

Academic Editor: Gennady Agre

Received: 13 June 2023
Revised: 13 July 2023
Accepted: 20 July 2023
Published: 21 July 2023

Copyright: © 2023 by the authors. Licensee MDPI, Basel, Switzerland. This article is an open access article distributed under the terms and conditions of the Creative Commons Attribution (CC BY) license (https://creativecommons.org/licenses/by/4.0/).

1. Introduction

In fact, assessment plays a very important role in the learning process [1]. Assessment is a process of evaluating knowledge, the ability to understand, and achievement of test takers' skills [2]. Assessment is used to measure students' abilities with the aim of selecting students for new admissions, measuring the level of understanding of post-learning material, and as a determinant of graduation. In addition, one of the benefits of conducting an assessment is as a reference for determining student learning flows. An example is the determination of material according to students' abilities [3] and determining the next material they need to study [4]. In addition, student assessments can streamline the allocation of resources needed to increase student learning competencies [5].

As test-takers, we often do not know whether these students' answers are valid or not, and whether they are taking it seriously or cheating. As students, we also sometimes come across questions that are very difficult, forcing us to answer to obtain the best grades even though we do not know the answers. This behaviour is called rapid-guessing behaviour. According to ref. [6], rapid-guessing behaviour occurs when test takers answer questions quicker than usual in a speeded test. However, assessment results can be invalid because students cheated or rapidly guessed the answer to the question [6]. Ref. [7] states that, therefore, to obtain the ideal assessment results, it is necessary to differentiate assessment results based on student behaviour, whether they answer by guessing (rapid-guessing behaviour) or answer seriously (solution behaviour). This rapid-guessing behaviour causes biased scores and unreliable tests, so it should be ignored.

Schnipke was the first to discover rapid-guessing behaviour when mapping the response times of the Graduate Record Examination Computer-Based Test (GRE-CBT). In her research, each question was mapped to its response time distribution as shown in

Figure 1. Response time is taken from how long it takes students to read to answer a question. In practice, to distinguish rapid-guessing behaviour and solution behaviour, we need to determine the threshold time.

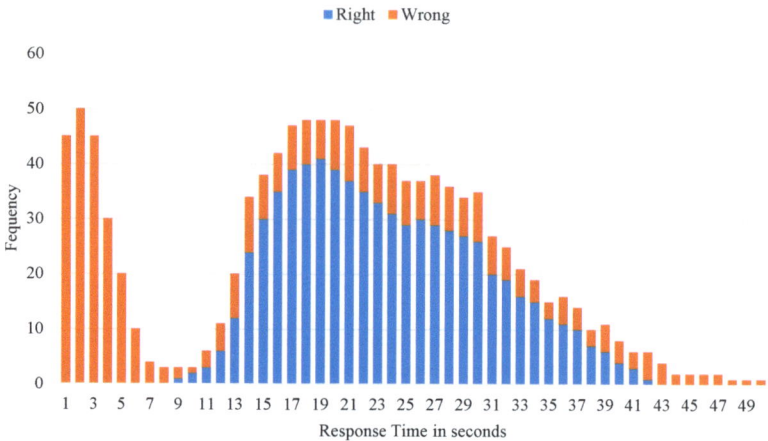

Figure 1. Example of RT distribution.

Several studies investigated how to determine the threshold time. Schnipke use visual inspection to determine threshold and distinguish both behaviours. A similar approach was carried out by DeMars [8], Setzer et al. [9], and Pastor et al. [10]. However, detecting rapid-guessing behaviour becomes more difficult using this approach when the RT distribution has the same RT peak. Students who answered by guessing and students who answered seriously both made overlapping response time distributions. Other researchers used the k-second method to determine the RT threshold, in which the fixed threshold value is generally set between three to five seconds [7]. K-second is the simplest threshold method. It does not require information about each item's surface features or response time distribution and is particularly useful with large item pools. Its one-size-fits-all nature, however, will often result in variations in misclassification across items [7–10].

Some other researchers use the surface features method to distinguish between the two behaviours. Surface features determine the RT threshold using several item features. Silm et al. [11] considered the test subject and item length in determining the RT threshold. Wise and Kong [12] considered the number of characters and whether there were tables or images. However, in both studies, the results of evaluating students' rapid guessing behaviour were not explicitly detailed. In contrast to methods that use time thresholds, Lin [13] processes the student's ability score (l) and item difficulty index (i) based on the Rasch model to determine guessing behaviour. They argue that if there is a large difference between the student's ability and the item difficulty index, then it is rapid-guessing behaviour.

This study aims to propose a correction to the determination of time thresholds as part of the identification of rapid-guessing behaviour in assessment. The correction we provide is that the determination of the threshold is not simply about how to choose the right number to be used as a threshold, but also needs to pay attention to how difficult the question and how the data processing technique is. We tried several data aggregation techniques such as sum, average, and maximum. We adopted the concept of k-seconds and combined it with the features of item response theory (IRT) to create a new approach in determining the time threshold for each item category. The questions were divided into three categories based on their difficulty according to IRT features. Data was obtained from online exams during lectures on campus. Response time is obtained from how long students work on questions (calculated from the time of opening to answering questions).

The expected benefit of this research is that the question maker can know which answers are given seriously by students and which are given fraudulently, so that the scores can be differentiated. This research is part of our larger research on computer adaptive assessment.

2. Related Works

Rapid-guessing behaviour is a phenomenon when students answer items rapidly without serious thought. In other words, students randomly guess the answers to the items. Rapid-guessing behaviour usually occurs in multiple choice tests. We discussed how rapid-guessing behaviour is detected in exams. There have been several variables used to detect rapid-guessing behaviour. The most popular approach is rapid-guessing detection based on response time (RT). Other variables include student ability, item difficulty, and response accuracy (RA). In the next section we discuss our proposed method and our contribution to rapid-guessing behaviour detection.

2.1. Detection Based on Response Time (RT)

Schnipke [6] is one of the first researchers that used RT thresholds as the basis for detecting rapid-guessing behaviour. Visual inspection was carried out on RT distributions of 17,415 students that took a computer-based Graduate Record Examinations Computer-based Test (GRE-CBT). The RT of correct and wrong responses for each item were separately plotted to visualize the distribution of RT of each item. In this study, rapid-guessing behaviour towards an item is indicated by a larger number of fast wrong responses in the RT distribution of the item. Figure 1 shows the distribution of two items, in which wrong responses are indicated by the red lines. In the first distribution, the RT for majority of the students is relatively short, and the number of wrong responses exceed the number of correct responses. While in the second distribution, the RT for majority of the students is relatively long and the number of correct responses exceeds the number of wrong responses. Therefore, the first distribution is classified as rapid-guessing behaviour and the second distribution is classified as solution behaviour (students fully consider the answer). Furthermore, in the second distribution that is classified as solution behaviour, the fastest RT of a correct response is five seconds; therefore, a RT under five seconds is rapid-guessing behaviour.

A similar approach was carried out by DeMars [8], Setzer et al. [9], and Pastor et al. [10]. However, detecting rapid-guessing behaviour becomes more difficult using this approach when RT distributions classified as rapid-guessing behaviour and solution behaviour possess similar peaks of RT. This is because the time needed to correctly answer items is indeed short.

Other researchers used the k-seconds method to determine the RT threshold, in which the fixed threshold value is generally set between three to five seconds [7]. The threshold value was then used to determine the response time effort (RTE) of the students. Wise [7] evaluated the proposed RTE model on students that were given mathematics and reading tests in varying times, days, seasons, and age groups. From the experimental results, it was indicated that RTE is influenced by several factors, namely gender, age, contents of an item, and time.

2.2. Detection Based on Combination of RT and Other Variables

Surface features is a method used to determine the RT threshold using several item features. Unlike the k-seconds method that sets the same RT threshold value to all the items, in the surface features method, each item is given an RT threshold based on its features. The features include the number of characters in an item, whether an item consists of tables and figures, and the subject being evaluated by the item. Several features that were used in previous studies and the resulting RT threshold values are shown in Table 1. Silm et al. [11] considered the subject of the test and item length in determining the RT threshold as shown in Table 1. Wise and Kong [12] took into consideration the number of characters and whether an item consisted of a table or figure in the determination of the RT threshold. However, in both studies, the results of rapid-guessing behaviour evaluation on the students were not explicitly detailed.

Table 1. Surface Feature Threshold.

Criteria	Threshold
Math/spatial reasoning problem	5 s
<200 characters	3 s
200–1000 characters	5 s
>1000 characters	10 s

Pastor et al. [10] used latent class analysis (LCA) to investigate whether there was a difference in solution behaviour patterns across three tests differing in content. They implemented the RT threshold value resulting from visual inspection of RT distributions into the LCA model. From the experiment that was carried out on undergraduate students, it was found that the results of the proposed method were similar to that of Wise et. al. [14], in which the solution behaviour pattern is consistent in all the tests differing in content. The experimental results were validated using the BCH approach (Bolck, Croon, and Hagenaars [15]), which involves performing a weighted ANOVA, with weights that are inversely related to the classification error probabilities [16].

Another study, proposed by Lee and Jia [17] combined RT and RA to determine the time threshold. Time thresholds were determined based on the participants' RTs for test 1 and test 2, as shown in Figure 2. The RT results of each test were then combined to be analysed manually using either common k-seconds or visual inspection of the RT distribution. The test was conducted on approximately 8400 junior high school students in mathematics with a composition of 40% students in a multistage test (MST) sample and 60% students in control sample. The proposed method is evaluated manually by the authors with expert inspection of the questions, such as the presence of tables or figures and the complexity of the questions.

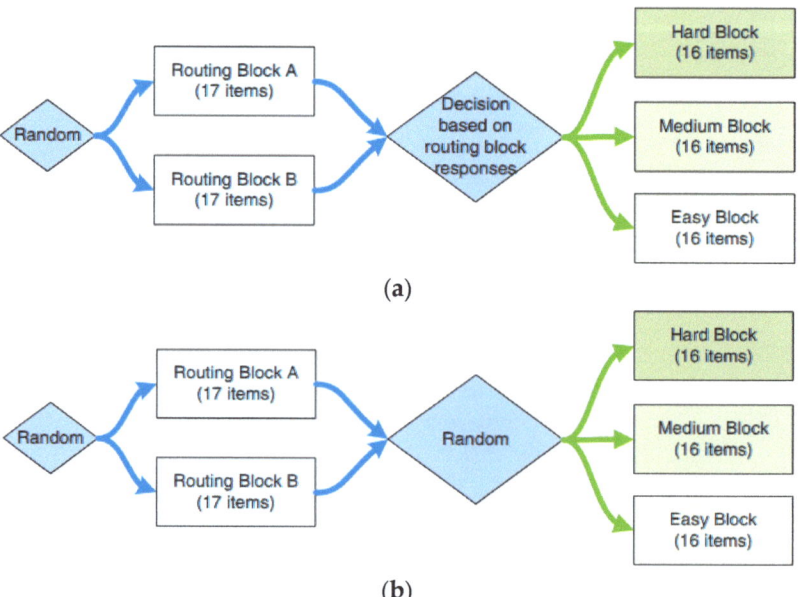

Figure 2. Test scenario using multistage test (MST) (**a**) and control test (**b**).

In contrast to the method that uses a time threshold, Lin [13] processed the value of the student's ability (l) and the difficulty index (i) based on the Rasch model to determine guessing behaviour. Student's ability (l) refers to the measure of how proficient a student is, and difficulty index (i) refers to the measure of how hard an item (question) is. They argue that if there is a big difference between the logit ability and the difficulty index of the question, it should be rapid-guessing behaviour. They classified the answers as rapid-guessing behaviour if $l - i \leq 2$. Answers that were classified as rapid-guessing behavior were removed from the dataset and used as the final test model on the language test of sixth-grade elementary school students. From the tests carried out, they found that the assessment of high-ability students had better precision.

Based on previous literature studies, no research has developed and corrected time threshold determination utilising IRT features and considering variations in data aggregation. Therefore, this study aims to combine the k-second method with IRT features to recognize the difficulty level of each question and utilise multiple data aggregation methods to distinguish rapid-guessing behaviour and solution behaviour. We compared the proposed method with previous methods such as the common k-second, surface, and normative. We pay attention to the data aggregation technique, because in the classification process it is not only about how to determine the right threshold value, but also the aggregation technique is also important. Some of the aggregation techniques we used include average, sum, and maximum. Then, the model is evaluated using accuracy, precision, recall, and F1 score parameters. The next section will describe this method in more detail.

3. Methods

This section details the methodology used for detecting rapid-guessing behaviour. As shown in Figure 3, this research consists of two main processes: a conventional test and rapid-guessing modelling.

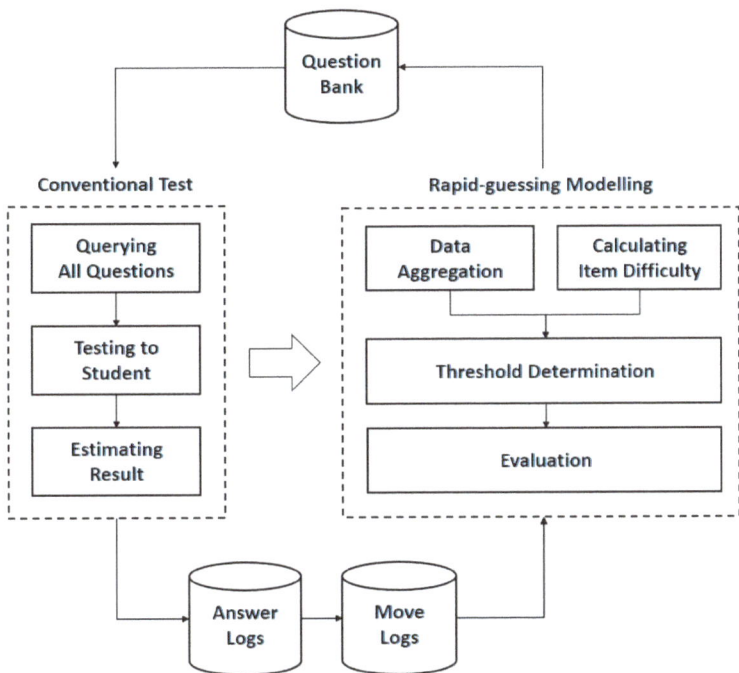

Figure 3. System design for detecting rapid-guessing behaviour.

3.1. Gathering Data from Conventional Test

One of the advantages of computer-based tests (CBTs) is that data on the student's activities from the start to the end of the test can be easily obtained. The data is accessible, provides meaningful information, and is unambiguous because every student has their own accounts and all activities of students are recorded. This study focuses on analysing user behaviour data from the system log, without taking into consideration demographic factors such as age, gender, and ethnicity of the students which may cause the proposed model to become biased towards these factors.

This study analyses student daily test data in a specific course. The examinees of the daily test are university students that are technologically literate. The students were first given a conventional test. The conventional test consisted of 40 items. The questions were multiple choice with one correct answer. All students worked on the same questions at the same time. This was so as to evaluate the comprehensive ability of the students in understanding the study material. Furthermore, the results of the comprehensive test are used to calculate item difficulty of each item in the test.

The platform used for this test is a web-based "i-assessment" software accessed through smartphones. The "i-assessment" software records student activity during the test and the answers of the students and stores the data in a database. The time a student accesses a question and the time the student answers the question is stored in the Answer Log table. Furthermore, the time a student navigates between questions is stored in the Move Log table. Every time a student gives an answer to each question, a pop up appears in the system asking, "Are you sure about your answer?". We use this data as a reference to distinguish answers that are guessing and not.

3.2. Conventional Test Information

The tests were administered to students of a widely recognized university in Indonesia. The students were given an end-of-semester daily test (quiz) by the lecturer. The detailed information is shown in Table 2. The test data was collected from two courses, namely software project management (SPM) and software engineering (SE). The SPM course is an undergraduate course, while the SE course is a postgraduate course. The duration of the conventional test was 90 min and consisted of 40 multiple-choice items, in which each item presented five answers to choose from. The average scores of each course showed that students in the SPM course had a fairly high score, as seen from the average score of 65.89. In contrast to students in SE courses, students have fewer high scores, as seen from the average score of 44.58. Even though the standard deviation of the SPM test scores was higher than that of the SE test scores, the minimum and maximum score were higher for the SPM test. However, these data alone are insufficient to adequately assess the educational evaluation process. Further analysis needs to be carried out with respect to the test items and other underlying factors of the students.

Table 2. Data summary.

Course and Duration	Class Member	Level	Score
SPM 90 min	45 students	Undergraduate	Mean = 65,89 Std = 15 Min = 38 Max = 93
SE 90 min	45 students	Graduate	Mean = 44,58 Std = 11,82 Min = 27,5 Max = 75

3.3. Rapid-Guessing Modelling

The first step in rapid-guessing modelling is data aggregation. This stage combines data from several tables into a single unit. Both the Answer Log table and the Move Log table possess a relationship with the Participant table and the Question table. The Answer Log table stores information on when students open a question, and when they answer the question. Meanwhile, the Move Log table stores information on when students moved from one question to another, regardless of when they answered the question. After gathering the relevant data, the Log Aggregation table is generated to store a summary of data of both the Answer Log and Move Log tables based on the key attributes of the log tables. This transformation process is called data aggregation. Data aggregation is the process of finding and gathering data and visualizing the data in a summarized format for an easier statistical analysis of the data. The Log Aggregation table possesses columns that are produced from the aggregation process, including sum, maximum, minimum, and average values as shown in Figure 4. The Log Aggregation table is then split with respect to the purpose of the data analysis based on questions, participants, and a combination of both.

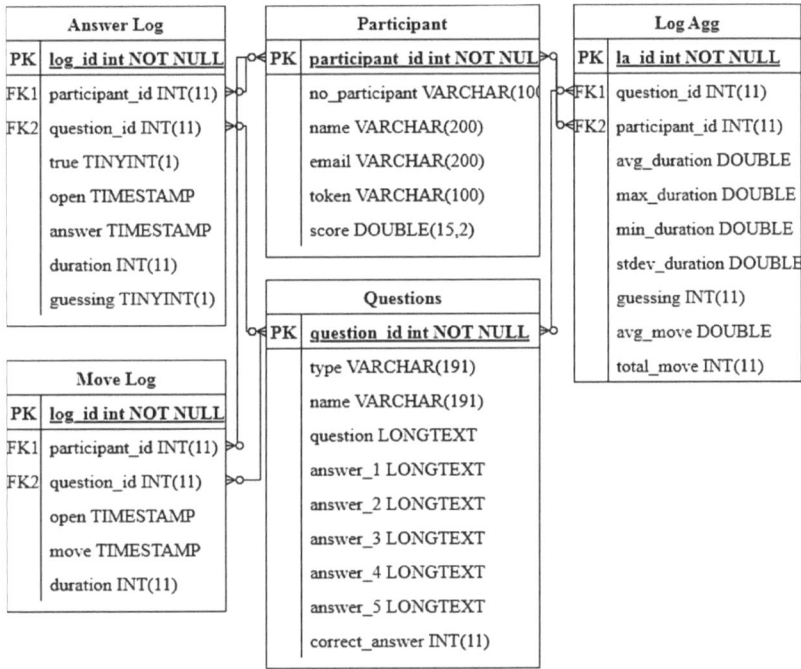

Figure 4. Aggregation table diagram.

The second step is calculating item difficulty. Item difficulty (b_i) is defined as the proportion of examinees that were able to correctly answer the item [18]. Item difficulty in item response theory(IRT) is derived from the z-score measurement method. Therefore, item difficulty is calculated by dividing the number of examinees that were unable deliver correct answer to item i (n_{fi}) by the total number of examinees that submitted a response item i (N_i) minus the number of examinees that were unable to submit (false answer) a response to item i (n_{fi}). The resulting value is then normalized using the natural logarithm to decrease the distribution value [19], as shown in Equation (1).

$$b_i = \ln\left(\frac{n_{fi}}{N_i - n_{fi}}\right) \qquad (1)$$

After that, from the question difficulty values, we categorised the questions into three labels, namely easy, medium, and difficult, based on the question difficulty parameters in

IRT. We labelled them using the fuzzy logic inference method. Figure 5 shows the member function of item difficulty. The y-axis shows the fuzzy inference value, while the x-axis value shows the item difficulty value. The range of item difficulty values is from −3 to 3. For this question, we directly divided it into three labels. The easy label is given if the item difficulty ranges from −3 to 0. Meanwhile, the medium label is given if the item difficulty ranges from −1 to 2. And finally, the difficult label is given if the item difficulty level is above 1. These three different labels are to categorise student responses, and then determine the threshold for each item label.

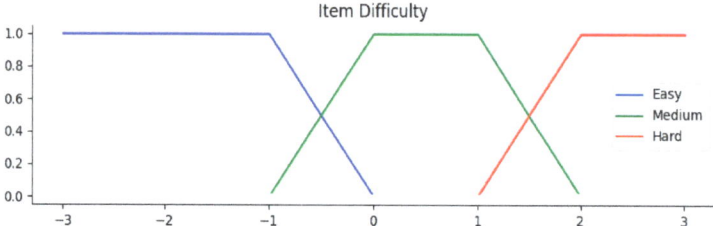

Figure 5. Item difficulty member function.

The third step is determining the threshold. We are inspired by the common k-seconds [7] method to determine the threshold. While the common k-seconds method sets all questions with the same threshold, we have a different approach. We categorize items into three labels based on their difficulty level, namely easy, medium, and hard. Each question label has its own threshold. The determination of the threshold is the same as the predecessor method, which is that we use a common value and then match with the dataset which value is the best. The value we agreed on was 3 s for questions with the hard label, and 2 s for questions with the easy and medium labels.

The last step is evaluation. We evaluate the model by calculating the evaluation matrix. We compare our proposed method with previous methods. In addition, we also compare various aggregation techniques, so that we can find out the effect of different aggregation techniques on the classification results.

4. Results and Discussion

We conducted experiments on students during lecture hours. Some of the steps were aggregating data, calculating item difficulty, determining threshold, and evaluation. The first step is to perform data aggregation. We collect data from the Answer Log and Move Log tables to be aggregated in an aggregation table according to the design. However, the parameters we use here are only time-related parameters, including avg_duration, max_duration, min_duration, stdev_duration, avg_move, and total_move. However, considering the processing time, we chose three main parameters to compare, namely avg_duration, max_duration, and total_move.

The second step is to calculate item_difficulty. We use the equation from IRT to calculate the item difficulty. Then, we assign labels to it using the inference method of fuzzy logic. Following the completion of the conventional tests by the students, the answer log was used in the RT-based guessing model. The guessing model proposed in this study only uses one parameter, namely time. Further analysis on the answer log data indicated several different behaviours exhibited by the students in giving responses to the presented items. These behaviours occurred due to the duration of the test (90 min), which is long for a multiple-choice test that consists of 40 items. The first behaviour exhibited by the students was that several students used the remaining time to reconsider doubtful responses after they had given responses to all the items. The second behaviour exhibited by the students was that several students spent a lot of time reading items that they deemed difficult, then they skipped the item without giving a response. After giving responses to the other items, the students then came back to the items they deemed difficult and gave a quick response.

Due to these exhibited behaviours, we investigated the use of several parameters to define RT in the proposed guessing model. The first parameter that we used to define RT was the time spent by the students to initially read an item and give a response, which we named duration. The second parameter was the accumulation of time spent on an item even after giving a response, which we named total move. The last parameter, named max time, was derived from the duration parameter, which was the longest time spent to initially read an item and give a response among all the students. We compared the performance of the guessing model with the use of these different parameters.

Table 3 shows the evaluation matrix of threshold determination for the SPM course and SE course. At a glance, the accuracy value of SE course is higher than that of SPM course. This difference is because the number of students taking the exam is not the same. There are more students in the SPM course compared to students in the SE course. This certainly affects the accuracy of the model. The more samples, the greater the potential for outlier behaviour. Therefore, outlier detection [20] is necessary to reduce bias.

Table 3. Evaluation of Threshold Determination Methods.

Couse	Parameter	Method	Accuracy	Precision	Recall	F1
Software Project Management (SPM)	avg_duration	Common k-second	66.57%	16.67%	5.27%	8.34%
		Surface	65.32%	16.67%	9.60%	13.78%
		Normative	67.01%	18.85%	4.33%	7.04%
		Modified k-second	68.36%	16.88%	2.44%	4.27%
	total_move	Common k-second	71.14%	-	0.00%	-
		Surface	71.03%	0.00%	0.00%	-
		Normative	71.14%	-	0.00%	-
		Modified k-second	71.14%	-	0.00%	-
	max_duration	Common k-second	69.62%	26.67%	03.01%	05.41%
		Surface	68.58%	28.44%	05.84%	09.69%
		Normative	70.11%	28.89%	02.45%	04.51%
		Modified k-second	70.87%	39.13%	01.69%	03.25%
Software Engineering (SE)	avg_duration	Common k-second	84.72%	16.67%	1.96%	3.51%
		Surface	84.72%	16.67%	1.96%	3.51%
		Normative	83.33%	17.86%	4.90%	7.69%
		Modified k-second	85.28%	16.67%	0.98%	1.85%
	total_move	Common k-second	85.83%	-	0.00%	-
		Surface	85.83%	-	0.00%	-
		Normative	85.83%	-	0.00%	-
		Modified k-second	85.83%	-	0.00%	-
	max_duration	Common k-second	85.83%	50.00%	0.98%	1.92%
		Surface	85.83%	50.00%	0.98%	1.92%
		Normative	85.13%	27.27%	2.94%	5.31%
		Modified k-second	85.83%	-	0.00%	-

Each table displays the evaluation matrix of our proposed methods compared to other threshold determination methods. In addition, each table is compared with various aggregation parameters. In general, the guessing model that used the modified k-seconds method to determine the RT threshold outperformed the other models in terms of accuracy. In the SPM course, using the avg_duration, the accuracy was 68% aggregation parameter, outperforming the other methods. Meanwhile, on the SE course, the accuracy was 85%, outperforming the other methods. Further analysis of modified k-seconds method revealed that the model performed better with the use of the total move and max time parameters. With the use of the total move parameter, the model achieved a higher accuracy. However, this model obtained a recall value of 0. This indicates that the model was unable to detect rapid-guessing behaviour. As a result of the recall metric having a value of 0, the precision and F1 score values were not able to be calculated.

Furthermore, the evaluation of the models based on the F1 score metric revealed that the guessing model that used the surface features method along with the guessing model that used the normative method to determine the RT threshold achieved the best performance. Further analysis of these two models revealed that the performance of both models was more stable with the use of the duration parameter.

Our experiments show that our proposed method, modified k-second, has superior accuracy compared to other methods in both courses. In addition, this study also proves that there is a difference in accuracy along with the difference in aggregation techniques. Aggregation using total_move has higher accuracy than using avg_duration or max_duration parameters. Therefore, further research needs to try other aggregation parameters, one of which is sum_duration. However, when viewed from the F1 score evaluation, the best method is the surface feature. Although in terms of accuracy, modified k-second recorded the highest value, this method has a very low recall value, because the count of students who guessed is very little (data imbalance). This causes the model to be biased, so that the model cannot properly accommodate class with little data [21]. For further research, several techniques need to be conducted to handle data imbalance, such as modifying preprocessing techniques, algorithmic approaches, cost sensitivity, and ensemble learning [21].

5. Conclusions

Assessment is used to measure students' abilities with the aim of selecting students for new admissions, measuring the level of understanding of post-learning material, and as a determinant of graduation. However, the results of the assessment may be invalid because the students cheated or rapidly guessed the answer to the question. Rapid-guessing behaviour is a phenomenon where students answer items rapidly without serious thought. Several researchers have conducted studies on how to detect rapid-guessing behaviour by analysing processing time with a certain threshold. However, existing methods have no developed and corrected time threshold determination utilising IRT features and considering variations in data aggregation. Therefore, this study aims to combine the k-second method with IRT features to recognize the difficulty level of each question and utilise multiple data aggregation methods to distinguish rapid-guessing behaviour and solution behaviour. We compared the proposed method and the data aggregation technique. Some of the aggregation techniques we used include average, sum, and maximum. Then, the model is evaluated using accuracy, precision, recall, and F1 score parameters.

This study proves that the correction of threshold determination that we proposed, modified k-second, succeeded in detecting guessing with an accuracy better than the other methods. In SPM courses, modified k-second has an accuracy of 68.36%, superior to other methods using the avg_duration parameter. This research also proves that the selection of aggregation techniques also greatly affects the level of accuracy. Total move is an aggregation parameter that has high accuracy. Meanwhile, average duration is an aggregation parameter that has lower accuracy. However, when viewed from the F1 score evaluation, the best method is the surface feature. Although in terms of accuracy, modified k-second recorded the highest value, this method has a very low recall value, because the count of students who guessed is very little (data imbalance). This causes the model to be biased, so that the model cannot properly accommodate class with little data. For further research, several techniques need to be conducted to handle data imbalance, such as modifying preprocessing techniques, algorithmic approaches, cost sensitivity, and ensemble learning.

Author Contributions: Conceptualization, U.L.Y. and A.N.P.B.; methodology, U.L.Y. and A.N.P.B.; software, A.N.P.B.; validation, A.N.P.B. and M.A.; formal analysis, M.A. and E.P.; investigation, A.N.P.B.; resources, A.N.P.B.; data curation, M.A. and E.P.; writing—original draft preparation, M.A.; writing—review and editing, E.P.; visualization, M.A.; supervision, U.L.Y. and E.P.; project administration, U.L.Y.; funding acquisition, U.L.Y. All authors have read and agreed to the published version of the manuscript.

Funding: This research was funded by Institut Teknologi Sepuluh Nopember (ITS) for WCP-Like Grant Batch 2, grant number 1855/IT2/T/HK.00.01/2022.

Data Availability Statement: The data that support the findings of this study are available from the corresponding author, M.A., upon reasonable request.

Acknowledgments: This work is part of the "i-assessment project", an adaptive testing-based test application.

Conflicts of Interest: The author declares no conflict of interest.

Abbreviations

	Notation and Acronym
RA	Response accuracy
RT	Response time
RTE	Response time effort
IRT	Item response theory
SE	Software engineering
SPM	Software project management
b_i	Item difficulty
N_i	The number of examinees that submitted a response item i
n_{fi}	The number of examinees that were unable to submit (false answer) a response to item i
l	Student's ability
i	Rasch model

References

1. *Scottish Qualifications Authority Guide to Assessment*; Scottish Qualifications Authority: Glasgow, UK, 2017; pp. 3–9.
2. Kennedy, K.J.; Lee, J.C.K. The changing role of schools in Asian societies: Schools for the knowledge society. In *The Changing Role of Schools in Asian Societies: Schools for the Knowledge Society*; Routledge: Oxfordshire, UK, 2007; pp. 1–228.
3. Hwang, G.J.; Sung, H.Y.; Chang, S.C.; Huang, X.C. A fuzzy expert system-based adaptive learning approach to improving students' learning performances by considering affective and cognitive factors. *Comput. Educ. Artif. Intell.* **2020**, *1*, 100003. [CrossRef]
4. Hwang, G.-J. A conceptual map model for developing intelligent tutoring systems. *Comput. Educ.* **2003**, *40*, 217–235. [CrossRef]
5. Peng, S.S.; Lee, C.K.J. *Educational Evaluation in East Asia: Emerging Issues and Challenges*; Nova Science Publishers: Hauppauge, NY, USA, 2009.
6. Schnipke, D.L. Assessing Speededness in Computer-Based Tests Using Item Response Times. Ph.D. Thesis, Johns Hopkins University, Baltimore, MD, USA, 1995.
7. Wise, S.L. Rapid-Guessing Behavior: Its Identification, Interpretation, and Implications. *Educ. Meas. Issues Pract.* **2017**, *36*, 52–61. [CrossRef]
8. Demars, C.E. Changes in Rapid-Guessing Behavior Over a Series of Assessments. *Educ. Assess.* **2007**, *12*, 23–45. [CrossRef]
9. Setzer, J.C.; Wise, S.L.; van den Heuvel, J.R.; Ling, G. An Investigation of Examinee Test-Taking Effort on a Large-Scale Assessment. *Appl. Meas. Educ.* **2013**, *26*, 34–49. [CrossRef]
10. Pastor, D.A.; Ong, T.Q.; Strickman, S.N. Patterns of Solution Behavior across Items in Low-Stakes Assessments. *Educ. Assess.* **2019**, *24*, 189–212. [CrossRef]
11. Silm, G.; Must, O.; Täht, K. Test-taking effort as a predictor of performance in low-stakes tests. *Trames* **2013**, *17*, 433–448. [CrossRef]
12. Wise, S.L.; Kong, X. Response time effort: A new measure of examinee motivation in computer-based tests. *Appl. Meas. Educ.* **2005**, *18*, 163–183. [CrossRef]
13. Lin, C.K. Effects of Removing Responses With Likely Random Guessing Under Rasch Measurement on a Multiple-Choice Language Proficiency Test. *Lang. Assess. Q.* **2018**, *15*, 406–422. [CrossRef]
14. Wise, S.L.; Ma, L.; Kingsbury, G.G.; Hauser, C. An investigation of the relationship between time of testing and test-taking effort. *Natl. Counc. Meas. Educ.* **2010**, 1–18. Available online: https://eric.ed.gov/?id=ED521960 (accessed on 12 June 2023).

15. Vermunt, J. Latent Class Modeling with Covariates: Two Improved Three-Step Approaches. *Political Anal.* **2017**, *18*, 450–469. [CrossRef]
16. Bakk, Z.; Vermunt, J.K. Robustness of stepwise latent class modeling with continuous distal outcomes. *Struct. Equ. Model.* **2016**, *23*, 20–31. [CrossRef]
17. Lee, Y.H.; Jia, Y. Using response time to investigate students' test-taking behaviors in a NAEP computer-based study. *Large-Scale Assess. Educ.* **2014**, *2*, 8. [CrossRef]
18. Ebel, R.L.; Frisbie, D.A. *Essentials of Educational Measurement*, 5th ed.; Prentice-Hall of India Private Limited: New Delhi, India, 1991; ISBN 0-87692-700-2.
19. Purushothama, G. Introduction to Statistics. In *Nursing Research and Statistics*; Jaypee Brothers Medical Publishers (P) Ltd.: New Delhi, India, 2015; p. 218. [CrossRef]
20. Singh, K.; Upadhyaya, S. Outlier Detection: Applications And Techniques. *IJCSI Int. J. Comput. Sci. Issues* **2012**, *9*, 307.
21. Ali, H.; Najib, M.; Salleh, M.; Saedudin, R.; Hussain, K. Imbalance class problems in data mining: A review. *Indones. J. Electr. Eng. Comput. Sci.* **2019**, *14*, 1552–1563. [CrossRef]

Disclaimer/Publisher's Note: The statements, opinions and data contained in all publications are solely those of the individual author(s) and contributor(s) and not of MDPI and/or the editor(s). MDPI and/or the editor(s) disclaim responsibility for any injury to people or property resulting from any ideas, methods, instructions or products referred to in the content.

Article

DEGAIN: Generative-Adversarial-Network-Based Missing Data Imputation

Reza Shahbazian *,† and Irina Trubitsyna *,†

Department of Informatics, Modeling, Electronics and System Engineering, University of Calabria, 87036 Rende, Italy
* Correspondence: reza.shahbazian@unical.it (R.S.); i.trubitsyna@dimes.unical.it (I.T.)
† These authors contributed equally to this work.

Abstract: Insights and analysis are only as good as the available data. Data cleaning is one of the most important steps to create quality data decision making. Machine learning (ML) helps deal with data quickly, and to create error-free or limited-error datasets. One of the quality standards for cleaning the data includes handling the missing data, also known as data imputation. This research focuses on the use of machine learning methods to deal with missing data. In particular, we propose a generative adversarial network (GAN) based model called DEGAIN to estimate the missing values in the dataset. We evaluate the performance of the presented method and compare the results with some of the existing methods on publicly available Letter Recognition and SPAM datasets. The Letter dataset consists of 20,000 samples and 16 input features and the SPAM dataset consists of 4601 samples and 57 input features. The results show that the proposed DEGAIN outperforms the existing ones in terms of root mean square error and Frechet inception distance metrics.

Keywords: machine learning; data cleaning; missing data; data imputation; generative networks

Citation: Shahbazian, R.; Trubitsyna, I. DEGAIN: Generative-Adversarial-Network-Based Missing Data Imputation. *Information* 2022, 13, 575. https://doi.org/10.3390/info13120575

Academic Editor: Peter Revesz

Received: 20 October 2022
Accepted: 8 December 2022
Published: 12 December 2022

Publisher's Note: MDPI stays neutral with regard to jurisdictional claims in published maps and institutional affiliations.

Copyright: © 2022 by the authors. Licensee MDPI, Basel, Switzerland. This article is an open access article distributed under the terms and conditions of the Creative Commons Attribution (CC BY) license (https://creativecommons.org/licenses/by/4.0/).

1. Introduction

Data cleaning is the process of fixing or removing incorrect, corrupted, incorrectly formatted, duplicate, or incomplete data within a dataset. There are many factors for data to be duplicated or mislabeled, especially when multiple data sources are combined. If data are incorrect, outcomes and algorithms are unreliable, or even the results are incorrect. The exact steps in the data cleaning process is highly dependent on the dataset; however, it is possible to establish a generalized conceptual data cleaning process [1,2] as described in the following:

1. **Removing duplicate or irrelevant data**: When datasets from multiple sources, clients, etc., are combined, the chance of duplicate data creation increases. Additionally, in some analyses, the irrelevant data could also be removed. Any information that does not pertain to the issue that we are attempting to solve is considered irrelevant.
2. **Fixing structural errors**: Structural errors occur when conventions, typos, or incorrect capitalization is observed due to the measurement or data transfer. For instance, if "N/A" and "Not Applicable" both appear, they should be analyzed as the same category.
3. **Filtering unwanted outliers**: If an outlier proves to be irrelevant for analysis or is a mistake, it needs to be removed. Some outliers represent natural variations in the population, and they should be left as is.
4. **Handling missing data**: Many data analytic algorithms cannot accept missing values. There are a few methods to deal with missing data. In general, missing data rows are removed or the missing values are estimated according to the existing data in the dataset. These methods are also known as data imputation.
5. **Data validation and quality assurance**: After completing the previous steps, it is needed to validate the data and to make sure that the data have sufficient quality for the considered analytics.

In this paper, we study the algorithms that are capable of handling the missing data as an important step of data cleaning. Different reasons can lead to missing values during the data collection phase, including, but not limited to, the faulty clinical data registration of patients [3], and sensor damages [4]. One basic method to handle this problem is to remove incomplete data. However, removing the data can diminish the number of samples when the dataset contains many samples with missing values [5]. Therefore, many researchers have tried to employ efficient and effective algorithms to handle the missing values. The effect of the missing data handling methods mainly depends on the missing mechanism. The missing data are categorized as follows [6]:

- **Missing completely at random (MCAR)**: It means that the probability of missing data does not depend on any value of attributes.
- **Missing at random (MAR)**: Meaning that the probability of missing data does not depend on its own particular value, but on the values of other attributes.
- **Not missing at random (NMAR)**: Meaning that the missing data depend on the missed values.

In general, the missing data handling methods could be categorized as data deletion, statistical methods, and machine learning (ML) based approaches. Among the ML based algorithms, generative adversarial networks (GANs) have attracted many researchers in recent years. The GAN has many applications, mostly with a focus on generating synthetic data. The missing value estimation could be considered synthetic data generation. Therefore, it is possible to use such networks in handling the missing data problem. However, the performance of the algorithms is dependent on many variables, such as data type, for instance, if the data belong to the category of an image, clinical dataset, energy dataset, etc. Accordingly, many variations of the GAN are introduced in the literature.

In this study, we conduct a literature review on missing data handling. We present the fundamentals of GAIN [7], a GAN-based algorithm and propose an improved version of GAIN called DEGAIN. In our method, we improve the GAIN by applying the idea of network deconvolution [8]. Convolutional kernels usually re-learn redundant data because of the strong correlations in many image-based datasets. The deconvolution strategy is proven to be effective on images; however, it has not been applied to the GAIN algorithm. DEGAIN is capable of removing the data correlations. We evaluate the performance of the proposed DEGAIN with the publicly available Letter Recognition dataset (Letter dataset, for short) and SPAM dataset. In particular, we use root mean square error (RMSE) and Frechet inception distance (FID) metrics and compare the performance of the proposed DEGAIN with the GAIN, the auto-encoder [9] and the MICE [10] algorithms.

The remainder of this paper is organized as follows: In Section 2, we shortly review the related works. In Section 3, we introduce the system model with mathematical relations and describe the proposed DEGAIN algorithm. The performance evaluation is presented in Section 4. Finally, Section 5 concludes the paper.

2. Related Works

In this section we start with a brief overview of incomplete information management perspectives. Next, we focus our attention on the techniques that replace missing values with the concrete ones and provide a brief review of the existing literature on the missing data handling problem.

2.1. Incomplete Information

Incomplete information arises naturally in many database applications, such as data integration, data exchange, inconsistency management, data cleaning, ontological reasoning, and many others [11].

In some applications, it is natural to allow the presence of incomplete data in the database and to support this circumstance using the proper approaches. The use of null values is the commonly accepted approach for handling incomplete data, and the databases containing null values are usually called *incomplete database*. Some recent proposals in this direction can be found in [11–16].

Intuitively, whenever the database has a structure defined a priori (as in the case of relational databases), some data (for instance, a fax number for a person) to be inserted could be missing. This situation occurs according to the following situations:

1. We are sure that the value exists but it is unknown for us;
2. We are sure that this value does not exist;
3. We do not know anything.

In the relational databases, the unique value, Null, is used in all three situation described before. However, in different applications null values are interpreted as unknown values. In cases like this, we will consider the null values as missing values.

The recent study of [17] that analyzed the use of null values in widely used relational database management systems evidenced that null values are ubiquitous and relevant in real-life scenarios; however, the SQL features designed to deal with them cause multiple problems. While most users accept the SQL behavior for simple queries (positive fragments of relational algebra), many are dissatisfied with SQL answers for more complex queries involving aggregation or negation.

For instance, in some circumscriptions, SQL can miss some tuples that should be considered answers (false negatives); in other cases, SQL can return some tuples that should not be considered answers (false positives). The first situation can be considered an under-approximation of the results and is acceptable in different scenarios. The second one is more critical, as the result might contain incorrect answers. The experimental analysis in [18] showed that false positive are a real problem for queries involving negation. In the observed situations, they were always present and sometimes they constituted almost 100% of the answers.

Theoretical frameworks allow multiple null values in incomplete databases, and the use of labeled nulls provides a more accurate depiction of unknown data. Certain answers, i.e., query answers that can be found from all the complete databases represented by an incomplete database, are a commonly accepted semantics in this paradigm. Unfortunately, the computation of certain answers is a coNP-hard problem [19], which restricts the practical usefulness. A possible solution is to use polynomial time evaluation algorithms computing a sound but possibly an incomplete set of certain answers [11,18,20]. The corresponding prototypes are described in [21,22].

In different applications, missing values cannot be tolerated and must be replaced by the concrete values. It should be noted that in some research works, the authors use the missing data as a feature for other decision makings such as error estimation. For instance the authors in [23] estimate the physical-layer transmission errors in cable broadband networks by considering the missing values. In one of the recent research studies, the authors proposed a new multiple imputation MB (MimMB) framework for causal feature selection with missing data [24]. However, in this paper, we assume that the missing value needs to be handled for further processing activities. The available strategies for handling missing data can be divided into traditional methods and ML-based algorithms, summarized in Figure 1 and described below.

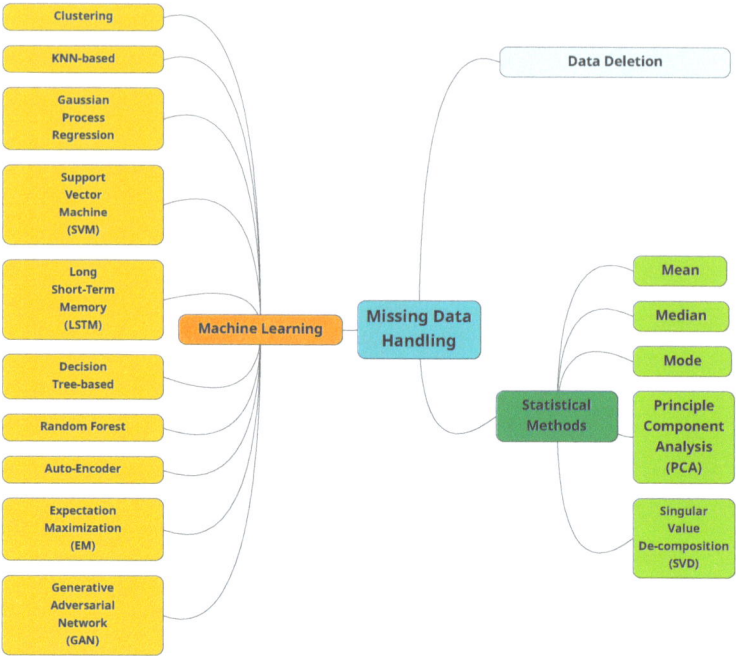

Figure 1. The categorization of traditional and machine learning based algorithms used for missing data handling.

2.2. Traditional Methods

Some of known traditional methods on missing data handling, including the case deletion, mean, median, mode, principal component analysis (PCA) and also singular value decomposition (SVD) are described in the following:

- **Case deletion (CD)**: In CD, missing data instances are omitted. The method has two main disadvantages [25]:
  1. Decreasing the dataset size;
  2. Since the data are not always MCAR, bias occurs on data distribution and corresponding statistical analysis.

- **Mean, median and mode**: In these methods, the missing data are replaced with the mean (numeric attribute) of all observed cases. Median is also used to reduce the influence of exceptional data. The characteristic of the original dataset will be changed by using constants to replace missing data, ignoring the relationship among attributes. As an alternative similar solution, we may use the mode of all known values of that attributes to replace the missing data [25]. Mode is usually preferred for categorical data.

- **Principal component analysis (PCA)**: This method is well-known in statistical data analysis and can be used to estimate the data structure level. Traditional PCA cannot deal with the missing data. Upon the measure of variance within the dataset, data will be scored by how well they fit into a principal component (PC). Since the data points will have a PC (the one that best fits), PCA can be considered a clustering analysis. Missing scores are estimated by projecting known scores back into the principal space. More details can be found in [26].

- **Singular value decomposition (SVD)**: In this method, data are projected into another space where the attributes have different values. In the projected space, it is possible to re-construct the missing data.

One of the main differences of the traditional methods and the machine learning based methods to handle the missing data is the capability of the optimization in ML. The ML-based methods follow an optimization process. ML-based methods can also extract the relation between data points, and therefore more precise estimation on the missing values.

2.3. Machine Learning Methods

Machine learning algorithms are categorized into supervised, semi-supervised and unsupervised [27]. Some of the main ML-based methods are clustering algorithms [28], k-nearest neighbors (KNN) [29], Gaussian process regression (GPR) [30], support vector machine (SVM) [31], long short-term memory (LSTM) [32], decision trees (DT) [33], random forests (RF) [34], auto-encoder (AE) [35], expectation maximization (EM) [36] and generative adversarial networks (GAN) [7].

- **Clustering**: These unsupervised learning algorithms group the samples with similar characters. To replace the missing value, the distance between the centroid of clusters with the sample is calculated, and the missing value of chosen cluster is replaced with the obtained value [28]. The minimum distance could be calculated by a variety of distance functions. One common function is l^2-norm [37]. The l^2-norm calculates the distance of the vector coordinate from the origin of the vector space.
- **k-nearest neighbors (KNN)**: This supervised algorithm replaces the missing value by the mean (or weighted mean) of the k nearest samples. These neighbor samples are identified by calculating the distance of the missing value with the available samples [28]. Currently, many variations of the original KNN have been proposed in the literature, including the SKNN, IKNN, CKNN, and ICKNN. KNN-based algorithms require heavy calculations to find the nearest neighbors [29]. Some of the common distance functions used in KNN are the Euclidean-overlap metric [38], Value Difference Metric [38] and mean Euclidean distance [38].
- **Gaussian process regression (GPR)**: GPR algorithms predict the output's variance based on non-linear probabilistic techniques. GPR-based algorithms estimate a probabilistic region for the missing values instead of point estimation. The performance of GRP-based algorithms is dependent on the used kernel function. This kernel is chosen according to the data type and the effective algorithms might use a combination of different kernels. Similar to KNN, the GRP-based algorithms also need heavy calculations, which is not the ideal case for large-scale datasets [30].
- **Support vector machine (SVM)**: SVM has applications in both classification and regression. SVM-based algorithms are non-linear and map the input to a high-dimensional feature space. SVM-based algorithms also use different kernels such as GPR [30].
- **Long short-term memory (LSTM)**: LSTM is a deep learning (DL) algorithm. As a subcategory of recurrent neural networks, DL shows improvement for time series compared with conventional ML algorithms. The training phase of the LSTM might be complex due to the vanishing gradient problem [32].
- **Decision tree (DT)**: DT-based algorithms partition the dataset into groups of samples, forming a tree. The missing data are estimated by the samples associated with the same leaves of the tree. Different variations of DT algorithms have been proposed by researchers, including the ID3, C4.5, CRAT, CHAILD and QUEST [39]. DT algorithms do not require prior information on the data distribution [33].
- **Random forest (RF)**: RF consists of multiple DTs in which the average value of the DT estimation is considered the missing value [34].
- **Auto-encoder (AE)**: As a class of unsupervised DL algorithms, AE learns a coded vector from the input space. The AE generally consists of three layers, including the input, hidden, and output. The objective in AE is to map the input layer to the hidden

layer and then reconstruct the input samples through the hidden vector. The elements of input attributes can be missed randomly in the training phase. Therefore, it is expected that the vectors in the input space consist of randomly missing vectors, and the output layer has the complete set of vectors. Performing this task, the AE will learn how to complete the missing data. Different versions of AEs have been introduced, such as VAE, DAE, SAE, and SDAE [35].

- **Expectation maximization (EM)**: EM algorithms are capable of obtaining the local maximum likelihood of a statistical model. These models entail latent variables besides the observed data and unknown parameters. These latent variables can be missing values among the data. The EM-based algorithms guarantee that the likelihood will increase. However, the price is slow convergence rate [36].

Generative Adversarial Networks

The supervised algorithms need labeled data for the optimization process. However, the data collection for these algorithms is complex. Generative adversarial networks (GANs) can produce synthetic data samples based on a limited set of the collected data. GANs are semi-supervised learning algorithms and generate synthetic data with a small set of collected data. The generated data are not the same as the collected data; however, they are very similar. GANs are among the foremost essential research topic within different research fields, including image-to-image translation, fingerprint localization, classification, speech and language processing, malware detection and video generation [27].

In this section, we introduce the main structure of the GAN, firstly proposed by Goodfellow et al. in 2014 [7]. The GAN consists of two components, the generator (G) and the discriminator (D) as shown in Figure 2. In the training phase, the noise and real data are the input and output of the G component. The dominant goal of the G is to alter the noise to the realistic data. The D component learns to discriminate the real and generated data.

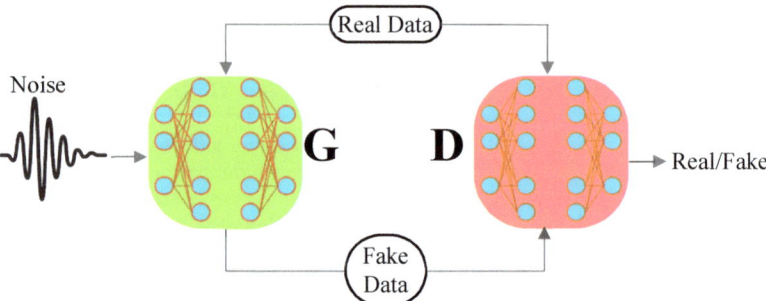

Figure 2. The general structure of GAN proposed by Goodfellow.

The training process of these two components are based on the pre-defined cost function as presented in Equation (1).

$$\min_G \max_D L(D,G) = E_{x \sim p_r(x)}[\log D(\mathbf{x})] + E_{z \sim p_z(z)}[\log(1 - D(G(\mathbf{z})))] \tag{1}$$

The G and D are two functions and can be denoted by a multi-layer perceptron (MLP). The G learns to alter the noise $\mathbf{z} \sim p_z(z)$ to the real data. It can be represented by $G(\mathbf{z};\theta_g)$ as a function, whose inputs are noise and outputs are generated synthetic data, in which θ_g indicates the parameters of the G component. The D component learns to discern the real and fake data. This component can be represented by $D(\mathbf{x};\theta_d)$, whose inputs are real and synthetic data and outputs are class labels, in which θ_d indicates the parameters of the D component. The cost function of the D and G components are presented in Equations (2) and (3), respectively:

$$L(\theta_d) = E_{x \sim p_r(x)}[\log D(\mathbf{x};\theta_d)] + E_{z \sim p_z(z)}[\log(1 - D(G(\mathbf{z};\theta_g)))] \tag{2}$$

$$L(\theta_g) = E_{z \sim p_z(z)} \left[\log(1 - D(G(\mathbf{z};\theta_g))) \right] \tag{3}$$

where (θ_d and θ_g) are updated until convergence is reached. When the real and synthetic data cannot be recognized by the D, the system has reached convergence. Mathematically specking, the convergence occurs when $D(\mathbf{x};\theta_d) = 0.5$. After the training phase, the G is ready to be utilized for producing the synthetic samples.

3. Proposed Method

3.1. System Model

A GAN-based data imputing method, called generative adversarial imputation nets (GAIN), was introduced in [7]. In GAIN, the generator component G takes real data vectors, imputes the missing values conditioned on the really observed data, and gives a completed vector. Then, the discriminator component D obtains a completed vector and tries to determine which element is really observed and which one is synthesized. To learn the desired distribution in the G component, some additional information is deployed for the discriminator D in the form of a hint vector. The hint vector shows the pieces of information about the missing quality of the real data to the D component, and D concentrates its heed on the quality of imputation for particular missing values. In other words, the hint vector assures the G component to be learned for generating data based on the actual data distribution [7].

Convolution, which applies a kernel to overlapping sections shifted across the data, is a crucial operation in many convolutional neural networks. Although utilizing CNN to synthesize images is not required in GANs, it is frequently done in order to learn the distribution of images [40]. The generator architecture is typically composed of the following layers:

- **Linear layer**: The noise vector is fed into a fully connected layer, and its output is reshaped into a tensor.
- **Batch normalization layer**: Stabilizes learning by normalizing inputs to zero mean and unit variance, avoiding training issues, such as vanishing or exploding gradients, and allowing the gradient to flow through the network.
- **Up sample layer**: Instead of using a convolutional transpose layer to up sample, it mentions using upsampling and then applying a simple convolutional layer on top of it. Convolutional transpose is sometimes used instead.
- **Convolutional layer**: To learn from up-sampled data, the matrix is passed through a convolutional layer with a stride of 1 and the same padding as it is up sampled.
- **ReLU layer**: For the generator because it allows the model to quickly saturate and cover the training distribution space.
- **TanH Activation**: TanH enables the model to converge more quickly.

Convolutional kernels are in fact relearning duplicate data because of the high correlations in real-world data. The convolution makes the neural network training difficult. In the following, we review the mathematical presentation of GAIN.

3.2. Problem Formulation

In a d-dimensional space $\mathcal{X} = \mathcal{X}_1 \times \cdots \times \mathcal{X}_d$ the $\mathbf{X} = (X_1, \ldots, X_d)$ is a random variable taking values in $\mathcal{X}$ with distribution $P(\mathbf{X})$. $\mathbf{M} = (M_1, \ldots, M_d)$ is a random variable in $\{0,1\}^d$. The $\mathbf{X}$ is called the data vector, and $\mathbf{M}$ is called the mask vector.

A new space $\tilde{\mathcal{X}}_i = \mathcal{X}_i \cup \{*\}$ is defined for $i \in \{1, \ldots, d\}$ where the start, $*$ does not belong to any $\mathcal{X}_i$, and represents an unobserved value. Defining $\tilde{\mathcal{X}} = \tilde{\mathcal{X}}_1 \times \cdots \times \tilde{\mathcal{X}}_d$ The variable $\tilde{\mathbf{X}} = (\tilde{X}_1, \ldots, \tilde{X}_d) \in \tilde{\mathcal{X}}$ is presented in Equation (4):

$$\tilde{X}_i = \begin{cases} X_i, & \text{if } M_i = 1 \\ *, & \text{otherwise} \end{cases} \tag{4}$$

where **M** indicates which components of **X** are observed. The **M** could be recovered from $\tilde{\mathbf{X}}$. In missing data re-construction, n independent and identically distributed copies of $\tilde{\mathbf{X}}$ are realized, denoted by $\tilde{\mathbf{x}}^1, \ldots, \tilde{\mathbf{x}}^n$ and defined in the dataset $\mathcal{D} = \{(\tilde{\mathbf{x}}^i, \mathbf{m}^i)\}_{i=1}^n$, where $\mathbf{m}^i$ is simply the recovered realization of **M** corresponding to $\tilde{\mathbf{x}}^i$. The goal is to estimate the unobserved values in each $\tilde{\mathbf{x}}_i$. The samples are generated according to $P(\mathbf{X}|\tilde{\mathbf{X}} = \tilde{\mathbf{x}}^i)$, that is, the conditional distribution of **X** given $\tilde{\mathbf{X}} = \tilde{\mathbf{x}}^i$ for each i to fill in the missing data points in $\mathcal{D}$.

The generator, G, takes as input $\tilde{\mathbf{X}}$, **M** and a noise variable **Z**, and outputs $\bar{\mathbf{X}}$, where $\bar{\mathbf{X}}$ is a vector of synthetic data. Let $G : \tilde{\mathcal{X}} \times \{0,1\}^d \times [0,1]^d \to \mathcal{X}$ be a function, and $\mathbf{Z} = (Z_1, \ldots, Z_d)$ be d-dimensional noise (independent of all other variables) [7]. The random variables $\bar{\mathbf{X}}, \hat{\mathbf{X}} \in \mathcal{X}$ are defined by Equations (5) and (6).

$$\bar{\mathbf{X}} = G(\tilde{\mathbf{X}}, \mathbf{M}, (\mathbf{1} - \mathbf{M}) \odot \mathbf{Z}) \tag{5}$$

$$\hat{\mathbf{X}} = \mathbf{M} \odot \tilde{\mathbf{X}} + (\mathbf{1} - \mathbf{M}) \odot \bar{\mathbf{X}} \tag{6}$$

where $\odot$ denotes element-wise multiplication. $\bar{\mathbf{X}}$ corresponds to the vector of estimated values and $\hat{\mathbf{X}}$ corresponds to the completed data vector.

The discriminator, D will be used to train G. However, unlike the standard GAN where the output of the generator is either real or synthetic, the output of GAIN is comprised of some components that are real and some that are synthetic. Rather than identifying that an entire vector is real or synthetic, the discriminator attempts to distinguish which are real (observed) or synthetic. The mask vector **M** is pre-determined by the dataset. Formally, the discriminator is a function $D : \mathcal{X} \to [0,1]^d$ with the i-th component of $D(\hat{\mathbf{x}})$ corresponding to the probability that the i-th component of $\hat{\mathbf{x}}$ was observed. The D is trained to maximize the probability of correctly predicting **M** and G is trained to minimize the probability of D predicting **M**. The quantity $V(D, G)$ is defined as presented in Equation (7).

$$V(D, G) = \mathbb{E}_{\hat{\mathbf{X}}, \mathbf{M}, \mathbf{H}} \left[\mathbf{M}^T \log D(\hat{\mathbf{X}}, \mathbf{H}) + (\mathbf{1} - \mathbf{M})^T \log \left(\mathbf{1} - D(\hat{\mathbf{X}}, \mathbf{H})\right) \right], \tag{7}$$

where log is an element-wise logarithm and dependence on G is through $\hat{\mathbf{X}}$. The goal of GAIN is presented in Equation (8):

$$\min_G \max_D V(D, G). \tag{8}$$

where the loss function $\mathcal{L} : \{0,1\}^d \times [0,1]^d \to \mathbb{R}$ is defined as presented in Equation (9):

$$\mathcal{L}(\mathbf{a}, \mathbf{b}) = \sum_{i=1}^d \left[a_i \log(b_i) + (1 - a_i) \log(1 - b_i) \right]. \tag{9}$$

3.3. Proposed Algorithm: DEGAIN

The proposed DEGAIN is originated from GAIN [7]. The main idea behind the DEGAIN is to use deconvolution in the generator and discriminator. Convolution applies a kernel to overlapping regions shifted across the data. However, because of the strong correlations in real-world data, convolutional kernels are in effect re-learning redundant data. This redundancy makes the neural network training challenging. The deconvolution can remove the correlations before the data are fed into each layer. It has been shown in [8] that the deconvolution can be efficiently calculated at a fraction of the computational cost of a convolution layer. The deconvolution strategy has proven to be effective on images; however, it has not been applied to GANs, including the GAIN.

Given a data matrix $X_{N \times F}$, where N is the number of samples and F is the number of features, the covariance matrix is calculated as $Cov = \frac{1}{N}(X - \mu)^T(X - \mu)$.

An approximated inverse square root of the covariance matrix could be calculated as $D = Cov^{-\frac{1}{2}}$ multiplied with the centered vectors $(X - \mu) \cdot D$. Accordingly, the correlation effects could be removed. If computed perfectly, the transformed data have the identity matrix as covariance: $D^T(X-\mu)^T(X-\mu)D = Cov^{-0.5} \cdot Cov \cdot Cov^{-0.5} = I$.

The process to construct X and $D \approx (Cov + \epsilon \cdot I)^{-\frac{1}{2}}$ is presented in Algorithm 1, where $\epsilon \cdot I$ improves the stability. The deconvolution operation is further applied via matrix multiplication to remove the correlation between neighboring pixels. The deconvolved data are then multiplied with w. The architecture of proposed method is depicted in Figure 3. When the training phase is completed, the G component is able to impute the dataset. There are two main loops for updating the parameters of the G and D components in Algorithm 1. First, batch samples from the noise and samples of real data are presented to the inner loop for updating the parameters of the D component. The cost function of the D component is then calculated by the given samples. Then, the D component's parameters are updated based on the initiated rate.

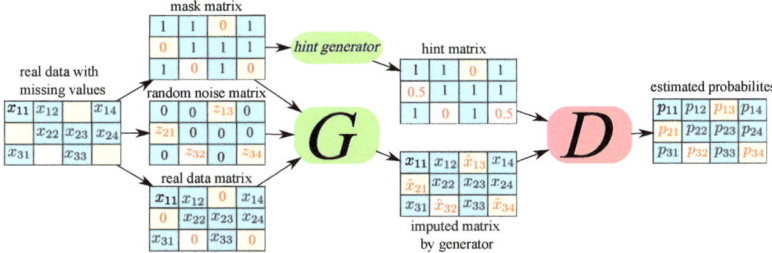

Figure 3. The general structure of DEGAIN.

Algorithm 1 (The DEGAIN algorithm: deconvolution and then training).

1: **Input:** N channels of input features $[x_1, x_2, \ldots, x_N]$, Number of epochs e; Number of Iteration of inner loop n; Updating rates (α_g and α_d);
2: **for** $i \in \{1, \ldots, N\}$ **do**
3:    $X_i = im2col(x_i)$
4: $X = [X_1, \ldots, X_N]$
5: $X = Reshape(X)$
6: $Cov = \frac{1}{M} X^t X \ \%[x_i]$ has M rows
7: $D \approx (Cov + \epsilon \cdot I)^{-\frac{1}{2}}$
   Training of G and D
8: **for** $(i=0; i<e; i++)$ **do**
9:    **for** $j=0; j<n; j++$ **do**
10:      batch samples from noise $Z \in R^{B \times L} \sim p_z(z)$;
11:      batch samples from noise $X \in R^{B \times M}$;
12:      $L(\theta_d) = \frac{1}{B} \Sigma_{b=1}^{B} log D(X_b, \theta_d) + log(1 - D(G(2Z_b), \theta_g))$;
13:      $\xi_d = \frac{\delta}{\delta \theta_d} L(\theta_d)$;
14:      $\theta_d^{j+1} = \theta_d^t + \alpha_d \xi_d$
15:    batch samples from noise $Z \in R^{B \times L} \sim p_z(z)$;
16:    $L(\theta_d) = \frac{1}{B} \Sigma_{b=1}^{B} log(1 - D(G(Z_b, \theta_g)))$
17:    $\xi_d = \frac{\delta}{\delta \theta_d} L(\theta_d)$;
18:    $\theta_g^{i+1} = \theta_g^t - \alpha_g \xi_g$;

4. Performance Evaluation

We evaluate the performance of the DEGAIN and compare the results with MICE [10], GAIN [7] and AE [9]. Multivariate imputation by chained equations (MICE) has emerged in addressing missing data. The chained equations approach can handle variables of varying types, for instance, the continuous or binary as well as complexities such as bounds. MICE is also a software package presented in R [10]. In multiple imputation algorithms such as

AE, multiple copies of the dataset are replaced by slightly different imputed values in each copy. In this method, the variability is modeled into the imputed values [9].

We perform each experiment 10 times, and within each experiment, we use 5-cross validations in terms of RMSE, which directly measures the error distance, and FID, which takes the distribution of imputed values into account. We use the Letter and SPAM datasets for comparing algorithms, and samples are missed with the rate of 20%.

4.1. Evaluation Metrics

In our experiments we consider the root-mean-square error (RMSE) and Frechet inception distance (FID) metrics to evaluate the performance of the proposed DEGAIN on handling the missing data. It should be noted that besides RMSE and FID, several other metrics are introduced in the literature including, but not limited to, mean absolute error (MAE), area under ROC curve or AUC score and F1-score. These metrics perform in different datasets or operations. For example, the F1-score is mostly used for binary classification. MAE is similar to RMSE; however, RMSE is more common in the literature. Comparing our method with GAIN, AE and MICE, we chose the related RMSE and FID metrics. Remember that RMSE is used for continuous values and measures the error between real values and imputed values for an incomplete dataset. FID converts a group of imputed samples to a feature space using a particular inception net layer. Assuming the converted layer as a continuous multivariate Gaussian distribution, the mean and covariance are predicted for both the imputed and real samples.

The mathematical presentation is shown in Table 1, where $\hat{N}$ is the number of missing values, y_i is the real missing value, and $\hat{y}_i$ is the imputed value. Additionally, the m_1 and m_2 denote the mean of the real and imputed data, respectively. C_1 and C_2 indicate the covariance of real and imputed data, respectively.

Table 1. Evaluation metrics used to assess the proposed algorithm's performance.

Metric	Formula
RMSE	$RMSE = \sqrt{\frac{1}{\hat{N}} \sum_{i=1}^{\hat{N}} (y_i - \hat{y}_i)^2}$
FID	$d^2((m_1, C_1)(m_2, C_2)) = \|m_1 - m_2\|_2^2 + \text{Tr}(C_1 + C_2 - 2(C_1 C_2)^{\frac{1}{2}})$

4.2. Dataset

Here we use the **Letter** and **SPAM** datasets. Letter is publicly available in the UC Irvine Machine Learning Repository and can be accessed through https://archive.ics.uci.edu/ (accessed on 20 July 2022). In this dataset, the objective is to identify each of a large number of black-and-white rectangular pixel displays as one of the 26 capital letters in the English alphabet. The character images are based on 20 different fonts and each letter within these 20 fonts is randomly distorted to produce a file of 20,000 unique stimuli. Typically, the first 16,000 items are used for training and then the resulting model is capable of predicting the letter category for the remaining 4000. The SPAM dataset consists of 4601 samples and 57 input features. In this dataset, the goal is to predict spam emails based on input features. SPAM is also publicly available at http://archive.ics.uci.edu/ml/datasets/Spambase/ (accessed on 10 August 2022). These datasets have no missing values and therefore, we use a 20% rate of missed samples.

4.3. Results

The evaluation are performed in Google Colab, on Python 3 with 12 GB of RAM. We used the base codes of GAIN [7] publicly accessible in https://github.com/jsyoon0823/GAIN (accessed on 10 July 2022). We modified the code and added the deconvolution to the Generator and Discriminator. The results are presented in Table 2 and illustrated in Figure 4. As can be seen in Table 2 and Figure 4, the performance of GAN-based algorithms, both

GAIN and DEGAIN perform much better than the auto-encoder (AE) [9] and MICE [10]. The DEGAIN is slightly better compared with the GAIN. The main advantage of the DEGAIN could be explored running on correlated large datasets of images. Therefore, as expected, the GAIN and proposed DEGAIN can be the most profitable in image datasets.

Table 2. Performance evaluations of proposed DEGAIN, GAIN [7], AE [9], MICE [10] in terms of RMSE and FID metrics.

	Proposed DEGAIN	**GAIN [7]**	**AE [9]**	**MICE [10]**
RMSE (Letter dataset)	0.096	0.101	0.142	0.166
Normalized FID (Letter dataset)	0.492	0.513	0.826	1
RMSE (SPAM dataset)	0.047	0.050	0.064	0.068
Normalized FID (SPAM dataset)	0.898	0.946	0.973	1

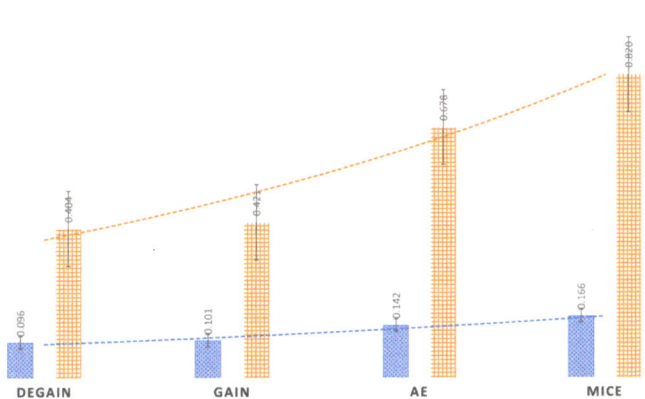

Figure 4. Illustration of evaluation results on RMSE and FID metrics for proposed DEGAIN, GAIN [7], AE [9] and MICE [10] on Letter dataset.

5. Conclusions

In this paper, we studied the traditional and machine learning based algorithms that could handle the missing data problem in data cleaning process. We reviewed the architecture of generative adversarial network (GAN) based models and their performance on missing data handling. We proposed an algorithm called DEGAIN to estimate the missing values in the dataset. The DEGAIN is based on the known GAIN algorithm that is already used in missing data imputation. We added deconvolution to remove the correlation between data. The evaluated performance of the presented method was performed on publicly available datasets, called Letter and SPAM. The RMSE and FID metrics on the results confirmed that the GANs are effective on re-constructing the missing values compared with the earlier auto-encoder or MICE algorithms. Additionally, the proposed DEGAIN performed well and improved the performance of GAIN. We believe that the main advantages of DEGAIN could be explored running on large image datasets, although it showed improvement even on our chosen datasets. This paper addresses one of many aspects of data quality: missing information. Inconsistent information, which could be handled using methods such as arbitration [41], is still an open issue that GAN-based methods and the proposed DEGAIN need to address in future works.

Author Contributions: Conceptualization, R.S. and I.T.; Methodology, I.T.; Software, R.S.; Supervision, I.T.; Validation, R.S. and I.T.; Visualization, R.S.; Writing—original draft, R.S. and I.T.; Writing—review and editing, R.S. and I.T. All authors have read and agreed to the published version of the manuscript.

Funding: This research was supported by MISE Project True Detective 4.0.

Data Availability Statement: The datasets used in this paper are publicly available at https://archive.ics.uci.edu/ml/datasets/letter+recognition (accessed on 20 July 2022) and http://archive.ics.uci.edu/ml/datasets/Spambase/ (accessed on 10 August 2022).

Conflicts of Interest: The authors declare no conflict of interest.

Abbreviations

The following abbreviations are used in this manuscript:

AE	Auto-Encoder
CD	Case Deletion
DT	Decision Tree
DL	Deep Learning
EM	Expectation Maximization
FID	Frechet Inception Distance
GRP	Gaussian Process Regression
GAN	Generative Adversarial Network
KNN	k-Nearest Neighbors
LSTM	Long Short-Term Memory
ML	Machine Learning
MAE	Mean Absolute Error
MAR	Missing At Random
MCAR	Missing Completely At Random
MLP	Multi-Layer Perceptron
NMAR	Not Missing At Random
PCA	Principal Component Analysis
RF	Random Forest
RMSE	Root Mean Square Error
SVD	Singular Value Decomposition
SVM	Support Vector Machine

References

1. Ilyas, I.F.; Chu, X. *Data Cleaning*; Morgan & Claypool: San Rafael, CA, USA, 2019.
2. O'Brien, A.D.; Stone, D.N. Yes, you can import, analyze, and create dashboards and storyboards in Tableau! The GBI case. *J. Emerg. Technol. Account.* **2020**, *17*, 21–31. [CrossRef]
3. Luo, Y. Evaluating the state of the art in missing data imputation for clinical data. *Briefings Bioinform.* **2022**, *23*, bbab489. [CrossRef] [PubMed]
4. Li, Y.; Bao, T.; Chen, H.; Zhang, K.; Shu, X.; Chen, Z.; Hu, Y. A large-scale sensor missing data imputation framework for dams using deep learning and transfer learning strategy. *Measurement* **2021**, *178*, 109377. [CrossRef]
5. Platias, C.; Petasis, G. A Comparison of Machine Learning Methods for Data Imputation. In Proceedings of the 11th Hellenic Conference on Artificial Intelligence, Athens, Greece, 2–4 September 2020; pp. 150–159.
6. Austin, P.C.; White, I.R.; Lee, D.S.; van Buuren, S. Missing data in clinical research: A tutorial on multiple imputation. *Can. J. Cardiol.* **2021**, *37*, 1322–1331. [CrossRef] [PubMed]
7. Yoon, J.; Jordon, J.; Schaar, M. Gain: Missing data imputation using generative adversarial nets. In Proceedings of the International Conference on Machine Learning, Stockholm, Sweden, 10–15 July 2018; pp. 5689–5698.
8. Ye, C.; Evanusa, M.; He, H.; Mitrokhin, A.; Goldstein, T.; Yorke, J.A.; Fermüller, C.; Aloimonos, Y. Network deconvolution. *arXiv* **2019**, arXiv:1905.11926.
9. Gondara, L.; Wang, K. Multiple imputation using deep denoising autoencoders. *arXiv* **2017**, arXiv:1705.02737.
10. Van Buuren, S.; Groothuis-Oudshoorn, K. mice: Multivariate imputation by chained equations in R. *J. Stat. Softw.* **2011**, *45*, 1–67. [CrossRef]
11. Greco, S.; Molinaro, C.; Trubitsyna, I. Approximation algorithms for querying incomplete databases. *Inf. Syst.* **2019**, *86*, 28–45. [CrossRef]
12. Calautti, M.; Console, M.; Pieris, A. Benchmarking approximate consistent query answering. In Proceedings of the 40th ACM SIGMOD-SIGACT-SIGAI Symposium on Principles of Database Systems, Virtual Event, China, 20–25 June 2021; pp. 233–246.

13. Calautti, M.; Caroprese, L.; Greco, S.; Molinaro, C.; Trubitsyna, I.; Zumpano, E. Existential active integrity constraints. *Expert Syst. Appl.* **2021**, *168*, 114297. [CrossRef]
14. Calautti, M.; Greco, S.; Molinaro, C.; Trubitsyna, I. Query answering over inconsistent knowledge bases: A probabilistic approach. *Theor. Comput. Sci.* **2022**, *935*, 144–173. [CrossRef]
15. Calautti, M.; Greco, S.; Molinaro, C.; Trubitsyna, I. Preference-based Inconsistency-Tolerant Query Answering under Existential Rules. *Artif. Intell.* **2022**, *312*, 103772. [CrossRef]
16. Calautti, M.; Greco, S.; Molinaro, C.; Trubitsyna, I. Querying Data Exchange Settings Beyond Positive Queries. In Proceedings of the 4th International Workshop on the Resurgence of Datalog in Academia and Industry (Datalog-2.0), Genova, Italy, 5 September 2022; Volume 3203, pp. 27–41.
17. Toussaint, E.; Guagliardo, P.; Libkin, L.; Sequeda, J. Troubles with nulls, views from the users. *Proc. VLDB Endow.* **2022**, *15*, 2613–2625. [CrossRef]
18. Guagliardo, P.; Libkin, L. Making SQL queries correct on incomplete databases: A feasibility study. In Proceedings of the 35th ACM SIGMOD-SIGACT-SIGAI Symposium on Principles of Database Systems, San Francisco, CA, USA, 26 June–1 July 2016; pp. 211–223.
19. Abiteboul, S.; Kanellakis, P.C.; Grahne, G. On the Representation and Querying of Sets of Possible Worlds. *Theor. Comput. Sci.* **1991**, *78*, 158–187. [CrossRef]
20. Libkin, L. SQL's three-valued logic and certain answers. *ACM Trans. Database Syst. (TODS)* **2016**, *41*, 1–28. [CrossRef]
21. Fiorentino, N.; Greco, S.; Molinaro, C.; Trubitsyna, I. ACID: A system for computing approximate certain query answers over incomplete databases. In Proceedings of the International Conference on Management of Data (SIGMOD), Houston, TX, USA, 10–15 June 2018; pp. 1685–1688.
22. Fiorentino, N.; Molinaro, C.; Trubitsyna, I. Approximate Query Answering over Incomplete Data. In *Complex Pattern Mining*; Springer: Berlin, Germany, 2020; pp. 213–227.
23. Hu, J.; Zhou, Z.; Yang, X. Characterizing Physical-Layer Transmission Errors in Cable Broadband Networks. In Proceedings of the 19th USENIX Symposium on Networked Systems Design and Implementation (NSDI 22), Renton, WA, USA, 4–6 April 2022; USENIX Association: Renton, WA, USA, 2022; pp. 845–859.
24. Yu, K.; Yang, Y.; Ding, W. Causal Feature Selection with Missing Data. *ACM Trans. Knowl. Discov. Data* **2022**, *16*, 1–24. [CrossRef]
25. Peng, L.; Lei, L. A review of missing data treatment methods. *Intell. Inf. Manag. Syst. Technol* **2005**, *1*, 412–419.
26. Folch-Fortuny, A.; Arteaga, F.; Ferrer, A. PCA model building with missing data: New proposals and a comparative study. *Chemom. Intell. Lab. Syst.* **2015**, *146*, 77–88. [CrossRef]
27. Mirtaheri, S.L.; Shahbazian, R. *Machine Learning: Theory to Applications*; CRC Press: Boca Raton, FL, USA, 2022.
28. Nagarajan, G.; Babu, L.D. Missing data imputation on biomedical data using deeply learned clustering and L2 regularized regression based on symmetric uncertainty. *Artif. Intell. Med.* **2022**, *123*, 102214. [CrossRef]
29. Emmanuel, T.; Maupong, T.; Mpoeleng, D.; Semong, T.; Mphago, B.; Tabona, O. A survey on missing data in machine learning. *J. Big Data* **2021**, *8*, 1–37.
30. Ma, Y.; He, Y.; Wang, L.; Zhang, J. Probabilistic reconstruction for spatiotemporal sensor data integrated with Gaussian process regression. *Probabilistic Eng. Mech.* **2022**, *69*, 103264. [CrossRef]
31. Camastra, F.; Capone, V.; Ciaramella, A.; Riccio, A.; Staiano, A. Prediction of environmental missing data time series by Support Vector Machine Regression and Correlation Dimension estimation. *Environ. Model. Softw.* **2022**, *150*, 105343. [CrossRef]
32. Saroj, A.J.; Guin, A.; Hunter, M. Deep LSTM recurrent neural networks for arterial traffic volume data imputation. *J. Big Data Anal. Transp.* **2021**, *3*, 95–108. [CrossRef]
33. Cenitta, D.; Arjunan, R.V.; Prema, K. Missing data imputation using machine learning algorithm for supervised learning. In Proceedings of the 2021 International Conference on Computer Communication and Informatics (ICCCI), Coimbatore, India, 27–29 January 2021; pp. 1–5.
34. Tang, F.; Ishwaran, H. Random forest missing data algorithms. *Stat. Anal. Data Mining: Asa Data Sci. J.* **2017**, *10*, 363–377. [CrossRef] [PubMed]
35. Ryu, S.; Kim, M.; Kim, H. Denoising autoencoder-based missing value imputation for smart meters. *IEEE Access* **2020**, *8*, 40656–40666. [CrossRef]
36. Nelwamondo, F.V.; Mohamed, S.; Marwala, T. Missing data: A comparison of neural network and expectation maximization techniques. *Curr. Sci.* **2007**, *93*, 1514–1521.
37. Eirola, E.; Doquire, G.; Verleysen, M.; Lendasse, A. Distance estimation in numerical data sets with missing values. *Inf. Sci.* **2013**, *240*, 115–128. [CrossRef]
38. Santos, M.S.; Abreu, P.H.; Wilk, S.; Santos, J. How distance metrics influence missing data imputation with k-nearest neighbours. *Pattern Recognit. Lett.* **2020**, *136*, 111–119. [CrossRef]
39. Rokach, L.; Maimon, O. Decision trees. In *Data Mining and Knowledge Discovery Handbook*; Springer: New York, NY, USA, 2005; pp. 165–192.
40. Benjdira, B.; Ammar, A.; Koubaa, A.; Ouni, K. Data-efficient domain adaptation for semantic segmentation of aerial imagery using generative adversarial networks. *Appl. Sci.* **2020**, *10*, 1092. [CrossRef]
41. Revesz, P.Z. On the semantics of arbitration. *Int. J. Algebra Comput.* **1997**, *7*, 133–160. [CrossRef]

MDPI AG
Grosspeteranlage 5
4052 Basel
Switzerland
Tel.: +41 61 683 77 34

Information Editorial Office
E-mail: information@mdpi.com
www.mdpi.com/journal/information

Disclaimer/Publisher's Note: The title and front matter of this reprint are at the discretion of the Guest Editor. The publisher is not responsible for their content or any associated concerns. The statements, opinions and data contained in all individual articles are solely those of the individual Editor and contributors and not of MDPI. MDPI disclaims responsibility for any injury to people or property resulting from any ideas, methods, instructions or products referred to in the content.

www.ingramcontent.com/pod-product-compliance
Lightning Source LLC
LaVergne TN
LVHW072326090526
838202LV00019B/2363